Sound Unseen

Sound Unseen

Acousmatic Sound in Theory and Practice

Brian Kane

UNIVERSITY PRESS

Oxford University Press is a department of the University of Oxford.
It furthers the University's objective of excellence in research, scholarship,
and education by publishing worldwide.

Oxford   New York
Auckland   Cape Town   Dar es Salaam   Hong Kong   Karachi
Kuala Lumpur   Madrid   Melbourne   Mexico City   Nairobi
New Delhi   Shanghai   Taipei   Toronto

With offices in
Argentina   Austria   Brazil   Chile   Czech Republic   France   Greece
Guatemala   Hungary   Italy   Japan   Poland   Portugal   Singapore
South Korea   Switzerland   Thailand   Turkey   Ukraine   Vietnam

Oxford is a registered trademark of Oxford University Press
in the UK and certain other countries.

Published in the United States of America by
Oxford University Press
198 Madison Avenue, New York, NY 10016

© Oxford University Press 2014

First issued as an Oxford University Press paperback, 2016

All rights reserved. No part of this publication may be reproduced, stored in
a retrieval system, or transmitted, in any form or by any means, without the prior
permission in writing of Oxford University Press, or as expressly permitted by law,
by license, or under terms agreed with the appropriate reproduction rights organization.
Inquiries concerning reproduction outside the scope of the above should be sent to the
Rights Department, Oxford University Press, at the address above.

You must not circulate this work in any other form
and you must impose this same condition on any acquirer.

Library of Congress Cataloging-in-Publication Data
Kane, Brian, 1973-
Sound unseen : acousmatic sound in theory and practice / Brian Kane.
      pages ; cm
Includes bibliographical references and index.
ISBN 978-0-19-934784-1 (hardback)
ISBN 978-0-19-934787-2 (online content)
ISBN 978-0-19-063221-2 (paperback)
1. Music—Acoustics and physics.   2. Musique concrete—History and criticism.
3. Music—Philosophy and aesthetics.   I. Title.
ML3805.K15 2014
781.2'3—dc23
            2013037100

For A

…la prise de conscience d'une situation d'écoute que, pour n'être pas nouvelle, n'avait jamais été repérée dans son originalité et baptisée d'un terme spécifique: la situation acousmatique.

…the awareness of a listening situation, not new, but whose originality had never been identified or given a specific name: the acousmatic situation.
—Michel Chion, *Guide des objets sonores*

CONTENTS

Acknowledgments xi

Introduction 1

PART ONE The Acousmatic Situation

1. Pierre Schaeffer, the Sound Object, and the Acousmatic Reduction 15

PART TWO Interruptions

2. Myth and the Origin of the Pythagorean Veil 45
3. The Baptism of the Acousmate 73

PART THREE Conditions

4. Acousmatic Phantasmagoria and the Problem of *Technê* 97
INTERLUDE. Must *Musique Concrète* Be Phantasmagoric? 119
5. Kafka and the Ontology of Acousmatic Sound 134

PART FOUR Cases

6. Acousmatic Fabrications: Les Paul and the "Les Paulverizer" 165
7. The Acousmatic Voice 180

Conclusion 223

Notes 227
Bibliography 277
Index 293

ACKNOWLEDGMENTS

Ever since I had the inkling to write a book about acousmatic sound numerous people have contributed in large and small ways to its completion. I want to thank Philipp Blume, Daniel Callahan, Amy Cimini, J. D. Connor, Joanna Demers, Ryan Dohoney, Emily Dolan, Joseph Dubiel, Will Eastman, Juliet Fleming, Walter Frisch, Milette Gaifman, Evelyne Gayou, Michael Gallope, Doug Gordon, Paul Grimstad, Marion Guck, David Gutkin, Andy Hamilton, Leigh Landy, Deirdre Loughridge, John Paulson, Jairo Moreno, Alexander Rehding, Margaret Schedel, Martin Scherzinger, Stephen Decatur Smith, Gary Tomlinson, Andrew V. Uroskie, and David Wessel for their conversations, questions, support, engagement, and enthusiasm.

Special thanks to my colleagues in the Department of Music at Yale and the graduate students in my seminar on acousmatic sound, in particular, Alexandra Kieffer for the acousmatic voice, Jennifer Chu for the nuns, Kevin Koai for Apollinaire, and Carmel Raz for the French gothic literature; to Seth Brodsky and Seth Kim-Cohen, my dear friends and interlocutors; to David E. Cohen and John Hamilton for help with all things Pythagorean; to Remi Castonguay for help negotiating 18th century French and 21st century library systems; to Gundula Kreuzer and Benjamin Piekut for their helpful comments on drafts at critical junctures; to James Hepokoski for his support, advice, editorial eye, sense of humor, and most of all, his friendship; to Whitney Davis, who has exemplified for me what it means to work in the humanities; to Ian Quinn, who has been the absolute best of colleagues; to Todd Cronan for countless conversations and shared projects over so many years; to my family, both the Kanes and the Handlers; to the editorial and production team of Jessen O'Brien, Mary Jo Rhodes, and Colleen Dunham for their fine work; and to Suzanne Ryan, whose steady hand steered this book from proposal to publication at the speed of sound.

Many ideas in this book were first presented in talks and lectures. I thank all those who raised a hand and asked a question. This book is stronger for all of your inquiry. For the invitation to speak about acousmatic sound, I thank Joseph Clarke and Kurt Forster at the Yale School of Architecture; Sherry Lee and the members of the Opera Exchange at the University of Toronto; Ron Kuivila and Wesleyan University; David Novak and the Columbia University Society of Fellows; Holly Watkins and the Butler School of Music at UT Austin; the graduate students at NYU for their lecture series, "Music, Language, Thought"; the Society for Music Theory and the Music and Philosophy Interest Group; UC Berkeley Art Research Center; Lydia Goehr and the participants in the aesthetics reading group; and the departments of music at the

University of Chicago, Cornell, Penn, New York University, and the University of Wisconsin-Madison.

Earlier versions of parts of this book have appeared elsewhere: "Jean-Luc Nancy and the Listening Subject," *Contemporary Music Review*, 31/5–6 (2012): 439–447, with permission from the Taylor and Francis Group; "Acousmate: History and De-Visualised Sound in the Schaefferian Tradition" *Organised Sound* 17/2 (Fall 2012): 179–188 and "L'objet Sonore Maintenant: Pierre Schaeffer, Sound Objects and the Phenomenological Reduction," *Organised Sound* 12/1 (Spring 2007): 15–24, with permission from Cambridge University Press; and "Acousmatic Fabrications: Les Paul and the Les Paulverizer," *Journal of Visual Culture* 10/2 (August 2011): 212–231, with permission from SAGE.

For permission to reproduce images, I thank Sean Roderick and Universal Music Enterprises, a Division of UMG Recordings, Inc., Annie Segan and the Arthur Rothstein Archive, The Yale University Library, Photofest NYC and Miranda Sarjeant at Corbis Images.

Additional research support was funded by the Mellon Postdoctoral Fellowship in Music at Columbia University, Yale University's Morse Fellowship and the Samuel and Ronnie Heyman Prize for Outstanding Scholarly Publication or Research.

And, as always, there is the amazing Adrienne. This book is dedicated to you.

Sound Unseen

Introduction

About an hour's drive from New Haven lies a small village near East Haddam with an unusual name, Moodus. Derived from the Wangunk term "Machemoodus," meaning "Place of Noises," Moodus possesses a peculiar soundmark.[1] Since the time of the Native Americans, residents of the area have keenly attended to the distinctive sound of tremors and underground rumblings that emanate from a cave located near Mt. Tom. These sounds have come to be collectively known as the "Moodus noises," and it is probably safe to say that no Connecticut phenomenon has inspired more curiosity, speculation, and marvel.

New Englanders have written many accounts of the Moodus noises. The earliest, from settlement days, reported that the Wangunks heard the voices of their gods in the rumblings and tremors. Mt. Tom was a site of divination and, according to folklorist Odell Shepard, those who lived in its proximity had "special access to the Divine." In the noises, the Wangunks "heard the immediate voice of the good spirit Kiehtan and also the rage of Hobbamock."[2] Rev. Stephen Hosmore, the first minister of East Haddam, confirms this view in a letter from 1728. "I was informed," Hosmore wrote, "that, many years past, an old Indian was asked, What was the reason of the noises in this place? To which he replied, that the Indian's God was very angry because Englishman's God was come here."[3] Hosmore's letter also describes the nature and duration of the noises. Not only had he heard them, their presence had been "observed for more than thirty years" by settlers in the region. His vivid description is worth repeating:

> Whether it be fire or air distressed in the subterraneous caverns of the earth, cannot be known; for there is no eruption, no explosion perceptible, but by sounds and tremors, which sometimes are very fearful and dreadful. I have myself heard eight or ten sounds successively, and imitating small arms, in the space of five minutes. I have, I suppose, heard several hundreds of them within twenty years; some more, some less terrible. Sometimes we have heard them almost every day, and great numbers of them in the space of a year. Often times I have observed them to be coming down from the north, imitating slow thunder, until the sound came near or right under, and then there seemed to be a breaking like the noise of a cannon shot, or severe thunder, which shakes the houses, and all that is in them.... Now whether there be anything diabolical in these things, I know not;

but this I know, that God Almighty is to be seen and trembled at, in what has been often heard among us."[4]

According to historian Richard Cullen Rath, both Native American and European settlers in early America understood natural sounds as corresponding to animate sources, "as bridges between visible and equally real invisible worlds."[5] The thundering sounds emanating from Moodus could be heard by both groups as "made by some great spiritual being." Although the religious beliefs of the Native Americans and European settlers led to divergent attributions concerning the divinities heard in the noises, Rath notes that this entire complex of animate natural sound was challenged by a shift in European-American soundways in the 18th century. Due, in part, to the rise of natural scientific inquiry, modern listeners, unlike their 17th-century counterparts, attributed many supernatural features of the natural soundscape to more worldly causes.[6]

Yet, superstition persisted alongside scientific inquiry well into the 18th century. Theories concerning the noises often embellished natural events with ominous forces and supernatural causes. One of the stranger accounts appears in the *Connecticut Gazette* from August 20, 1790. It describes the legend of Dr. Steel, a mysterious visitor drawn to Moodus by the dark enigma of the noises:

> Various have been the conjectures concerning the cause of these earthquakes or Moodus noises, as they are called. The following account has gained credit with many persons—It is reported that between 20 and 30 years ago, a transient person came to this town, who called himself Dr. Steel, from Great Britain, who having had information respecting these noises, made critical observations at different times and in different places, till at length he dug up two pearls of great value, which he called Carbuncles…and that he told people the noise would be discontinued for many years, as he had taken away their cause; but as he had discovered others in miniature, they would be again heard in process of time. The best evidence of the authenticity of this story is that it has happened agreeably to his prophecy. The noises did cease for many years, and have again been heard for two to three years past, and they increase—three shocks have been felt in a short space, one of which, according to a late paper, was felt at New London, though it was by the account much more considerable in this and the adjacent activity.[7]

Like the removal of a growth or tumor that would improve the health of a patient, removal of the carbuncles would make the noises cease, but only temporarily. As fate would have it, a large tremor shook the region the following May. Its effects were felt as far as New York and Boston.[8] While the tremor helped to legitimate Dr. Steel's prophecy in the minds of credulous residents, the validity of his reasoning became far less important than the entertaining story itself. The alchemical Dr. Steel and his discovery of mysterious carbuncles became a standard part of local folklore, inspiring numerous legends, stories, poems, and ballads.[9]

In contrast to Dr. Steel's carbuncular theory, less eccentric explanations of the Moodus noises vied for legitimacy during the 19th century. In 1836, John Warner Barber, the Connecticut historian, offered a theory: "the cause of these noises is explained by some to be mineral or chemical combinations, at a depth of many thousand feet beneath the surface of the earth."[10] Barber's explanation was likely influenced by chemical experiments with reactive sulfur and iron mixtures that were common at the time.[11] In addition, Barber included a description of the sound, reported to him by a local resident: "It appeared to this person as though a stone of a large body fell, underneath the ground, directly under his feet, and grated down to a considerable distance in the depths below."[12] Soon after, in 1841, Barber's theory was challenged by a group of Wesleyan professors. Arguing that an explosion of mineral compounds could not produce an agitation large enough to generate tremors of the size associated with the Moodus noises, they proposed a different hypothesis: electricity. Interruptions in the natural flow of electricity in the Earth's crust would be adequate to produce the devastations and tremors associated with the noises. If electrical forces could rattle the skies in the form of thunder and lightning, the same phenomenon could perhaps explain the noises' powerful underground rumblings and quakings.[13]

By the turn of the 20th century, William North Rice, a professor of geology at Wesleyan, attributed the noises to minor seismic activity: "The noises are simply small earthquakes, such as are frequent in many regions of greatly disturbed metamorphic strata.... The rocks are apt to be in a state of strain or tension, which will from time to time produce such slight vibratory movements as are heard and felt in the Moodus noises."[14] Professor Rice's successor at Wesleyan, Wilbur Garland Foye, developed Rice's theory, attributing the cause of the small quakes to ongoing readjustments of Connecticut's rocky crust after the glacial period.[15] The slow process of glacial retreat released the pressure and strain on the crust underneath, leading to occasional shifts of rock masses near the surface. Foye's hypothesis received some confirmation during the construction of Route 11, about 20 miles east of Moodus. Holes drilled into the bedrock revealed shifting and thrusting consistent with Foye's explanations, although here the release of stress was due to manmade disturbances of the bedrock, not natural forces.[16]

A far cry from the dreadful rage of Hobbamock, "The Moodusites of today listen to the noises with greater equanimity," writes C. F. Price, "because science has solved the mystery."[17] And yet, when they rumble, they still have the power to shake such confidence. While I was writing this book, the noises sounded again. On the evening of March 23, 2011, a loud boom rattled East Haddam and shook houses in the vicinity of Moodus. The sound was thought by many to be an explosion. More than 30 firefighters searched the village and its surroundings for the source of the blast, but found nothing. Craig Mansfield, East Haddam's emergency management director, began to suspect that it was simply the Moodus noises. "You hear old-timers talk about feeling their house shake and hearing loud groans," said Mansfield, "but in all my 23 years with the town, I've never experienced anything like this." The next morning, the U.S. Geological Survey confirmed that a 1.3 magnitude earthquake had struck the region. No damage was reported.[18]

The Moodus noises are acousmatic sounds. Strictly speaking, they fit the standard definition of the term, cited by Pierre Schaeffer and others: "Acousmatic, adjective: a sound that one hears without seeing what causes it."[19] The cause of the noises—whether seismic, chemical, carbuncular, or divine—remains unseen; its sound is an audible trace of a source that is invisible to the listener. Yet, aside from a few vivid descriptions of the sounds themselves, discussions of the Moodus noises (mine included) tend to emphasize two notable features. First, most accounts focus on speculation concerning the source of the sound through the causal ascriptions that historical listeners have made. Because the source of the noises remains obscure, the desire to uncover it generates much of the interest in the sounds themselves. Second, many accounts focus on the various effects the noises have wrought on auditors. Terror, curiosity, bemusement, awe, theophany, wonder—listeners have experienced these feelings, and more, before the noises.

Yet, these two aspects of the Moodus noises—concern with the source of the sound and its effects on the listener—do not squarely align with the use to which Pierre Schaeffer, and the tradition of those directly influenced by him, deployed acousmatic sound. While the canonical account of acousmatic sound is presented in Schaeffer's massive *Traité des objets musicaux*, Michel Chion summarized many of Schaeffer's findings in his authorized *Guide des objets sonores*. The *Guide* provides a synoptic and perspicuous account of Schaeffer's thought by reorganizing his ideas into topical headings. Here is Chion's very first entry:

1) Acousmatic: a rare word, derived from the Greek, and defined in the dictionary as: adjective, indicating a sound that one hears without seeing what causes it.
The word was taken up again by Pierre Schaeffer and Jérôme Peignot to describe an experience which is very common today but whose consequences are more or less unrecognized, consisting of hearing sounds with no visible cause on the radio, records, telephone, tape recorder etc.
Acousmatic listening is the opposite of direct listening, which is the "natural" situation where sound sources are present and visible.
The acousmatic situation changes the way we hear. By isolating the sound from the "audiovisual complex" to which it initially belonged, it creates favorable conditions for reduced listening which concentrates on the sound for its own sake, as sound object, independently of its causes or its meaning (although reduced listening can also take place, but with greater difficulty, in a direct listening situation).[20]

There are three important aspects to consider. First, Chion notes a relationship between the acousmatic experience of sound, which is "very common today," and the ubiquitous presence of modern forms of audio technology, in particular those designed for sound transmission, inscription, storage, and reproduction. Unlike the rare Moodus noises, acousmatic sound is here described as an everyday phenomenon, a result of our immersion in sound flooding from elevators, radios, cars, computers, and stereos. However, it would be incorrect to claim that the acousmatic experience of sound *originates* in modern audio technology. Schaeffer argued that the originary experience of acousmatic sound could be traced back to the school of Pythagoras. Etymologically, the term "acousmatic" refers to a group of Pythagorean disciples known as the *akousmatikoi*—literally the "listeners" or "auditors"—who, as legend has it, heard the philosopher lecture from behind a curtain or veil. According to Chion, Pythagoras used the veil to draw attention away from his physical appearance and toward the meaning of his discourse.[21] The central role of the Pythagorean veil in Schaefferian tradition blocks the *causal* identification of acousmatic experience with modern audio technology in order to make a more striking claim. Modern audio technology does not create acousmatic experience; rather, acousmatic experience, first discovered in the Pythagorean context, creates the conditions for modern audio technology. Radio, records, the telephone, and the tape recorder exist within the horizon first opened by the Pythagorean veil.

Second, Chion emphasizes the relationship between acousmatic experience and the partition of the sensorium. Acousmatic experience entails the "isolation of sound from the 'audiovisual' complex." Hearing is separated from seeing (and the rest of the sensory modalities) and studied for its own sake. This separation is, as Schaeffer calls it, "anti-natural." It requires effort to divide the sensorium given its "natural" intersensorial condition, where multiple senses simultaneously and cooperatively provide information about the environment. According to Chion, one "effect of the acousmatic situation" is that "sight and hearing are dissociated, encouraging listening to sound forms for themselves (and hence, to the sound object)."[22] The acousmatic experience of sound allows a listener to attend to the sound itself, apart from its causes, sources, and connections to the environment. Just as the Pythagorean veil directed the attention of the disciples onto the meaning of the master's discourse, the isolation of hearing from seeing in the acousmatic experience directs the listener's attention toward the sound as such. It allows a listener to grasp the sound itself as a "sound object," a term about which I have much to say in the following chapters. It also affords a special mode of listening that is focused entirely on the sound object, known as "reduced listening."

Third, Chion claims that reduced listening is facilitated when the source of the sound is unseen or hidden, what he calls "indirect listening." However, reduced listening can also take place, albeit with more difficulty, in a "situation where sound sources are present and visible," called "direct listening." What is the relationship between reduced listening and the acousmatic experience (or "indirect listening")? Although there has been much confusion about the precise relationship of these two terms in the discourse on acousmatic sound, the two should be distinguished. By separating a sound from its (visible) source or cause, the acousmatic experience of sound facilitates reduced listening; however, reduced listening can occur in situations where sources are visible and present—situations that are not, strictly speaking, acousmatic. This distinction is not always preserved, and many writers on acousmatic experience treat the terms synonymously. For instance, the philosopher Roger Scruton, in his *Aesthetics of Music*, claims that when listening to music, "we spontaneously detach the sound from the circumstances of its production, and attend to it as it is in itself: this, the 'acousmatic' experience of sound, is fortified by recording and broadcasting, which completes the severance of sound from its cause that has already begun in the concert hall.... The acousmatic experience of sound is precisely what is exploited by the art of music."[23] Even in the direct listening situation of the concert hall, Scruton claims that we are not attending to the source of the sound (*this* clarinet, *that* trumpet). Rather, the musical listener listens to an order that is distinct from the material world of causes, a reduced listening to musical tones irreducible to their sources.

Chion's Schaefferian account of acousmatic sound can be summarized under three headings: technology, the division of the sensorium, and reduced listening. The acousmatic experience is encountered in certain forms of ancient and modern technology (Pythagorean veil, architectural screen, tape recorder, loudspeaker, etc.) that divide hearing from the rest of the sensorium. This division encourages reduced listening—a way of attending aesthetically to sounds as such, apart from their worldly causes. The purpose of the acousmatic experience in the Schaefferian tradition is, as Chion says, to "change the way we hear," to draw attention away from the source of the sound (whether visually present or not) and onto its intrinsic audible properties. The source of the sound is severed from its audible effects, so that the latter can be studied separately, placed into morphological categories or systematically integrated into musical compositions. The separation of the senses is purposive, a way of discarding the sonic source in order to orient attention toward aesthetically appreciated sonic effects alone.

While the Moodus noises are acousmatic sounds, according to the standard definition of the term, they are not typically listened to for their aesthetic properties. If anything, their sonic properties, like those described in Hosmore's letter, are ultimately used to provide clues about the potential source of the sound. The natural conditions at Moodus make the source of the noises invisible to a listener and thus aid in splitting the sensorium, creating an experience of hearing without seeing. Yet, the aesthetic orientation toward the sound of the noises does not follow. In a case like the Moodus noises, an aesthetic orientation toward their sound is not the relevant mode of audition. Yet, I would argue that the acousmatic character of the sounds matters, in that the enigma of their source—its invisibility and uncertainty—is a central feature of the experience.

There are many cases, like the Moodus noises, where such sounds are neither heard primarily as aesthetic objects, nor capable of being made intelligible in aesthetic terms. For instance, the aesthetic orientation cannot address the central role that acousmatic sounds have played in Judeo-Christian religion, from the invisible voice of the Jewish God to the Catholic confessional. Nor can this aesthetic orientation account for the role played by acousmatic sounds in psychoanalysis; the psychic effect of the disembodied, acousmatic voice has not only contributed to the spatial arrangement of the analytic session (where, in the famous photos of Freud's office, the analyst is tucked away from the analysand's view), the position of the superego in Freud's late topographical models, and the development of psychoanalytic technique, but, in the form of the "sonorous envelope" (the prenatal experience of sound in the mother's womb), is claimed to play a central role in the formation of the subject.[24] Nor can the aesthetic orientation deal with the full employment of acousmatic sounds in the *production and performance* of music. This would include not only the positioning of singing nuns behind grilles and grates during the era of *clausura*, the setting of offstage voices and instruments in opera to produce divine effects, or the use of darkened auditoriums to produce quasi-religious effects in German *Dunkelkonzerte* or Georg Friedrich Haas's contemporary compositions, but also the application of architectural techniques to hide the orchestra at Bayreuth and elsewhere.[25] Nor can the aesthetic orientation make sense of the way that acousmatic sounds have come to be a topic of concern in the humanities: from Chion's invention of the cinematic *acousmêtre*; to Carolyn Abbate's analysis of acousmatic sounds that conjure unheard, ineffable, metaphysical events; to Mladen Dolar and Slavoj Žižek's investigation of the acousmatic voice as a form of social interpellation; to recent work in literary studies—topics like "acousmatic blackness" or the use of acousmatic sound in Ralph Ellison's work—which sit at the nexus of critical race theory and sound studies.[26]

To understand the significance of acousmatic sound in these various domains, we must listen to them anew. While, undoubtedly, the aesthetic orientation toward acousmatic sound has contributed to historical and current practices of musical listening, exclusive focus on it limits our ability to consider acousmatic listening—that is, the experience of acousmatic sound—as a cultural practice. In this book, I attempt to theorize acousmatic sound differently. As an alternative to the aesthetic approach to acousmatic sound, I take the position that acousmatic listening is a shared, intersubjective practice of attending to musical and nonmusical sounds, a way of listening to the soundscape that is cultivated when the source of sounds is beyond the horizon of visibility, uncertain, underdetermined, bracketed, or willfully and imaginatively suspended. The term "acousmatic listening" should be understood as a rubric intended to capture a set of historically situated strategies and techniques for listening to sounds unseen. Thus, there is a double entendre in the subtitle of this book. Instead of a book that describes the use of acousmatic sound in compositional and aesthetic terms (i.e., a book on the theory and practice of "acousmatic music"), this book is written to develop a theory of acousmatic listening as a historical and cultural practice, one with clearly defined characteristics.[27]

To begin moving in this direction, I present a few central propositions that I develop in the chapters that follow. While I am aware that I must forgo offering the kind of substantial evidence and argumentation for these propositions provided later in the text, I hope that they will serve as points of orientation for the reader and offer clear contrasts with the Schaefferian tradition of acousmatic sound.

1. I work with a model of sound that has three necessary components: source, cause, and effect. Sounds, as we know, only occur when one object activates or excites another. For instance, a rosined bow is rubbed against a string or cymbal; air is forced across a cane reed or a vocal tract (then shaped by a mouth, tongue, and teeth); or a raindrop collides with a windowpane. The interaction of a source (cymbal, string, reed, vocal tract, or windowpane) with a cause (rosined bow, moving air, raindrop) produces an audible effect. We can formulate this as a proposition: Every sonic effect is the result of the interaction of a source and a cause. Without this interaction, there is no emission of sound.[28]
2. Just because a sonic effect is the result of an interaction of a source and a cause does not entail that a listener is certain about the source and cause based on hearing the sonic effect alone. Typically, the environmental situation will aid in determining the source and cause of the sound.[29] But, there might be cases where I cannot determine the source from the effect, or the effect is ambiguous. For instance, as I walk across a college campus, I might hear various chirps from above, which I know come from birds in the trees, although I may not be able to see them through the foliage. Normally, I have no worries about making such inferences. But, perhaps, as I walk past the art school, I hear a chirp that sounds slightly amiss. It is possible that the chirp I hear is not from a bird, but from a little electronic circuit hidden in a tree—a clever piece of sound art—designed to imitate the sound of a bird.[30] If I spot a loudspeaker tucked away in the tree, near the location from which the sound emanates,

I will likely be satisfied that I have discovered the source of the sound. The same thing happens if I see a bird suddenly fly away from the location of the sound. Although one might be tempted to treat this example hyperbolically, as a case of global skepticism—and thereby immediately assert that I cannot know anything about the world because it may always turn out to be otherwise than what I expected—I draw a much more humble conclusion. An auditory effect, apart from the environmental situation in which it is located and our ability to explore that situation with all our senses, is often insufficient for determining its cause. To formulate a second proposition: The sonic effect, by itself, underdetermines its source and cause.

3. The underdetermination of source and cause motivates a reification of the sonic effect. By bracketing an effect from its source and cause, I transform a sound from an event into an object. The autonomy of a sonic effect is constituted only when the gap between the effect and its source or cause is disregarded. In the aesthetic orientation of acousmatic sound, that is precisely the point. The autonomous sound, bereft of its source, is then integrated into the virtual world of musical composition; shedding its source, it can fully participate in the virtual connection of tone to tone, in the metaphorical gravity of tonal-harmonic organization, or in the expressive analogies of musical sound with emotional states. The autonomous sonic effect becomes a sound object. At the same time, there is a countervailing tendency, perhaps ineluctable, to find a source for autonomous sonic effects.[31] Steven Connor argues that "human beings in many different cultural settings find the experience of a sourceless sound uncomfortable, and the experience of a sourceless voice intolerable... the disembodied voice must be habited in a plausible body." The autonomous sound or voice is supplemented with an imaginary body. Connor refers to this phenomenon as the *vocalic body*, the idea of "a surrogate or secondary body, a projection of a new way of having or being a body, formed and sustained out of the autonomous operations of the voice."[32] For example, we could say that the Wangunks gave a *vocalic body* to the Moodus noises by imagining them to be the voices of Kiehtan or Hobbamock. Furthermore, we can move away from the voice and generalize Connor's term by referring to the production of a *sonic body* elicited by acousmatic sound. A third proposition: Acousmatic sounds encourage the imaginative projection of a *sonic body*.

4. If acousmatic listening is a practice, then it should be possible to trace its history. In the Schaefferian tradition, there have been attempts to talk about the history of acousmatic sound before, yet they have all foundered on the same methodological problem: The history of acousmatic sound has been mistaken for a history of the word "acousmatic." Given the rarity of this word, one ends up with only a piecemeal and diffuse historical account. One reason against privileging the presence of the word "acousmatic" as a central criterion for a historical account is that, I would argue, historical agents have not often recognized the extent to which they employed the practice of acousmatic listening. Acousmatic listening, while *audible*, has not, in all cases, reached a level of explicit *audibility*, in the sense that it is not always recognized as part of a culture or style of listening.[33] Yet, if one considers acousmatic listening as a practice—that is, a way of listening to the soundscape that is cultivated when the

source of the sounds is beyond the horizon of visibility, uncertain, underdetermined, bracketed, or willfully and imaginatively suspended—it is surely the case that acousmatic listening was alive and well, even in eras when the term "acousmatic" did not exist.[34] To write a history of acousmatic listening would then mean to gather significant instances of privileging hearing over seeing, of cultivating situations where sounds are detached from their causal sources, and of techniques for listening to sounds unseen in order to tell a story about how such practices have affected views about music, the senses, philosophy, and ourselves. Making acousmatic listening explicit should be a priority of any history or theory of the topic. A fourth proposition: The history of acousmatic sound is not a history of the word "acousmatic." It is a history of the practice of acousmatic listening.

5. If acousmatic listening is a practice, one should be able to talk about its meaning or the way that it conceptually articulates the audible world of those who employ it. While one can indeed talk about the meaning of the practice, one should be careful not to treat its meaning like an essence. The meaning of the practice cannot be specified apart from the actual context and use to which it is employed. However, this is where its history becomes pertinent—for it allows us to track the replications and propagations of the practice from agent to agent, and thus, to find central cases, norms, deviations, and patterns. One central, replicated feature of acousmatic listening appears to be that underdetermination of the sonic source encourages imaginative supplementation. In many cases, the *sonic body* projected onto acousmatic sound is taken to be transcendent. Acousmatic listening is often deployed in order to grant auditory access to transcendental spheres, different in kind from the purely sonic effect—a way of listening to essence, truth, profundity, ineffability, or interiority. However, we cannot specify *precisely* the kind of transcendence heard in the sound or the exact meaning of the practice without appealing to the specific context, culture, and experience of the agents involved. In fact, there is no guarantee that any numbers of agents in the same contexts employ the practice in identical ways. Thus, although we can articulate some basic conditions for hearing a sound acousmatically, at a certain point, a theory of acousmatic sound must give way to the social and historical agents who employ it as a practice. This leads to my fifth proposition: The meaning of the practice of acousmatic listening cannot be defined in abstraction from those who employ it.

Finally, a few words about the organization of this book. As the subtitle indicates, it is conceived as a *theory and practice* of acousmatic sound. While various chapters may emphasize the history of the practice more than the theory, or vice versa, the book was written in the form of a continuous argument, and I think it is best understood in that manner.

In part 1, "The Acousmatic Situation," I offer a philosophical reading of Schaeffer's concept of *l'acousmatique* and its special relationship with *l'objet sonore*, the sound object. I outline the development of Schaeffer's thinking, from the initial moments of *musique concrète* to the mature project of the *Traité*. My focus is on Schaeffer's employment of Husserlian phenomenology—in particular his use of the transcendental reduction, or *epoché*—to define the acousmatic reduction and its privileged object, the sound object. After presenting Schaeffer's theory, I raise a set of objections to be explored in the subsequent chapters: (1) I argue that Schaeffer's method does not allow him to adequately characterize the history of acousmatic listening; driven by a phenomenological account of history as "originary experience," Schaeffer's thinking about music, sound, and technology is ahistorical and mythic; (2) I argue that Schaeffer's theory does not give adequate consideration to the cause, the source, or even the production of sound; thus, Schaeffer offers a "phantasmagoric" view of musical material, one that occludes its manner of production; (3) I argue that there is an ontological problem in Schaeffer's characterization of the sonic effect as a sound object.

I revisit the history of acousmatic sound in part II, "Interruptions," by way of a critique of its standard historiography. As noted above, the history of acousmatic sound in the Schaefferian tradition is often confounded with a history of the word "acousmatic." This places undue weight on two contexts: the Pythagorean school, with its veil and *akousmatikoi* (or silent disciples), and the rare French term *acousmate*. In chapter 2, I investigate the first context by posing a simple question: When and where did the Pythagorean veil first emerge? A patient investigation into the ancient sources reveals a history of the veil (and the acousmatic disciples who sat opposite it) that cannot be reconciled with the Schaefferian account. This exposes the mythic use to which the veil has been employed in the Schaefferian tradition and disallows any phenomenological claims about the veil as initiating the originary experience of acousmatic sound. In chapter 3, I investigate the second context, the word *acousmate*. Again, I pose a simple question: Where did this word come from and how does it relate to the Pythagorean tradition? After describing the context in which the word was coined and first used, I demonstrate that the word originally had nothing to do with the Pythagorean school. Then, by tracing its dissemination into various contexts—medical, psychological, and literary—I pinpoint the moment when Pythagorean and French contexts were first associated.

In part III, "Conditions," I move beyond the Schaefferian tradition and start to sketch an alternative historical account of acousmatic listening as a practice (chapter 4). In particular, I argue that the history of modern acousmatic listening is sutured to a lineage of musical phantasmagoria that reaches fruition with the birth of Romanticism, the aesthetics of absolute music, and Wagnerian architectural reforms of the concert hall. I consider a wide variety of evidence: from Schopenhauer's use of bodily techniques designed to ready the listener for the experience of music's disclosure of profound metaphysical truths; to architectural projects (realized and unrealized, from the grilled galleries of Italian churches to the hidden orchestra at Bayreuth) for performance spaces where the musician's body would be partially or entirely obscured in order to preserve music's transcendental nature from contamination by empirical sources; to literary and philosophical fantasies where absorbed listeners shut their eyes in order to disclose and experience music's transcendental power. In all cases, the auditory effect is separated from its source; the latter is phantasmagorically occluded so that the former can be taken as transcendent, as manifesting a virtual or spiritual world separate from the mundane. These claims form the basis of a set of theoretical conditions about the production of acousmatic phantasmagoria.

I also argue that these conditions are prolonged in Schaeffer's phenomenology of the sound object and in his works of *musique concrète*. But must all *musique concrète* be phantasmagoric? In the interlude between chapters 4 and 5, I consider an internal critique of Schaeffer's work by his pupil, the composer Luc Ferrari. Through a discussion of his piece *Presque rien*, I trace how Ferrari breaks the grip of Schaefferian phantasmagoria by self-consciously emphasizing the materiality of recorded sound and producing an aesthetic situation that encourages reflection upon the affordances of recording devices used in the production of *musique concrète*.

In chapter 5, I develop an alternative theory of acousmatic sound, by way of a close reading of Kafka's tale "The Burrow." Extrapolating from Kafka, I advance arguments for some of the central propositions mentioned above. This chapter is the theoretical core of the book. It attempts to rethink the terms of acousmatic sound apart from the ontology of the sound object. In the final sections of chapter 5, I test these arguments and develop their implications against a variety of examples from music, literature, film, sound studies, and philosophy.

In part IV, "Cases," I continue to test the theory developed in part III. Chapter 6 is a case study of guitarist Les Paul. The personal motivations behind Paul's overdubbed recordings, his unusual production and studio techniques, his relentless invention of electronic gadgets, and his challenges with live performance provide a matrix wherein many of the book's central themes intersect: technology, recording, subjectivity, identity, underdetermination, the uncanny, the Pythagorean veil, and the separation of the senses. I argue that Paul's career was shaped by his encounters with acousmatic sound and hone in on a central problem: How can one perform acousmatic music live while maintaining the underdetermination of the source by the effect that is the hallmark of acousmatic sound?

In chapter 7, I focus on the acousmatic voice. Taking my cue from Slavoj Žižek and Mladen Dolar, who explicitly describe the Lacanian "object voice" as acousmatic, I argue that the acousmatic voice has played an unacknowledged but crucial role in Husserl and Heidegger's philosophical theories of the voice. Lacanian theorists like Žižek and Dolar prolong this tradition. By closely reading Dolar's *A Voice and Nothing More* in terms of the theory put forth in part III, I expose a set of critical problems and inconsistencies in the Lacanian treatment of the voice. In particular, I argue that Dolar reifies the acousmaticity of the voice, making it into a permanent condition, and that his treatment of the acousmatic voice is phantasmagoric, masking the technique at play in the psychoanalytic session.

PART ONE

The Acousmatic Situation

1

Pierre Schaeffer, the Sound Object, and the Acousmatic Reduction

IMPROVISED ONTOLOGY

In 1948, working in the studios of Radiodiffusion Française, Pierre Schaeffer began to keep a set of journals describing his attempt to create a "symphony of noises."[1] These journals, published in 1952 as *A la recherche d'une musique concrète*, portray Schaeffer's initial experiments as anything but systematic. Anxious to compose a concrete music yet perpetually dissatisfied, he roves from conventional instruments to unconventional tools, from the recording studio to the booth, percussing and sounding object after object to find a suitable candidate. The list of objects is a veritable *abécédaire*: alarm clocks, bicycle horns, birdcalls, bits of wood, clappers, coconut shells, klaxons, organ pipes, rattles, vibrating metal strips, whirligigs; then recordings of bells, buffer stops, orchestras, piano improvisations, trains, xylophones, and zanzis. Throughout his experiments, Schaeffer remains in the grip of two recurrent desires: a compositional desire to construct music from concrete objects—no matter how unsatisfactory the initial results—and a theoretical desire to find a vocabulary, solfège, or method upon which to ground such music. In those early days, Schaeffer's improvised compositional techniques were indissociable from an improvised ontology, not only in search of a concrete music but a basic theoretical unit upon which to compose such music.

One constant in Schaeffer's work, from the early days of *A la recherche* to his mature *Traité des objets musicaux* and *Solfège de l'objet sonore*, was his fixation on the word "object." Alhough the objects kept shifting, the term "object" persisted like an *idée fixe*. The trajectory of the term over the course of 1948 is revealing. In March, an "object" refers to a physical-material thing—a source for the production of sound: "Back in Paris I have started to collect objects. I have a 'Symphony of noises' in mind...."[2] By April, the object has acquired a modifier. Now a "sound object," it still refers to the physical-material source and not the effect of the sound: "I am amongst the turn-tables, the mixer, the potentiometers...I operate through intermediaries. I no longer manipulate sound objects myself. I listen to their effect through the microphone."[3]

In early May, now working with recordings of trains made at the Batignolles station in addition to stock recordings, the sound object is supplemented with a new term, the "sound fragment." Unlike the physical-material sound object, a sound fragment designates a bit of recorded sound, the "effect" emitted from a sound object and engraved into a spiral groove:

> I lower the pick-up arm as one rhythmic group starts. I raise it just as it ends, I link it with another and so on. How powerful our imagination is! When in our minds we pick out a certain rhythmic or melodic outline in a sound fragment like this, we think we have its musical element.[4]

A few days later, Schaeffer exploits the infinite repeatability of the fragment to distinguish it from the sound object, the physical-material cause. "Repeat the same sound fragment twice: there is no longer event, but music."[5] Repetition musicalizes the sound fragment by removing the dramatic and anecdotal traces of its original causal context.

Identifying the sound fragment was an important step in breaking the grip of the physical-causal source. The recorded fragment, not the physical source, acquired the plasticity of compositional material. By removing the attack from a recording of a bell, Schaeffer noted, "the bell becomes an oboe sound."[6] Similarly, "if I compensate for the drop in intensity with the potentiometer, I get a drawn-out sound and can move the continuation at will."[7] The transformations of the "cut bell," or *cloche coupée*, produced unexpected auditory results and revealed the potential of recorded sounds when considered separately from their physical-causal sources.

By May 15th, Schaeffer's work with both bells and trains led to a generic conclusion about the sound fragment. "For the 'concrete' musician there is no difference between the cut bell and the piece of train: they are 'sound fragments.'"[8] The sound fragment reduces the specific difference of the physical-material source to the generality of the sample. Yet, in the very same journal entry, the "sound object" returns, albeit transformed, to reassert its priority. Schaeffer writes:

> I have coined the term *Musique Concrète* for this commitment to compose with materials taken from "given" experimental sound in order to emphasize our dependence, no longer on preconceived sound abstractions, but on sound fragments that exist in reality, and that are considered as discrete and complete sound objects, even if and above all when they do not fit in with the elementary definitions of music theory.[9]

The sound object can no longer be understood as the material-physical cause of the sound in distinction to the effect captured on disc. In Schaeffer's improvised ontology, the sound object has leapfrogged over the fragment to assume a new significance. More than simply a sample or bit of recorded sound, the sound object now suggestively appears to designate something "discrete and complete," the fruit of a mode of "considering" or listening to the fragment torn from the whole. It seems to be the disclosure of a minimal unit of heard sound upon which to ground the project of *musique concrète*, a novel discovery that cannot be assimilated into "music theory."

Despite the constant drift of Schaeffer's terminology over the course of 1948, the claim that a sound object is inassimilable into music theory is intriguing. The earliest works of *musique concrète* were made using special phonograph discs with a locked or closed groove (*sillon fermé*) to create repetitive units. Eventually, these discs were replaced with magnetic tape, where small loops or isolated segments would undergo electronic processing or be repeated without variations. These small pieces of recorded sound, which often underwent elaborate manipulation, filtering, signal processing, and editing, acted as stand-ins for the motive, the smallest musical gestalt deployed to organize the surface of musical works. However, these bits of sound were radically discontinuous with the motive, in that composition with sound objects displaced the note or tone as music's fundamental ontological unit.

By choosing to identify his practice as *musique concrète*, Schaeffer was trying to differentiate his approach from traditional practices of musical composition bound to the note.[10] Abstract music, which Schaeffer contrasted with *musique concrète*, was music that began with the note, organized its musical thinking in terms of the note, and then draped it in the guise of acoustic or electronic sound. Abstract music gave the ideal note a sonorous body through the realization of scores by performers or engineers. It began silently in the head and ended in the vibrating garment of sound. German *elektronische Musik*, a child of serialism, was just such an abstract music. With its "rules...formulated like an algebra," Schaeffer disparagingly referred to it as "music *a priori*."[11] Concrete music was to be the exact opposite—a music that began with sounds recorded from the world and sought to perceive in them (and abstract from them) musical values. The emphasis was placed on listening; the ear would have to train itself to hear these new musical values unique to the sonic materials deployed.

The "sound object," first conceived in the improvised ontology of Schaeffer's experiments in the spring of 1948, would continue to undergo modification and explication for the next two decades—but it always retained the features of discreteness and completeness that characterized its initial leap ahead of the sound fragment.[12] In explicating and clarifying his theory of the sound object, Schaeffer introduced the concept of the acousmatic. "The sound object," Schaeffer tersely states, "is never revealed clearly except in the acousmatic experience."[13] In what follows, I try to show why this is indeed the case. To do so, I will explicate Schaeffer's mature theory of acousmatic experience, the sound object, and reduced listening (*écoute réduite*) as presented in the *Traité des objets musicaux*. This theory is cast in explicitly phenomenological terms, and I argue that Schaeffer's phenomenology is much closer to Husserl than it is to Schaeffer's French contemporary, Maurice Merleau-Ponty.[14] For without a good understanding of the Husserlian preoccupations of Schaeffer's work, one cannot adequately characterize the relationship between acousmatic experience, the sound object, and reduced listening. Once those various parts of Schaeffer's mature theory have been distinctly separated, the theory and practice of acousmatic listening—the real focus of interest in this book—can begin to be addressed.

DOING AND KNOWING

In 1966, after 15 years of work, Pierre Schaeffer published the *Traité des objets musicaux*. The first draft, which was stolen along with Schaeffer's luggage in Turin, had been rewritten four times in those 15 years. According to Schaeffer, the text had become a veritable "thinking machine."[15] The *Traité*, which is extraordinarily broad in scope, represents the summation of Schaeffer's research into *musique concrète*, containing reflections on the aesthetics of music, views on the nature of musical instruments and electronic studio tools, psychoacoustic findings, typologies and morphologies of sound, pedagogical recommendations, anthropological and ethnological considerations of the origins of music, and a bevy of other topics too numerous to mention here. Throughout the text, phenomenology is employed not merely as a method, but also, and more importantly, as a kind of commitment that may have indeed been present from the very beginning, only coming into focus slowly and patiently.

In the *Traité*, Schaeffer writes, "For years, we often did phenomenology without knowing it, which is much better than... talking about phenomenology without practicing it."[16] Rather than simply clinching Schaeffer's avocation for phenomenology, this suggestive sentence opens a series of questions about the relationship between phenomenology and Schaeffer's work as a theorist and composer. If doing phenomenology is distinct from knowing it, how did Schaeffer's actions compare with his method? When Schaeffer began to realize that what he was doing was phenomenology, in what ways did this realization alter his practice? Considering the varieties of phenomenology available to Schaeffer—Husserlian, Heideggerian, Sartrean, Merleau-Pontian—it is not trivial to inquire about what kind of phenomenology Schaeffer was unknowingly doing. More provocative than illuminating, Schaeffer's tantalizing sentence needs further qualification.

Makis Solomos argues that Schaeffer's style of phenomenology is much closer to Merleau-Ponty than Husserl, that the *Phenomenology of Perception* played a "capital role" in introducing Schaeffer to phenomenology, "offering an immediate and quasi-poetic introduction" to the discipline.[17] Indeed, striking similarities can be found between the *Phenomenology* and the *Traité*. For example, Merleau-Ponty writes, "To return to the things themselves is to return to that world which precedes knowledge"; was this not also the world to which Schaeffer sought to return, a world of concrete sounds prior to the signification and sense that such sounds accrued through musical and cultural usage?[18] Do we not hear the echo of Merleau-Ponty's view that "we shall find in ourselves, and nowhere else, the unity of and the true meaning of phenomenology" in Schaeffer's claim that "man describes himself to man, in the language of things"?[19] Like the gestalt figures that litter the pages of Merleau-Ponty's text, are we not supposed to find in Schaeffer's explorations of the locked groove (*sillon fermé*) and the cut bell (*cloche coupée*) small figurations of a much larger field—namely, a field of listening understood not simply as the physiological response to an auditory stimulus but as a field of sound objects intentionally constituted by the subject through various modes of listening? Even in Schaeffer's tantalizing sentence, one can sense him standing in the footprints of Merleau-Ponty, who famously proclaimed "that phenomenology can be practiced and identified as a manner or style of thinking, that it existed as a movement before arriving at complete awareness of itself as a philosophy."[20]

The striking similarities cannot be discounted. But such similarities do not tell the whole story of Schaeffer's relationship to phenomenology. Although Solomos is correct to assert that *Phenomenology of Perception* introduced Schaeffer and others of his generation to phenomenology, perhaps even giving it a general orientation and persuasive power, I would argue that Schaeffer's reading of Husserl's texts—*Ideas, Formal and Transcendental Logic*, and the *Cartesian Meditations*—was more significant in influencing his actual phenomenological practice. Throughout the *Traité*, Schaeffer remains quite close to the letter of Husserlian phenomenological orthodoxy, often calling upon it when trying to articulate his views on the sound object, reduced listening, and the acousmatic field. Consistently, Schaeffer deploys techniques that are Husserlian in character: the transcendental-phenomenological reduction, the eidetic reduction, imaginative free variation, and the reactivation of originary experience.[21]

In chapter 15 of the *Traité*, entitled "Objects and Structures," Schaeffer explicitly claims his phenomenological inheritance. The tantalizing sentence about knowing and doing phenomenology opens a subsection on "The transcendence of the object." The transcendent object addressed here is, indeed, the sound object, first identified in the early days of *musique concrète*. But after encountering Husserl's theory of objects, Schaeffer realized that the sound object, first introduced in the improvised ontology of *In Search of a Concrete Music*, could be systematically defined. According to Schaeffer, "only after the fact did we recognize a conception of the object which has been presupposed by our research, [a conception] circumscribed by Edmund Husserl with a heroic demand for precision to which we are far from claiming."[22] Explicitly retrospective, the theory of the sound object presented in the *Traité* is intended to show how Schaeffer's first intuitions about the sound object were congruent with the phenomenological theory of objects.

So what is Husserl's conception of the object? An object must not be mistaken for an entity. In everyday parlance, these two terms are often synonymous, but there is an important distinction to be made. An entity refers to an externally existent thing. An object only comes into being when it is cognized, when it is something capable of being apprehended by a subject. One could imagine a world full of material entities, but it would lack objects unless a subject was to be conscious of them. Moreover, if the consciousness of the subject were simply a stream of distinct experiences, each unsynthesized and disconnected from the last, no object would emerge from that manifold.[23] Husserl is interested in the necessary conditions under which objects are possible. One necessary condition is that there be a subject available to cognize them.

A factor motivating Husserl's theory of objects is his desire to find a single ontology that covers not only objects presented in sensuous perception but also logical and mathematical objects, which cannot be directly, sensuously perceived. A centaur, a proposition, or a formula is as much of an object for Husserl as a horse or a man. The difference between the centaur and the horse depends on its mode of presentation. While the former cannot be given in the mode of direct sensuous presentation—we cannot vividly see a real centaur as we can see a horse standing in a field—it is available to the subject in imagination. Perception, desires, memory, fantasy, and imagination are all considered different modes of presenting objects to a subject. Whether I perceive a man walking across the street, recollect it later that day, or use him as an example in a mathematical word problem, the objectivity of the object remains the same, regardless of the object's mode of presentation. Moreover, whether or not entities exist—they might indeed—is irrelevant to Husserl's investigation.[24]

After citing Husserl's "conception of the object," Schaeffer poses a question about the objectivity of the object. "What are the conditions which permit us, as well as others, the recognition of *objectivity*?"[25] Schaeffer is really asking two questions. First, under what conditions can an object be identified? Second, how is it that objects can be intersubjective? In regard to the first question, Schaeffer cites "a well-known passage" from *Ideas*, where Husserl uses the example of a table to explain the difference between an entity and an object.[26]

> Let us start with an example. Constantly seeing this table and meanwhile walking around it, changing my position in space in whatever way, I have continually the consciousness of this one identical table as factually existing "in person" and remaining quite unchanged. The table-perception, however, is a continually changing one; it is a continuity of changing perceptions. I close my eyes. My other senses have no relation to the table. Now I have no perception of it. I open my eyes; and I have the perception again. *The* perception? Let us be more precise. Returning, it is not, under any circumstances, individually the same. Only the table is the same, intended to as the same in the synthetical consciousness which connects the new perception with the memory.... The perception itself, however, is what it is in the continuous flux of consciousness and is itself a continuous flux: continually the perceptual Now changes into the enduring consciousness of the Just-Past and simultaneously a new Now lights up, etc. The perceived thing in general, and all its parts, aspects, and phases...are necessarily transcendent to the perception.[27]

The table is always seen from some particular perspective. From no single point are all parts of the table visible simultaneously. In order to see those parts that are invisible from *this* location, I must circle around to the backside of the table; but from *that* location, I can no longer behold what I saw from *this* location. Perceptually, I am presented with a stream of various perspectives, each unique and distinct from the last. Husserl refers to this stream of perspectival views as a series of "adumbrations" (*Abschattungen*).

How is the table ever known as the same? If we take the series of adumbrations as such, we have only a series of multiple acts of consciousness—each registering the look of the table at a particular place and time, from a certain vantage point, or under specific illumination. Yet, nothing immanent to the stream of adumbrations unifies them. Husserl argues that the identity of the object is provided through an act of consciousness, which synthesizes the stream of adumbrations. As each new percept is connected to the one just past and grasped as a whole, an object emerges that can be identified as the same across a variety of acts of consciousness. I can perceive the table-object, but I can also imagine it, narrate a story about it, or hold various beliefs about its provenance. In each of these acts, the object I intend is the same. If we hold ourselves to the flow of adumbrations, we have only a series of perceived *qualities*, but through the synthesis of these qualities, we are able to posit the *identity* of the object, as something that transcends the stream of adumbrations. The act of mental synthesis, which Husserl refers to as a *noesis*, is correlated with an intended object, a *noema*, irreducible to any single adumbration.

When Schaeffer speaks of the "transcendence of the object," he means it in the sense that the object I intend is not immanently contained in the stream of perceptual adumbrations. Schaeffer writes, "lived particulars [i.e., adumbrations] are the multiple visual, audible, tactile impressions which succeed one another in an incessant flux, across which I tend towards a certain object, I 'intend' it [*je le 'vise'*] and the diverse modes according to which I connect myself to this object: perception, memory, desire, imagination, etc."[28] This is Husserl's theory in miniature. Schaeffer describes the flow of adumbrations, an object that comes into being from out of this flow due to its constitution by a subject's intentionality, and the possibility that these objects may appear under various modes of presentation. Yet, the main focus is on the transcendence of the object. To clarify the nature of this transcendence, Schaeffer compares a perceptual object, like a table, with an "ideal object" like "a mathematical theorem."[29] In either case, the object can be recollected after some interval of time, and my original experience and memory will refer to one and the same object, whether table or theorem; but the mathematical theorem is not individualized in time and space, because the theorem is not dependent upon having been encountered at some particular spatiotemporal location crucial for establishing identity. All instances of the theorem are necessarily identical in a way that all instances of a table are not. The transcendence of objects, whether ideal or perceptual, is demonstrated by the fact that the subject can refer to them again and again, in various modes of presentation and at different times. The objectivity of an object depends on this kind of repeatable reference.

There is a classic example involving a transposed melody that is often deployed to illustrate this point.[30] Take, for instance, a melody played on a violin and a transposed version in which none of the pitches are the same as in the original. Despite the transposition and its wholly new set of adumbrations, the two melodies are recognizable as "the same." The object in both cases is identical. Thus, an object is not the same as a physical-material entity, which, from a scientific perspective, causes my perceptions. Being the correlate of an act of synthesis on my part, an object is irreducible to any particular adumbration, or even all of them taken together. To grasp the difference between the stream of auditory adumbrations and an object, an appeal must be made to the manner in which the auditory event is experienced. Melodies, cries, harmonic progressions, samples, or other sonic events are not experienced as a discrete array of auditory perceptions; rather, according to the phenomenologist, they are experienced as transcendent objects possessing distinct boundaries, durations, identifying qualities, and properties. The phenomenologist of sound does not deny that there is a stream of auditory adumbrations; rather, the focus is on how parts of the stream are primordially grasped as a unity—as a constituted object, or set of objects, transcending any particular adumbration. The transcendent object grounds the possibility of hearing *the same thing* across the multiple acts of listening by a single subject, despite variations in location, attentiveness, knowledge, or fluctuations in the acoustic signal. The transcendental unity of the sound object is a noematic correlate to a synthetic noetic act by the listening subject. This is the background to Schaeffer's claim that "it is *in my experience* that this transcendence *is constituted.... To each domain of objects corresponds thus a type of intentionality. Each of their properties depends on acts of consciousness that are 'constitutive,' and the object perceived is no longer the cause of my perception. It is 'the correlate.'*"[31]

Schaeffer's second question concerns the problem of intersubjectivity: Given that the object is constituted by an intentional act of the subject, how is it that multiple subjects can intend the same object? Rather then provide an argument, Schaeffer asserts that the transcendence of the object is shared by multiple subjects, thus presupposing a shared objective world:

> The object transcends not only the diverse moments of my individual experience but the whole set of these individual lived experiences: it [the object] is placed in a world that I recognize as existing for all. If I direct myself towards a mountain, it appears to me as the same...across the multiplicity of my points of view; but, I also admit that the companion who marches at my side is directed towards the same mountain as I am, while I have reason to think that he has a different view of it. The consciousness of an *objective world*...is presupposed.[32]

What holds good for perceptual objects (like mountains) also holds for the sound object, but exactly how the sound object grounds the agreement and coordination between different listeners remains unexplained.[33] Schaeffer collapses two situations into one: He conflates (1) the condition that allows one listener to hear the same object several times and identify it as the same with (2) the condition that allows various subjects to correlate their experiences of one and the same object. The noetic-noematic constitution of objects is conflated with intersubjective agreement. Because an object transcends the stream of adumbrations from which it is constituted, and thus is bound to no specific empirical fact, Schaeffer treats this nonempirical foundation as if it adequately guarantees, or at best permits, the sharing of objects between multiple subjects. Yet, this is a very slender basis upon which to ground an account of the intersubjective experience of objects.

Although Schaeffer lacks a thorough account of intersubjectivity, one should not be too critical, for he lacked a good model to emulate. Husserl too was unable to give an adequate account of the nature of intersubjectivity. In the last of his *Cartesian Meditations*, Husserl argues for the importance of empathy as a capacity whereby the subject is able to imagine or project an inner life onto the minds of others. Yet, Husserl spends little time addressing the question of how the objects of others can be brought to a "higher psychic sphere" where shared cultural products and ideas can be shared by a community.[34] The problem of intersubjectivity preoccupied Husserl in his later writings and became one of the great bugbears for phenomenological philosophers. Husserl's famous notion of the lifeworld, or *Lebenswelt*, can be seen as an attempt to deal with the problem of intersubjectivity; but one can also find this problem treated in Heidegger's (and later Merleau-Ponty's) emphasis on the primordiality of "being-in-the-world," where, in order to defeat solipsism, consideration of the subject begins by being placed, always already, into a shared world.

THE ACOUSMATIC REDUCTION AND THE *EPOCHÉ*

After addressing the "transcendence of the object," Schaeffer turns toward the "naïve thesis of the world, the *epoché*."[35] This is significant because, without the *epoché*, there can be no discussion of "acousmatic experience." Just as the Husserlian theory of the object allows Schaeffer to define the sound object, the phenomenological *epoché* allows for a definition of *l'acousmatique*. The two moments are sutured together. Recall Schaeffer's statement: "The sound object is never revealed clearly except in the acousmatic experience." To show why this is the case, we must now investigate the relationship between acousmatic experience and the phenomenological *epoché*.

The reduction of the natural standpoint, also known as the phenomenological *epoché*, is one of the most famous procedures in Husserlian phenomenology.[36] Husserl identifies the natural standpoint (or attitude) with a commonsense view of the world: a world immediately available or "on hand," where I am surrounded by objects and things of which I have immediate knowledge; where I operate habitually and often without reflection; where things possess significance and utility in relation to my interests and goals; a world that has spatial and temporal extension, and to which I am bound through everyday involvement.[37] For Husserl, the natural attitude is the given. But to become aware of the natural attitude, there must be some way of holding it at bay, so that it can be examined. Husserl, borrowing the term *epoché* from ancient skepticism, suggests that we should employ an act of "bracketing" or "suspending," an act of refraining from judgment about the exterior world in order to experience it anew. In Dan Zahavi's description, "Our investigation should be critical and undogmatic... it should be guided by what is actually given, rather than what we expect to find."[38] Rather than committing to the external world by positing it to be factually given, the *epoché* is a method for suspending the posited world and observing it as a startling phenomenon. Husserl often describes the suspension of the natural attitude as a new, presuppositionless beginning in philosophical method.[39] Yet, it could also be characterized as a return to the original impulse of philosophy, as identified by Aristotle: "For it is owing to their wonder that men both now begin and at first began to philosophize."[40]

The *epoché* has implications for philosophy's relation to natural science. For the phenomenologist, the natural sciences remain bound to the natural standpoint in that they are predicated on an unexamined *belief or faith* in the exterior world. In Schaeffer's words, "The elaborate discourse of science is founded on this initial act of faith."[41] This is not to criticize the results of science as useless or mistaken. In fact, as Husserl writes, "to know it [the external world] more comprehensively, more trustworthily, and more perfectly than the naïve lore of experience is able to do... is the goal of the *sciences of the natural standpoint*."[42] However, classical scientific method has minimized the contribution made by the observer to this knowledge. As Merleau-Ponty was quick to note, science often reduces phenomena to the effects of stimuli upon an organ, yet finds itself unable to explain how those phenomena are experienced. His use of gestalt figures, visual illusions, and phantom limbs was intended to illustrate how perceptual phenomena were irreducible to collections of individual stimuli, like retinal arrays of light or patterns of activation in the nervous system. "When we come back to phenomena we find, as a basic layer of experience, a whole already pregnant with an irreducible meaning, not sensations with gaps between them."[43]

For Schaeffer, the natural standpoint must be overcome if musical research is ever to disclose the grounding of musical practice. By bracketing out the physically subsisting fact-world, by barring judgments in relation to it, and by leaving us only with auditory phenomena, hearing can no longer be characterized as a subjective deformation of external things.[44] The *epoché* "*completely shuts me off from any judgment about spatiotemporal factual being. Thus I exclude all sciences relating to this natural world* no matter how firmly they stand there for me, no matter how much I admire them... *I make absolutely no use of the things posited in them*... [nor] *the propositions belonging to them... none is accepted by me; none gives me a foundation.*"[45] Listening becomes a sphere of investigation containing its own immanent logic, structure, and objectivity—one that is irreducible to the physical science of acoustics.

The introduction of the acousmatic reduction is modeled on Husserl's *epoché*. By barring visual access to the source of the sound, it is intended to draw our attention to the sound's immanent properties and objectivity. Schaeffer, following the definition given in Larousse, defines the word *acousmatic* as an adjective, "*referring to a sound that one hears without seeing the causes behind it.*"[46] The term derives from the ancient Greek word *akousmatikoi*, the name given to the disciples of Pythagoras who listened to the master's lectures through a curtain. According to legend, the physical body of Pythagoras was hidden from the *akousmatikoi*, leaving them with only the sound of their master's voice.[47] Schaeffer, working in the years after World War II, felt the new technologies of recording, telecommunications, and radio to be continuous with the ancient acousmatic experiences of the Pythagorean students. Schaffer writes, "In ancient times, the apparatus was a curtain; today, it is the radio and the methods of reproduction, with the whole set of electro-acoustic transformations, that place us, modern listeners to an invisible voice, under similar circumstances."[48]

In Schaeffer's application of the *epoché*, the spatiotemporal causes of sounds are bracketed in order to distinguish them from the sound itself, grasped as a transcendent object. The *epoché* is deployed to distinguish an acousmatic field of listening from the field of acoustics. Schaeffer explicitly contrasts the acousmatic situation with the natural attitude, which is presupposed by the field of acoustics:

> In acoustics, we started with the physical signal and studied its transformations via electro-acoustic processes, in tacit reference to... a listening that grasps frequencies, durations, etc. By contrast, the acousmatic situation, in a general fashion, symbolically precludes any relation with what is visible, touchable, measurable.[49]

Although the acousmatic experience of sound still allows for the possibility of speculating upon or inferring causal sources, it bars direct access to visible, tactile, and physically quantifiable assessments as a means to this end. The translation or transcription of sounds by scientific instruments is barred. The acousmatic experience reduces sounds to the field of pure listening, "*la pure écouter.*"[50] By shifting attention away from the physical cause of my auditory perception toward the *content* of this perception, the goal is to become aware of precisely what it is in my perception that is given with certainty, or "adequately."[51] After the reduction, only the acousmatic field remains.

The distinction is clearer in French, where Schaeffer contrasts *l'acoustique* with *l'acousmatique*. (In English, we might say that *acoustics* gives way to *acousmatics*—if we could pardon the neologism.) More than a methodological distinction, Schaeffer demonstrated the practical difference between acoustics and acousmatics in his work with magnetic tape. Certain sounds, when transposed or edited, would maintain expected perceptual features. For instance, doubling the tape speed would produce a perceived transposition by an octave. In other cases, the auditory results were unpredictable. Doubling the tape speed on certain inharmonic sounds would produce transpositions at intervals other than the octave. The intentionality of the ear, and its divergence from scientific hypotheses tied to acoustic experimentation, demonstrated the primacy of the subjective constitution of the sound object, and the difference between the acoustic and acousmatic fields.[52]

Part of Schaeffer's originality was to see a profound affinity between the phenomenological *epoché* and the role played by the Pythagorean veil. Schaeffer conceived of modern sound reproduction technologies like radio, the loudspeaker, and the tape recorder as participating in the "actuality of an ancient experience," originally opened by the Pythagorean veil.[53] Just as Husserl deploys the *epoché* to bracket any claims about spatiotemporal factuality in order to grasp phenomena in their objectivity, Schaeffer understands the Pythagorean veil (and its perpetuation in the form of modern audio technology) as a tool for bracketing the spatiotemporal factuality of the sonic source. This encourages two fundamental changes: First, the objectivity of sound is grasped as a phenomenon, and second, attention is redirected to the particular essential characteristics of a given sound.

This change in listening does not occur immediately upon encountering acousmatic technologies like the Pythagorean veil, tape recorder, or loudspeaker. Schaeffer offers an illustration. When auditioning a recording of a horse galloping across the pampas, visible clues are no longer available to help in the reconstruction of the source.[54] Naturally, a competent listener *recognizes* the sound as a horse galloping and treats it as an index, pointing back toward its source. The veil would appear, at first encounter, to encourage curiosity about what lies behind it. The acousmatic reduction by itself does not dismiss this possibility—it still allows for the identification of sources and causes—but it bars access to visual and tactual means to satisfy this goal. *Indexical listening* is still available as a possible modality. However, the acousmatic reduction disorients and redirects listening by *reducing sounds to the field of hearing alone*. "Often surprised, often uncertain, we discover that much of what we thought we were hearing was in reality only seen, and explained, by the context."[55] This is a significant realization. For Schaeffer, the acousmatic experience of sound opens up the possibility of identifying modes of listening more *essential* than those that depend primarily on context. Sound is always in danger of being apprehended as something other than itself—of possessing what Timothy Taylor calls "residual signification."[56] Take, for example, the recording of the galloping horse. When treated indexically, "... there is no sound object: there is a perception, an auditory experience, through which I intend [*je vise*] another object."[57] A sound object only truly emerges when a sound no longer functions *for another* as a medium, but rather is perceived *as such*.

The emergence of the sound object from the acousmatic situation is precarious. However, the tenuousness of the situation is bolstered by the fact that recordings can be repeated over and over without variation. Counteracting the overwhelming curiosity evoked by the encounter with the veil, mechanical repetition overrides desire and offers a solid footing for the experience of the sound object.

> In fact, Pythagoras' curtain is not enough to discourage our curiosity about causes, to which we are instinctively, almost irresistibly drawn. But the repetition of the physical signal, which recording makes possible, assists us here in two ways: by exhausting this curiosity, it gradually brings the sound object to the fore

as a perception worthy of being observed for itself; on the other hand, as a result of ever richer and more refined listenings, it progressively reveals to us the richness of this perception.[58]

Schaeffer's experience with locked-groove recordings and, later, tape loops, was foundational for stabilizing the emergence of the sound object from within the acousmatic situation. Schaeffer writes, "In order to retrieve this fervor of listening, this fever of discovery, it is necessary to have lived through those instants, which any interested person can personally experience, when sound imprisoned on tape repeats itself endlessly identical to itself, isolated from all contexts."[59] The locked-groove recording or tape loop, like a word spoken over and over again, halts the flow of signification and promotes, through repetition, the hearing of sounds as such. The "fervor of listening" is inversely proportional to a sound's function as an index or sign. Thus, "the better I understand a language, the worse I *hear* it."[60]

MODES OF LISTENING

Schaeffer understood the acousmatic reduction as more than simply a *theoretical* prescription to withhold presuppositions. Rather, it promoted an *art* of listening. The acousmatic experience of sounds is a concrete, lived experience, operating at the perceptual level. It must be heard. From the very inception of *musique concrète*—before he articulated his project in terms of acousmatics—Schaeffer's desideratum was to articulate an art of listening appropriate to his compositions, a way of conveying to others how to listen to *musique concrète*. In this respect, Schaeffer's journals are revealing, especially those written while he was working on his initial *concrète* piece, the *Étude aux chemins de fer*. Schaeffer writes,

> As soon as a record is put on the turntable a magic power enchains me, forces me to submit to it, however monotonous it is. Do we give ourselves over because we are in on the act? Why shouldn't they broadcast three minutes of "pure coach" telling people that they only need to know how to listen, and that the whole art is in hearing? Because they are extraordinary to listen to, provided you have reached that special state of mind that I'm now in.[61]

How can one articulate that "special state of mind" and instill it in others? It is an understatement to say that listening is a challenging field to theorize, for there is no direct material artifact produced by listening. It is often extraordinarily challenging to convey to others what is being heard in some stretch of sound such that they can reproduce the intended experience. Again, Schaeffer looks to phenomenology for guidance. Like Husserl, who lavishes attention on describing the relationship between objects and the various modes of presentation in which they appear, Schaeffer dedicated many pages in the *Traité* to the sound object and the various modes of listening that one employs when auditioning it. To put it schematically, Schaeffer addresses two dimensions of listening sorted along typical Husserlian lines: the noetic and the noematic. His famous categorization of the four basic modes of listening falls on the noetic side of this project; his theory of the sound object falls on the noematic side.

Four verbs are used to divide up the field of listening: *écouter, entendre, comprendre*, and *ouïr*. Each of Schaeffer's verbs indicates a distinct mode (*fonction*) of listening.[62] Each mode must be understood as a unique type of intentional noetic act—a sense-giving act of listening—correlated with a particular type of auditory object.

Ouïr, which is often simply translated as "perception," is the most primordial mode of listening. According to Schaeffer, "Strictly speaking, I never cease to perceive [*d'ouïr*]. I live in a world which does not cease to be here for me, and this world is sonorous as well as tactual and visual."[63] I am always already in-the-world, and this world is perceptually manifested *for me*. From this perspective, *ouïr* is the most basic mode in which the auditory manifestation of the world is apprehended. It constitutes the "*fond sonore*" shared by all other modes of listening or ways of attending to the sonorous world.[64] However, this foundation remains hidden in our everyday attentiveness to the source and meaning of sounds. Here, Schaeffer's thinking strongly echoes Merleau-Ponty, who often reflected on the rediscovery of the primordial world of perception. However, unlike Merleau-Ponty, Schaeffer spends little time investigating this *fond sonore*, preferring to focus on other modalities. *Ouïr* provides that which is *passively* "given to me in perception," but it must be contrasted with other, more active forms of attentiveness and intentionality.[65]

Comprendre, which is sometimes translated as "comprehending" or "understanding," refers specifically to the reception of sounds mediated by sign systems or languages—a type of listening aimed at getting the message from an utterance or proposition. Michel Chion, in his guide to Schaeffer's *Traité*, glosses the term: "Comprehending means grasping a *meaning*, values, by treating the sound as a sign, referring to this meaning through a language, a code...."[66] *Comprendre* extends beyond linguistic utterances to systems like music that employ quasi-linguistic auditory signs. Much of what gets taught in elementary harmony classes institutes this kind of listening, showing students how to compose, evaluate, and understand a well-formed tonal phrase, one that demonstrates the requisite musical grammar, proper use of musical *topoi*, or correctly reproduces a given musical style.

The two remaining verbs, *entendre* and *écouter*, are commonly used to describe the active and passive modes of listening that translate into the English equivalents "to hear" and "to listen." For Schaeffer, *écouter* designates a mode of listening that is securely bound to the natural attitude, where sounds are heard immediately as indices of objects and events in the world. *Écouter* situates sounds in the surrounding sonorous milieu, grasps their distance and spatial location, and identifies their source and cause on the basis of sonic characteristics.[67] It is an information-gathering mode in which sounds are used as indices for objects and events in the world. For example, if we are crossing the street and suddenly hear the sound of squealing tires, our information-gathering listening mode could mean the difference between life and death. In this mode, "sounds are an index to a network of associations and experiences; we are concerned with causality; it is a question of living and acting in the world, ultimately of survival."[68] *Écouter* is active, situated, positional, and indexical. It is also unreflective. When we are in the natural attitude, we immediately posit the objects presented to us perceptually as really existing—there is no reflection on the manner in which the objects are intentionally constituted or upon the variety of their modes of givenness. When listening in this mode, "[I am] directed towards the event, I hold onto my perception, I use it without knowing it."[69]

Écouter has often played a problematic role in Schaeffer's aesthetics of *musique concrète*. For Schaeffer, the "musical" as such begins only when the source of sounds has been eliminated. Schaeffer consistently uses the term "anecdotal" to describe a mode of listening fixated on sonic sources or causes, a mode clearly captured under the heading of *écouter*. While working on the *Étude aux chemins de fer*, Schaeffer often despaired that his experiments were falling prey to simple anecdotalism:

> Isn't the noise of [train] buffers first and foremost anecdotal, thus antimusical? If this is so, then there's no hope, and my research is absurd.[70]

> My composition hesitates between two options: dramatic or musical sequences. The dramatic sequence constrains the imagination. You witness events; departures, stops. We observe. The engine moves, the track is empty or not. The machine toils, pants, relaxes—anthropomorphism. All this is the opposite of music. However, I've managed to isolate a rhythm and contrast it with itself in a different sound color. Dark, light, dark, light. This rhythm could very well remain unchanged for a long time. It creates a sort of identity for itself, and repeating it makes you forget it's a train.[71]

The categorical divide between the musical and the anecdotal is presented without argument, and many composers before and after Schaeffer would dispute this rigid division. Yet, he strongly maintained this view for his entire career. That aside, it should be noted that the historical and critical popularity enjoyed by the *Étude aux chemins de fer* as an exemplary piece of *musique concrète* is a bit surprising when viewed in the light of Schaeffer's own aesthetics. For the study is hardly about trains at all; rather, it uses trains to generate contrasting rhythms and tone colors. The "trainness" of the sounds, when heard in the way Schaeffer intends, is separated from their purely musical values. In other words, the étude studies rhythms, not trains.[72]

The final mode, *entendre*, must be contrasted with *écouter*. *Entendre* is the mode of listening to a sound's morphological attributes without reference to its spatial location, source, or cause; we attend to sounds as such, not to their associated significations or indices. *Entendre* shares the Latin root *intendere*, with the central phenomenological concept of intentionality. Schaeffer is absolutely clear about this connection; he writes, "For *entendre*, we retain the etymological sense, 'to have an intention.' What I hear [*j'entends*], what is manifested to me, is a function of this intention [*intention*]."[73] This connection is lost when *entendre* is translated as "hearing," obscuring the close association between this mode of listening and Schaeffer's phenomenological preoccupations.

"Reduced listening" is Schaeffer's name for the audible act of attending to the sound apart from its source.[74] This is perhaps an unfortunate choice on Schaeffer's part, because of the confusion it causes: Reduced listening (*écoute réduite*) falls under the mode *entendre*, not *écouter*. It is as if *écouter* becomes *entendre* when the indicative or communicative signification of sounds is reduced.[75] *Entendre* (or "reduced listening") emerges when *écouter* and *comprendre* are barred. When sounds are auditioned under the mode *entendre*, "I no longer try, through its intermediary, to inform myself about some other thing (an interlocutor or his thoughts). It is the sound itself that I intend [*je vise*], that I identify."[76] In reduced listening, sound no longer appears as a medium or placeholder for "some other thing."

Entendre plays a central role in two halves of Schaffer's work, his musical research and his composition. Concerning his research, *entendre* is the mode of listening that forms the basis for his *Programme de la Recherche Musicale* (PROGREMU). John Dack describes PROGREMU as Schaeffer's "ultimate ambition...to discover the basic foundations of musical structure and meaning and that this could only be achieved once the sounds were freed from their causal origins."[77] In order to attain this end, Schaeffer encouraged musicians "to learn a *new solfège* by systematic listening to all sorts of sound objects."[78] Chion describes this new *solfège* as "a kind of becoming aware of the new materials of music while distrusting preconceived ideas and relying first and foremost upon what *one hears [on entend]*."[79] Through the selection and appreciation of sonic attributes, it is possible to construct a taxonomy of sounds, capable of organizing and classifying not only the typical sounds of instrumental music, but "the entire sound universe."[80]

On the compositional side, *entendre* is the mode of listening identified with Schaeffer's aesthetic preference for reduced listening. In this regard, the titles of Schaeffer's various études are revealing. Unlike the *Étude aux chemins de fer*, which identifies the source in the title of the work, the later studies remain wholly abstract: *Étude aux objets, Étude aux allures, Étude aux sons animés*. Rather than identify the source, these later works derive their material from a variety of sources, and then organize it in order to bring out some shared aspect, such as its grain, its duration, its register, or its timbre. These features of the sound object are afforded by *entendre*. Sounds are not employed as indicative or communicative signs; rather, the object is used to focus the listener on some intrinsic feature of the sound, regardless of its worldly reference. If Schaeffer initially worried about the difference between anecdotal and musical sequences in *musique concrète*, the later studies have effaced all traces of this worry by excising the former. The musical sequence alone is promoted. In the *Étude aux objets*, Schaeffer even deploys a plan that is based on traditional musical forms. The opening movement, "*Objets exposés*," smartly indicates its musical function as an exposition of musical materials. The first phrase, for left loudspeaker alone, concatenates eight sound objects of various character, only to be followed by a "counter-theme" for the right speaker, also formed of eight different objects.[81] The rest of the movement sequences and superimposes material taken from the phrases in a manner that is loosely fugal in character. The other movements in the étude are also based on the opening material, developing and drawing connections between sounds through the use of overlapping, mixing, and montage. The final movement, "*Objets rassemblés*," is also described as a "stretto."[82]

Unlike the purely musical études of Schaeffer in the late 1950s that efface all traces of *écouter* for the sake of *entendre*, a work like Pierre Henry's *Variations pour une porte et un soupir* thematizes the alternation of the musical and the anecdotal in an elegant manner.[83] The title of Henry's work is telling; it is a set of variations *for*—not *on*—a door and a sigh. Thus, it is conceptually closer to a work that names its instrumental forces, like Messiaen's *Theme and Variations for Violin and Piano*, than to a work like Reger's *Variations and Fugue on a Theme of Mozart*. The latter is based on a well-known musical theme and (existing in both orchestral and piano versions) is perhaps conceived as indifferent to its instrumental forces. In Henry's short movement entitled "*Étirement*," all sounds come from a recording of a creaking door hinge. In the left and right speakers, Henry begins with—pun intended—opening gestures. The referentiality of the sounds is brought to the fore by the sharp, percussive stridulations of the creaking hinge. The listener can sense the size and heft of the door, and the physical force required to move it. In the course of one minute, Henry transmogrifies these creaking doors into a musical duet by editing and sequencing passages that bring out dramatic melodic profiles, layered to create overlapping and unexpected entrances. The doors lose their characteristic "doorness" and are metamorphosed into flatulent tubas, rumbling contrabasses, or honking baritone saxophones. At the moment when the sounds are most continuous, having reached a crescendo, Henry ends the piece by slowly letting the doors creak shut in a closing *rallentando* where each snap of the hinge is distinctly and clearly articulated. The ear of the listener hovers between anecdotal reference and musical autonomy—oscillating between *écouter* and *entendre*.

THE EIDETIC REDUCTION AND THE END OF IMPROVISED ONTOLOGY

If the only concern of this chapter were to introduce Schaeffer's concept of *l'acousmatique*, I could stop right here. We have seen how the acousmatic reduction is modeled on Husserl's phenomenological *epoché*. We have also seen how the acousmatic reduction brings various modes of listening to the attention of the listener. By defamiliarizing everyday practices of listening, the acousmatic reduction makes these modes perspicuous. (Or, to use less visually centered terms, we could say that the acousmatic reduction brings these modes of listening into audibility.) If Schaeffer prefers *entendre* or reduced listening to other modes, this is not a valuation that *necessarily* follows from the acousmatic reduction. It must be noted that all modes are available within the acousmatic situation. The acousmatic situation is not a constraint on modes of listening; it is a way of bringing those modes into focus.

Although this is an important point, one that has been generally underappreciated in the reception of Schaeffer's work, there is still more to say about the role of the acousmatic reduction in Schaeffer's project if we want to explain the reasons behind his claim that "the sound object is never revealed clearly except in the acousmatic experience." If the acousmatic reduction brings the variety of modes of listening to the fore without preference for one of them, what is the special relationship between the sound object and acousmatic reduction?

As I argued earlier, by barring our access to visual, tactile, and measurable causes of sounds, the acousmatic reduction reduces sounds to the field of hearing alone. The listener is directed away from the physical object that causes a perception, toward the *content* of that perception. This shift is useful not only for bringing modes of listening into audibility, but also for establishing a few negative claims about what cannot constitute the sound object.[84] Once the *content* of some auditory perception is distinguished from its source or cause—once a split between the sonic source and its auditory effect has been established—then it is no longer possible to think of the sound object as determined by some physical thing. This is why Schaeffer claims that "the sound object is not the instrument that was played," nor is it reducible to "a few centimeters of magnetic tape."[85] These negative definitions might lead one to assume that, if the sound object is irreducible to some physical thing, then it must be reducible to some subjective state. Schaeffer anticipates this line of thought:

> To avoid confusing [the sound object] with the physical cause of a "stimulus," we seemed to have grounded the sound object on our subjectivity. But...the sound object is not modified...by the variations in listening from one individual to another, nor with the incessant variations in our attention and our sensibility. Far from being subjective...[sound objects] can be clearly described and analyzed.[86]

The challenge or "ambiguity" of the sound object is to realize that it is indeed "an objectivity linked to a subjectivity."[87] So what constitutes the objectivity of this ambiguous object?

To answer this question, Schaeffer supplements the acousmatic reduction or *epoché* with a second reduction, known in phenomenology as the *eidetic reduction*. The motivation behind Schaeffer's use of the eidetic reduction is simple; if the sound object is intended to ground the identification of sounds across multiple acts of listening and among multiple listeners, then the basis for its objectivity must be explained. The use of sound objects in *musique concrète* may help us to perceive and appreciate specific qualities of sound objects, but a piece of *musique concrète* is not a philosophical argument for the objectivity of sound objects generally. This is where the eidetic reduction comes into play. The eidetic reduction is a technique deployed by Husserl intended to reveal an object's *essential* features. Starting with some particular object, Husserl encourages the philosopher to detach it from its real situation and treat it as an "arbitrary example" that acts as a guiding model, "a point of departure for the production of an infinitely open multiplicity of variations."[88] By producing a series of "free variants," each of which is also imagined, "it then becomes evident that a unity runs through this multiplicity." In the act of producing a set of free variations, "an *invariant* is necessarily retained...according to which all the variants coincide: a *general essence*." The essence of an object "proves to be that without which an object of a particular kind cannot be thought," or in other words, an essence discloses the very condition of the possibility of some object's identity.[89] For Husserl, such essences form the basis of an object's objectivity; for without an a priori grasp of an object's essence, we could not identify and re-identify particulars.

In the *Cartesian Meditations*, Husserl returns to his example of a table in order to show how "imaginative free variation" operates as a technique for disclosing essences:

> Starting from this table-perception as an example, we vary the perceptual object, table, with a completely free optionalness, yet in such a manner that we keep perception fixed as perception of something, no matter what. Perhaps we begin by fictionally changing the shape or color of the object quite arbitrarily....In other words: Abstaining from acceptance of its being, we change the fact of this perception into a pure possibility, one among other quite "optional" pure possibilities— but possibilities that are possible perceptions. We, so to speak, shift the actual perception into the realm of non-actualities, the realm of the as-if.[90]

Three aspects of the process are worth noting: (1) Imaginative variation reveals *invariant* properties of the transcendent object. By imagining the table in a variety of changing contexts (changing its shape, color, structure, etc.), the essence of the phenomenon comes to be grasped and understood. Variation is a technique for revealing *essence*. (2) By undergoing the reductive test of the *epoché*, by bracketing all theses dependent upon the external world, imaginatively varied intentional objects are freed from all bonds to the external world. Thus, *the distinction between fiction and reality becomes moot*. In the lectures on the *Idea of Phenomenology*, Husserl explains that when considering essences, "perception and imagination are to be treated exactly alike," because any "suppositions about existence are irrelevant."[91] (3) Since existential questions are irrelevant, it is no longer possible to argue that transcendent objects are merely subjective fictions. For Husserl and Schaeffer, the contents of our mental acts possess a special type of objectivity. Schaeffer writes: "No longer is it a question of knowing how a subjective hearing interprets or deforms 'reality,' to study reactions to stimuli; hearing itself becomes the origin of the phenomenon to study."[92] Hearing, whether imagined or real, presents us with indubitable evidence or data. Based on such indubitable evidence, intentional objects are both *ideal and objective* or, in Husserl's terminology, "ideal objectivities."[93]

In a section of the *Solfège de l'objet sonore* entitled "The objectivity of the object," Schaeffer relies upon variation and eidetic reduction to clarify the objective character of the sound object.[94] In each of his examples, Schaeffer takes the same recording and gives it a variety of electronic variations. By taking a sound and using electronic means to alter its qualities, Schaeffer *pedagogically produces* a set of variations with the aim of disclosing the sound object's *invariant* and *essential* features. The sound of a gong gently rolled with soft mallets is played twice, followed by variants: by adjusting the potentiometers, the envelope of the object is varied; by using low and high pass filters, the mass and grain of the object are varied; subtle shifts in volume create an object with more *allure*, or internal beating; and finally, a combination of techniques produces another variant. As a listener, not only do we recognize the different variations as variations, we also hear them as one and the same sound object. The objectivity of the sound object is intended to emerge across its various instances.

No two instantiations are exactly the same: From an acoustician's point of view, the signal would contain measurable differences in each case; from the phenomenological point of view, each variant differs in aspect from the last. Schaeffer concludes,

> we must therefore stress emphatically that [a sound] object is something real [i.e., objective], in other words that something in it endures through these changes and enables different listeners (or the same listener several times) to bring out as many aspects of it as there have been ways of focusing the ear, at the various levels of attention or intention of listening.[95]

While employing these examples to demonstrate the objectivity of the sound object, Schaeffer wants to defend against any reduction of this demonstration to a set of studio tricks. "The purpose of these manipulations, these technical tricks, is purely pedagogical. It is an anticipation of the way in which the ear becomes increasingly alert, the more often one listens to the same object."[96] By emphasizing the pedagogical use of these "manipulations," Schaeffer is also noting that there is nothing specifically technical about the objectivity of the sound object. It could have been demonstrated otherwise than with mechanical means; one could have simply imagined such variations for oneself. *Once Schaeffer commits to the eidetic reduction, there can be no essential difference between imagined hearing and actual hearing.* The "mode of givenness" may change, but the "central core" remains the same.[97]

Many of the techniques developed for producing *concrète* works depend upon variation. The composer subjects recorded sounds to filtration, editing, looping, reverberation, and changes in speed or direction. The results of such processes must be auditioned again and again to determine whether these variations present us with "the same" sound object or a new sound object entirely. Each variation is an investigation into the objectivity of the sound object. Although Schaeffer clearly incubated his ideas about the sound object from within the *concrète* context, one must not treat his *solfège* as simply a method for learning *musique concrète*. The point of his phenomenological project is to identify an object capable of grounding both acoustics and our musical practices, be they concrete or abstract. Schaeffer's desideratum is to systematize what first began as an improvised musical ontology.

ORIGINARY EXPERIENCE AND THE PROBLEM OF HISTORY

After the eidetic reduction, Schaeffer's musical ontology is much clearer. A sound object is disclosed as a particular type of transcendental object, the typing of a sonic token defined by the possession of certain invariant features. Each empirical token of a sound object is identifiable and re-identifiable based on noetic synthetic acts, just like any other kind of object; each sound object, as a type capable of having many tokens or instances, is identifiable based on the recognition of a set of invariant, essential features. At the level of the token, it makes no difference under what mode the sound object is heard. We need not be attending to a sound's immanent morphological features to grasp it as an object; we could just as well be listening to it for its source or cause. But this is not the case when we talk about a sound object as a type. Here, all non-immanent properties have been stripped away. After undergoing the test of the *epoché*, after being acousmatically reduced and heard via "reduced listening," we can start to imaginatively vary a sound object in order to disclose its essential, invariant properties. Those invariant properties, which are always morphological for Schaeffer, identify a sound object as a specific kind or type. A sound object in this sense is an ideal object; it inhabits an order of essences (in the phenomenological sense) that guarantees repetition without difference. It insures ascriptions of identity to sounds across a variety of contexts, and thus also governs ascriptions of difference and variation, which are so central to musical composition. A sound object, in its fullest sense, is to be ontologically distinguished from the realm of empirically sounding events in that its ideal "being" guarantees infinite empirical identification and re-identification without divergence.

The eidetic reduction also clarifies the relationship between the sound object and technology. For Schaeffer, the empirical repetition afforded by technologies of recorded sound is simply a consequence of the ideality and repeatability of the sound object. Technology may be important, but it would be a misunderstanding of Schaeffer's thinking to assume that the sound object is in any way the result of modern sound technology. The Pythagorean veil or the loudspeaker, both of which encourage the acousmatic reduction and recognition of sounds distinct from their causes, find their condition of possibility in the ideal objectivity of the sound object.

In contrast to Schaeffer's claim from May of 1948 that the sound object "does not fit in with the elementary definitions of music theory," the ideal objectivity of the sound object is perfectly *music theoretical*. It follows upon and re-inscribes the ideality that was previously attached to the note: It defines a class, a type possessing tokens. Each sound object is a specific essence, an ideal objectivity posited as the ground that guarantees its repeatability. But as an ideality, this sound object does not exist in the world. It is heard *in sounds*, but must also be distinguishable from the actual sonorousness of sounds. The sound object is not itself sonorous. In the silence of imagined sound, where there is nothing actually vibrating, one can perform intentional acts that depend on the sound object's ideal stability, such as conceiving, comparing, composing, and distinguishing sounds.

The ontological grounding offered by the sound object challenges the claims of acoustics, or any science bound to the natural attitude. From Schaeffer's perspective, the acoustician is mistaken to take the signal as primary. Nowhere is Schaeffer more explicit on this point than when he writes, "One forgets that *it is the sound object, given in perception, which designates the signal to be studied*, and that, therefore, it should never be a question of reconstructing it on the basis of the signal."[98] This is an orthodox phenomenological strategy: By grounding the acoustician's signal upon the sound object, Schaeffer considers his investigation to be more originary, since it provides an ontological foundation to the merely empirical (or ontic) conclusions of acoustics. Compare this strategy with Heidegger's description of phenomenological reduction from *The Basic Problems of Phenomenology*. "Phenomenological reduction means leading phenomenological vision back from the apprehension of a being, whatever may be the character of that apprehension, to the understanding of the Being of this being."[99] This ontological understanding—"the Being of beings"—consistently resists the habitual tendency to gather our ontological terms from the natural attitude.

Not only is the hidden foundation of the acoustician's signal revealed as grounded upon the sound object, the sound object also underlies and determines our own subjectivity. According to Schaeffer,

> I must *free myself from the conditioning* created by my previous habits, by passing through the test of the *epoché*. It is never a question of a return to nature. Nothing is more *natural* than obeying the dictates of habit. [Rather,] It is a question of an *anti-natural* effort *to perceive what previously determined my consciousness without my knowing it*.[100]

The process of phenomenological reduction lends to the sound object a strange trajectory: Methodologically, one discloses the sound object only at the end of the investigation, after a series of interlocked reductions; but ontologically, the sound object is absolutely first, a priori. The priority of the sound object is evinced when Schaeffer writes, "I must re-visit the auditory experience, to re-grasp my impressions, to re-discover through them information about the sound object."[101] Due to the danger of continually losing the sound object to habit, one must constantly become reacquainted with it. But one can only be *reacquainted* with something with which one was already familiar. Perhaps the strangeness of this trajectory becomes less mysterious, less portentous, when we realize that it is simply teleological.

Yet, only through incessant revisiting, re-grasping, and rediscovering is the sound object revealed as the "originary experience" of phenomenological investigation.[102] In the phenomenological literature, an originary experience designates something quite specific; it marks the discovery of some transcendental region or field of inquiry (such as geometry, logic, technology, etc.) by a founding (noetic) act, which discloses a horizon containing all future investigations of that region. Through reactivation, an originary experience is available to all inquirers at all times. It is an inquiry into the propagation of essences, into the sense and structure that make some region of experience or thought possible, not into the factual circumstances or engagements of particular historical individuals or modes of apprehension. To explicate this concept, it is useful to compare Husserl's introduction of the originary experience of geometry, as presented in *The Origin of Geometry*, with Schaeffer's use of the concept. Husserl writes,

> The question of the origin of geometry...shall not be considered here as the philological-historical question, i.e., the search for the first geometers who actually uttered pure geometrical propositions, proofs, theories...or the like. Rather, our interest shall be the inquiry back into the most original sense in which geometry once arose, was present as the tradition of millennia, is still present for us, and is still being worked on in a lively forward development.[103]

This "regressive inquiry" or *Rückfrage* avoids anything that could be called historical.[104] The question of origins replaces the question of beginnings. Although Schaeffer first discovers the sound object by means of a material engagement with real technical devices in the studios of Radiodiffusion Française, in his mature theory, the revelations that emerged from the *cloché coupée* and the *sillon fermé* no longer constitute new phenomena. Hearkening back to the time of Pythagoras and echoing Husserl's own analyses of the origin of geometry, the disclosure of the sound object from within the acousmatic and eidetic reductions is less a historical phenomenon than the rediscovery of an originary experience first disclosed in ancient Greece and reactivated by the technology of sound reproduction.

An analogy can be drawn between the geometer and the electronic musician. Husserl writes,

> The geometer who draws his figures on the board produces thereby factually existing lines on the factually existing board. But his experiencing of the product, qua experiencing, no more *grounds* his geometrical seeing of essences and eidetic thinking than does his physical producing. This is why it does not matter whether his experiencing is hallucination or whether, instead of actually drawing his lines and constructions, he imagines them in a world of phantasy.[105]

The same could be said of the sound object. Whether a sound is locked in a groove, looped on a tape, or hallucinated in fantasy, the contingent and constantly varied experience of sound cannot provide a foundation for its qualitative, indicative, or communicative aspects. The geometrical drawing, with all of its crooked lines, is akin to the acoustician's signal—empirical, inessential, and contingent. As a vehicle to arrive at the sound object, *the empirical phenomenon "does not matter."* However, in the drive to locate a secure grounding for aural experience, *experience itself falls away*. Husserl says as much:

> [The] *pure eidetic sciences*... are pure of all positings of matters of fact; or, equivalently: *in them no experience, as experience*, that is, as a consciousness that seizes upon or posits actuality, factual existence, *can take over the function of supplying a logical ground*. Where experience functions in them it does not function *as* experience.[106]

Experience remains curiously ungrounded in phenomenology's eyes and must be supplemented after the fact with an ideal objectivity. Experience becomes secondary to its role of providing evidence for disclosing essences. Through a sleight of hand, phenomenology covertly places its ontology prior to experience, and then subsequently discloses the ontological horizon *as if* it were always already present—as if *its* ontology made experience possible in the first place.

In the Husserl passage just cited, this is made explicitly clear; the "pure eidetic sciences," if they want to remain free of the vulgar contingency of history, causality, or culture, *must* remain free of the "positings of matters of fact." Such vulgar positings (i.e., history, biography, culture, fact, contingency, chance, etc.) might sully the immaculate purity of philosophy as a rigorous science. In Husserl's privileged domain of geometry, the ethical imperative to avoid contingency at all costs is clearly demonstrated where the "originary experience" of geometry cunningly displaces any kind of material-historical investigation into its beginnings. The phenomenological necessity to end-run contingency, to remove the historical from history, is a self-imposed blind spot, an act of hardheaded idealism.

OBJECTIONS

The motivation for this chapter was to clarify the statement "the sound object is never revealed clearly except in the acousmatic experience." By rehearsing Schaeffer's argument and articulating how he models his research in the *Traité* upon Husserlian phenomenology, I have tried to show the precise relationship between the sound object and the experience of the acousmatic reduction. They are not the same. The acousmatic reduction restricts listening to the field of hearing alone, by bracketing visual, tactile, and other sensory means of assessing sounds. The acousmatic reduction is Schaeffer's version of the phenomenological *epoché*. Within the acousmatic reduction, various prominent modes of listening emerge. Some modes are indicative and communicative, where the sounds are used as signs to direct the listener's attention to physical-causal sources or linguistic meanings; others are self-reflexive, directing attention toward the intrinsic qualities and characteristics of sounds. *Entendre* and reduced listening are of the latter variety, *écouter* and *comprendre* of the former. The habitual everydayness of *écouter* and *comprendre* is disturbed after undergoing the acousmatic reduction. As for the sound object, it underlies all the various modes of listening, for a sound object is the basic ontological unit in Schaeffer's account. If it is only clearly revealed in the mode *entendre*, this is because Schaeffer thinks that the additional signification added to a sound when treated as an indicative or communicative sign can be reduced away without essentially changing the ontology of the sound. The essential qualities of the sound object are revealed in a process of imaginative variation, or eidetic reduction. This further reduction brings out the invariant features of an object and discloses these features as constituting the object's ideal objectivity.

I reiterate this account because the relationship among the acousmatic reduction, the sound object, and reduced listening is not always clearly understood. Even some of our finest writers on 20th-century sound, music, and technology occasionally miss these distinctions. Frances Dyson writes,

> Pierre Schaeffer, for instance, taking an essentially phenomenological approach, argued for "acousmatics"—a reduced listening that would bracket sounds from their musical and cultural origin and focus listening on sounds "in themselves" without recourse to their visual or material source.[107]

The imprecision in this sentence—which glosses acousmatics by identifying it with reduced listening and places an emphasis on sounds in themselves and the separation of the senses—may appear insignificant. Yet, without a precise distinction between these various parts of Schaeffer's project, we cannot really subject acousmatic experience to a thorough, honest, and clear-sighted assessment. Indeed, "the sound object is never clearly revealed except in the acousmatic experience," but it does not follow that acousmatic experience is *necessarily* beholden to the theory of the sound object.[108] Nor do we need to understand the acousmatic experience of sounds according to the phenomenological approach of Schaeffer. In fact, there is much left to be said about acousmatic experience *in distinction to* Schaeffer's affirmation of the sound object and reduced listening, and *apart from* his phenomenological method.

The chapters that follow investigate acousmatic experience in other terms than those proposed by Schaeffer. But before moving on to those investigations, I will quickly present three objections to Schaeffer's theory, with the acknowledgment that each objection functions as a starting point for investigations of acousmatic experience in the chapters that immediately follow. My three objections concern (1) the phantasmagoric effacement of technology in Schaeffer's thinking; (2) the mythic use of the Pythagorean veil; and (3) the ontological problem that emerges when sounds are conceptualized as sound objects that reify sonic effects, rather than events that bind source, cause, and effect together. These three objections are further developed in parts II and III.

The Ontological Problem

By positing the sound object as the ontological grounding of musical experience, Schaeffer commits himself to an ahistorical view about the nature of musical material. Of course, for Schaeffer, that is precisely the point; the sound object must be defined in a purely objective manner in order to ground subsequent research. Schaeffer employs phenomenology in the same way that Husserl did, as a rigorous science that veers away from the naturalistic grounding of the physical sciences. However, one might object that the severe reduction required to "disclose" the sound object is not worth the effort, since it sacrifices all ties of the sound object to its context and history. Despite Schaeffer's goals, the method used to disclose a sound object as an essence ends up denaturing the object and thus distorting the resulting essence.

This objection follows along the lines first proposed by Theodor Adorno. In the late 1920s, Adorno argued that "the cognitive character of art is defined through its historical actuality."[109] The compositional act is engaged, from the very beginning, in a historical dialectic presented *in the form of* musical material. Adorno writes,

> It is the material which provides the stage for progress in art, not individual works. And this material is not like the twelve semitones with their physically patterned overtone relationships, interchangeable and identical for all time. On the contrary, history is sedimented in the figurations in which the composer encounters the material; the composer never encounters the material separate from such figurations.[110]

The equivocal term "figuration" is intended to capture this dialectic of material and history: Sounds and notes do not simply constitute a realm of essence detachable from their moment, sites of production, or reception. Rather, they need to be recognized as a sedimentation of historical and social forces.

But such figurations are precisely the *disjecta membra* cast aside by reduced listening. The indicative and communicative sign is dismissed as inessential to Schaeffer's foundational project. In order to have an existence in the domain of the musical work, indicative and communicative signs must be reconstructed on the basis of the sound object. This style of reconstruction is hardly value-neutral. In fact, it reveals a bias that is manifest in the phenomenological method itself, despite its claims to be merely a descriptive science. As Adorno once wrote, "The form of phenomenological description borrowed from the sciences, which is supposed to add nothing to thought, changes it in itself."[111] This change is made in the name of securing an a priori ontological foundation, but the benefits of such a foundation are attained at the expense of historically sedimented "residual signification." Schaeffer, unwilling to see his own composing and theorizing as historically conditioned, deludes himself into describing a sonic material that necessarily stands outside history. What Adorno writes about Husserl also holds of any foundational musical ontology: "ostensible original concepts...are totally and necessarily mediated in themselves—to use the accepted scientific term—'laden with presuppositions.'"[112]

Although the acousmatic reduction does not bar the possibility of hearing sounds in relation to their source, when combined with the eidetic reduction, it changes the way sounds are conceptualized. They become audible phenomena, understood as ontologically distinct from their causal sources. Either we hear through the sound object to its source or attend to it for its own intrinsic features—but in either case, the sound object, taken as a phenomenon, has priority. This phenomenalization of sound, which is part and parcel of Schaeffer's acousmatic *epoché*, encourages the listener to understand sounds as objects, not as events. An event-based ontology of sounds is not congruent with a Husserlian emphasis on intentional, transcendent objects or noema. Unlike an event-based ontology, where the effect of a sound is not conceptually distinguished from its source or cause, Schaeffer's theory assumes a split from the outset. This authorizes a reification of the sonic effect and makes it impossible to accurately determine the ontological relation of effect to source and cause within a Schaefferian framework.

Even theorists who claim allegiance to Schaeffer have not accepted his reification of the sound object. For instance, Michel Chion, who often praises Schaeffer's work, challenges the strict separation of source and effect when he introduces the figure of the *acousmêtre* in *The Voice in Cinema*.[113] For Chion, the magical powers of the *acousmêtre*—the strange cinematic figure of an audible voice without a clearly visible body—depend on "whether or not the *acousmêtre* has been seen."[114] The *acousmêtre* is never an essence, ontologically indifferent to the source from which it is emitted; rather, the gap that separates the voice from its source generates the *acousmêtre*'s strange potency. Never suspended, never bracketed, the *acousmêtre* depends on the *paradox* of the effect without a cause—a paradox that has been reduced in Schaeffer's eidetic theory of the sound object. In chapter 5, I will return to Chion and the *acousmêtre*, along with literary examples from Kafka and Poe, to show how an auditory effect always underdetermines its source and cause; and how the strange potency attached to such underdetermined sounds challenges any kind of eidetic reduction.

Phantasmagoria

Schaeffer maintains an essentialist view of technology. Rather than theorize the acousmatic reduction in its specific relationship to modern audio technology, Schaeffer conceives of it as the reactivation of an ancient telos, an originary experience presupposed and retained in our practices, yet always available to be re-experienced in its fullness. He writes,

> The acousmatic situation, in a general fashion, symbolically precludes any relation with what is visible, touchable, measurable. Moreover, between the experience of Pythagoras and our experiences of radio and recordings, the differences separating direct listening (through a curtain) and indirect listening (through a speaker) *in the end become negligible*.[115]

Instead of capitalizing on this difference and distinguishing the manner in which new forms of technology produce *historically unique affordances or opportunities*, Schaeffer conjures technology into an archetype, disclosing a realm of essence that is always already present—and thus essentially ahistorical. Phantasmagorically, Schaeffer masks the technical specificity and labor involved in the production of the sound object, in order to present an autonomous realm of sonic effects without causes. In the "fervor of listening," Schaeffer effaces the historical and material specificity of the locked groove (*sillon fermé*) in the name of the disclosure of an eidetic sound object. In other words, acousmatic experience is treated like a horizon of possibility that underlies certain kinds of experiences epitomized in modern audio, rather than as a field constituted through material engagement with various forms of technology, both visual and auditory.

Carlos Palombini has convincingly argued for the explicit connection between Heidegger's and Schaeffer's views on technology.[116] In particular, both Heidegger and Schaeffer conceive of the technological domain as distinct from its particular cultural and social manifestations. According to Heidegger, "Technology is not equivalent to the essence of technology."[117] This is no anodyne claim; Heidegger assumes a split between the factual and the essential. Instead of negotiating with technology in its concrete, material manifestations, it must be reconceived as an ontological perspective, a new form of understanding or disclosing the world. Heidegger writes, "Technology is therefore no mere means. Technology is a way of revealing."[118]

Schaeffer would agree. Materially and historically specific forms of technology (magnetic tape and its possibilities of editing, splicing, and playback; the *phonogéne*, the morphophone, analogue filters, and artificial reverberators) may have afforded the conditions for developing *musique concrète*, but Schaeffer views technology as something far greater than the sum of such material conditions. More than just a prosthesis for the senses, technology discloses a "way of revealing." Schaeffer writes,

> The age of mechanism, denounced wrongly by Pharisees of spiritualism, is the age of the most inordinate human sensibility. It is not solely a question of machines for making, but of machines for feeling which give to modern man tireless touch, ears and eyes, machines that he can expect to give to him to see, to hear, to touch what his eyes could never have shown him, his ears could never have made him hear, to touch his what his hands could never have let him touch. As this enormous puzzle, which knowledge of the exterior world is, composes itself, strengthens itself, verifies itself and finally "sets" into shape, man recognizes himself in it: he finds in it the reflection of his own chemistry, his own mechanisms.[119]

But what is ultimately revealed? The celebration of new possibilities for feeling and sensation is superseded by man's recognition of himself, where "man" is characterized wholly abstractly. This is no account of historically specific persons involved in artistic or critical engagements with the technological means at hand; rather, Schaeffer presents a picture of ahistorical, existential man discovering himself within a teleological horizon. What modern technology reveals for Schaeffer is little more than an abstract glimpse into an ancient originary experience. Where "man describes himself to man, in the language of things," the "voice" of technological things is silenced. In chapter 4, I will revisit the relationship of acousmatic listening and phantasmagoria and present the historical context for their close affiliation. Additionally, I posit a set of philosophical conditions that underlie cases of musical phantasmagoria and propose a more productive model for understanding the role of technology, broadly construed, in the production of acousmatic experience.

Myth

Roland Barthes once said, "Myth deprives the object of which it speaks of all history."[120] In the Schaefferian discourse, the sound object is indeed the object that has exchanged its history for myth. The terms of that myth are well defined: The experience of the electronic musician in the studio reactivates the ancient originary experience of the Pythagorean disciples who heard the master speak from behind a veil.

This claim is not authorized by a patient historical account, but simply by an act of mythic identification.

The mythic identification between Pythagoras and the composer of *musique concrète*, initiated by Schaeffer, is prolonged in his students' work. This is clearly evinced in François Bayle's writings on acousmatic music. Bayle offers a standard account of the history of the term "acousmatic," tracing its origins to the legendary accounts of Pythagoras lecturing to his disciples from behind a veil. But two extra features are added: First, he writes that the Pythagorean disciples were placed in the dark; second, he writes that the *akousmatikoi* developed a special technique for concentrated listening.[121] As I will show in chapter 2, neither of these features has sufficient historical evidence to support it. Rather, they resemble Bayle's own modifications to and prescriptions for the practice of *musique concrète*. Bayle has been instrumental in developing darkened spaces for the performance of acousmatic music, in which an engineer at a mixing console spatially projects sounds. Thus, a specious identification is produced between the ancient acousmatic situation of the Pythagorean disciples and Bayle's own practice. Just as Pythagoras announced his teachings to his pupils in the dark from behind a veil, so too does the acousmatic music composer project his discourse into a darkened hall while remaining obscure. The loudspeaker and mixing console prolong the Pythagorean veil. As for the second feature, one can imagine that these special listening techniques foreshadow Schaeffer's *écoute réduite*, various kinds of sonic *solfège*, or even the link between the Husserlian technique of phenomenological *epoché* and the acousmatic reduction. To say the least, historical accuracy does not motivate Bayle's account. When the distance between our technological devices and the veil of Pythagoras becomes negligible, sadly, we are in the presence of ideology. As Marx wrote, "...we must pay attention to this history, since ideology boils down to either an erroneous conception of this history, or a complete abstraction from it."[122]

PART TWO

Interruptions

2

Myth and the Origin of the Pythagorean Veil

In 1977, three university psychologists performed a simple experiment. On three separate occasions, separated by intervals of two weeks, subjects were presented with a list of plausible statements culled from reference works on general topics: history, politics, sports, biology, current affairs, the arts, geography, and such. Some statements were true, and others were false, but none were likely to be known by their subjects—college students. In each session, 60 statements were presented; 40 were new each time, and 20 were repeated on all three occasions. The subjects were asked to rate how confident they were that the statements encountered were true or false. Over the course of the three sessions, the repeated statements ranked progressively higher in terms of confidence in their truthfulness than the new statements, which remained at a constant ranking throughout. Psychologists refer to this phenomenon as *the truth effect*: "The repetition of a plausible statement increases a person's belief in the referential validity or truth of that statement."[1]

When we study the discourses on acousmatic sound, we see the truth effect at work. For whenever this strange word, "acousmatic," is used by composers, theorists, artists, media scholars, and musicologists, a set of statements follows in tow—statements about the origin, etymology, transmission, and meaning of the term. Some statements invoke the figure of Pythagoras: They may speak of his technique of lecturing from behind a curtain or veil, oftentimes in the dark; or of the division of his school into exoteric and esoteric disciples; or of obscuring his appearance in order to develop techniques of concentrated listening in his pupils; or of his secret understanding of the distinct epistemologies of the eye and the ear. Others articulate the preservation of this Pythagorean tradition in modern forms of sound reproduction and radio transmission. These repeated claims take on the solidity of truth—grounding claims, organizing conceptual schemata, and shaping practices.

In the previous chapter, I raised the objection that the Schaefferian tradition persists in a mythic identification of the composer of *musique concrète* with Pythagoras. How best can one demonstrate the pervasiveness of this myth? There is no single central text describing the founding, meaning, and transmission of the term "acousmatic" from Pythagoras to the present day. Rather, there are multiple partial accounts that circulate in various discourses on acousmatic sound. The whole is nowhere directly presented; like a landscape composited from multiple snapshots, it is only revealed by overlaying the various pieces into a complete image.

Since writings on acousmatic sound are usually intended for specialized and distinct audiences (composers, film theorists, opera scholars, theorists of vocality, etc.), I cannot assume that the reader is familiar with all of the pieces that comprise that image. Thus, I present a set of statements selected from the writings on acousmatic sound before offering a synthesis of these statements into a "key myth"—to borrow a phrase from Levi-Strauss.[2] These statements contain a collection of mythemes: individual, repeated units of mythological narrative, of various sizes and comprehensiveness, deployed for a variety of purposes. The mythemes are not directly stated as such but discerned through replication. Their presence is marked by the fact that they are repeated again and again. I ask the reader to focus on both the variations and similarities among the mythemes.

To facilitate my construction of the key myth, I introduce a notational convention: Each mythic statement (or source from which I construct the key myth) is prefaced with the letter "M" and a number, so that I can refer to them explicitly and efficiently later in the text. The first set of statements contains overviews intended as schemata or thumbnail guides for tracing a comprehensive account of the origin and transmission of the term "acousmatic."

M1. Michel Chion, *The Voice in Cinema*:

Let us go back to the original meaning of the word acousmatic. This was apparently the name assigned to a Pythagorean sect whose followers would listen to their master speak *behind a curtain*, as the story goes, so that the sight of the speaker wouldn't distract them from the message.... The history of the term in interesting. The French word *acousmate* designates "invisible" sounds. Apollinaire, who loved rare words, wrote a poem in 1913 entitled "Acousmate," about a voice that resonates in the air. The famous *Encyclopédie* of Diderot and d'Alembert (1751) cites the "Acousmatiques" as those uninitiated disciples of Pythagoras who were first obliged to spend five years in silence listening to their master speak behind the curtain, at the end of which they could look at him and were full members of the sect. It seems that Clement of Alexandria, an ecclesiastic writing around 250 B.C., may be the sole source of this story, in his book *Stromateis*. The writer Jérôme Peignot called this term to the attention of Pierre Schaeffer.[3]

M2. François Bayle, *Musique acousmatique*:

a. acousmatic—Situation of pure listening, without attention being diverted or reinforced by visible or foreseeable [*prévisible*] instrumental causes.[4]

b. Pythagoras (6th cent. B.C.) invented an original device [*dispositif orginal*] for attentive listening, by placing himself behind a curtain [*rideau*] when lecturing to his disciples, in the dark, and in the most rigorous silence. *Acousmatic* is the word used to designate this situation—and the disciples who thereby developed their technique of concentration. Moreover, this philosopher, mathematician and musician left no writings.[5]

c. During the birth of the first "*musiques de bruit*," described by Schaeffer in his first methodological treatises, writer and poet Jérôme Peignot declared in 1955 in *Musique animée*, a broadcast of the *Groupe de musique concrète*: "*What words could designate this distance that separates sounds from their origin....Acousmatic sound means (in the dictionary) a sound that one hears without revealing* [déceler] *the causes. Ah, good! Here is the very definition of the sound object* [l'objet sonore], *this element at the base of musique concrète, the most general music there is....*"[6]

d. In his *Traité des objets musicaux*...P. Schaeffer reclaimed the term "acousmatic" by attaching it to the phenomenological reduction or *epoché*, and to reduced listening.[7]

M3. Anonymous [Francis Dhomont], "New Media Dictionary," *Leonardo* 43(3):

Acousmatics—Derived from the name of a disciple of Pythagoras who listened to his lessons from behind a curtain so he would not be distracted by the physical presence of the master and could give his full attention to the content of the message. In the early twentieth century, the French word "*Acousmate*" (a noun from the Greek *Akouma*, "that which can be heard") could still be found in the two-volume *Larousse pour tous*. There it is defined as "an imaginary noise, or a noise for which no cause or author can be found." But in 1955, when *musique concrète* first appeared, writer and poet Jérôme Peignot started using the French adjective *acousmatique* to mean "the distance that separates sounds from their origin," referring to the impossibility of penetrating the speakers to reconstitute visual elements that could be related to the sounds. In 1966, Pierre Schaeffer considered calling his *Traité des objets musicaux "Traité d'acousmatique."* Finally, around 1974, to distinguish between and avoid any confusion with electroacoustical performances or transformed instruments (Ondes Martenot, electric guitars, synthesizers, real-time audio-digital systems), François Bayle introduced the French expression *musique acousmatique* to refer to music "that is shot, that is developed in the studio, that can be projected to an audience, like film."[8]

M4. Francis Dhomont, "Is there a Québec Sound?":

By shrouding "behind" the speaker (a modern Pythagorean partition) any visual elements that could be linked to perceived sound events (such as instrumental performers on stage), acousmatic art presents sound on its own, devoid of causal identity, thereby generating a flow of images in the psyche of the listener.[9]

In these overviews, the founding, history, and transmission of the term "acousmatic" divide into two periods. The first, spanning the school of Pythagoras through his reception in the Greek, Roman, and early Christian world, focuses on the origins and meaning of the word "acousmatic." It touches on various mythemes of the Pythagorean school and its divisions, the invention and deployment of the Pythagorean veil, and the separation of the eye and the ear as a pedagogical technique. The second period spans from the time of Diderot to Apollinaire's poetry, then to Schaeffer, Peignot, and the age of *musique concrète*, and finally to François Bayle and *musique acousmatique*. It also encompasses the expansion of the term into a modern variant, "*acousmate.*" Maintaining these periods as thematic, I present another set of statements grouped in the first period, exclusively addressing the Pythagorean origins and meaning of the term.

M5. Diderot, *L'encyclopédie*:

To understand the Acousmatics, it is necessary to know that the disciples in the school of Pythagoras were divided in two separate classes by a veil [*voile*]; that the first class, the most advanced, who underwent five years of silence without having seen their master at the rostrum, and having always been separated from the others the entire time by a veil, were finally admitted into the space of the sanctuary where they could hear and see him face to face; they are called Esoterics. The others who remained behind the veil were called as Exoterics or Acousmatics.[10]

M6. Mladen Dolar, *A Voice and Nothing More*:

[Acousmatic] has a precise technical meaning: according to Larousse, "acousmatic" describes "the noise which we hear without seeing what is causing it." And it gives its philosophical origin: "The Acousmatics were Pythagoras' disciples who, concealed by a curtain, followed his teaching for five years without being able to see him." Larousse follows [the *Life of Pythagoras* of] Diogenes Laertius (VIII, 10): "[His pupils] were silent for the period of five years and only listened to the speeches without seeing Pythagoras, until they proved themselves worthy of it." The Teacher, the Master behind a curtain, proffering his teaching from there without being seen: no doubt a stroke of genius which stands at the very origin of philosophy—Pythagoras was allegedly the first to describe himself as a "philosopher," and also the first to found a philosophical school. The advantage of this mechanism was obvious: the students, the followers, were confined to "their Master's voice," not distracted by his looks or quirks of behavior, by visual forms, the spectacle of presentation, the theatrical effects which always pertain to lecturing; they had to concentrate merely on the voice and the meaning emanating from it.[11]

M7. Jérôme Peignot, "Musique concrète":

To try and finish in good time with the expression "musique concrete" why not use the word "acousmatic," taken from the Greek word akousma, which means "the object of hearing." In French, the word "acousmatique" already describes those disciples of Pythagoras who, during five years, only heard his lessons hidden behind a curtain, without seeing him, and keeping a rigid silence. Pythagoras was of the view that a simple look at his face could distract his pupils from the teachings that he was giving them. If one gives the word an adjectival form, acousmatic, it would indicate a sound that one hears without being able to identify its origin.[12]

M8. Beatriz Ferreyra:

a. The term "acousmatic" comes from Pythagoras and his method of teaching: he taught behind a curtain so that his pupils could only hear his voice without seeing him, without visual support, without recognizing the sound source.[13]

b. [Pierre Schaeffer] also introduced the idea of "reduced hearing," a sort of acrobatic exercise that pushed the composer to hear the sound without examining the cause of its production, to hear the sound out of its context. This abstraction of causality is now one of the foundations of electroacoustic musical composition. That's why our music is called acousmatic music. It comes from Pythagoras's technique of teaching philosophy to his students behind a screen, so his students listened to his voice without seeing him. They were called the Acousmates.[14]

M9. Carolyn Abbate, "Debussy's Phantom Sounds":

Schaeffer invoked the figure of Pythagorus, recalling how one of the cults of disciples surrounding the great mathematician listened to him teach from behind a curtained arras, the better to focus their thoughts on the content of his speech, not be distracted by his body or his gestures. Power accrues to the utterance and not the person; words are also freer, something more than the speech of a human being; they point not merely to Pythagorus and his earthly form, but become symbols that detach entirely from an agent of utterance to take on other meanings.[15]

M10. Pierre Schaeffer, *Traité des objets musicaux*:
Acousmatic, the Larousse dictionary tells us, is the: "*Name given to the disciples of Pythagoras who, for five years, listened to his teachings while he was hidden behind a curtain, without seeing him, while observing a strict silence.*" Hidden from their eyes, only the voice of their master reached the disciples. It is to this initiatory experience that we are linking the notion of acousmatics, given the use we would like to make of it here. The Larousse dictionary continues: "*Acousmatic, adjective: is said of a noise that one hears without seeing what causes it.*" This term...marks the perceptive reality of a sound as such, as distinguished from the modes of its production and transmission. The new phenomenon of telecommunications and the massive diffusion of messages exists only *in relation to* and *as a function of* a fact that has been rooted in human experience from the beginning: natural, sonorous communication. This is why we can, without anachronism, return to an ancient tradition which, no less nor otherwise than contemporary radio and recordings, gives back to the ear alone the entire responsibility of a perception that ordinarily rests on other senses. In ancient times, the apparatus [*dispositif*] was a curtain; today it is the radio and the methods of reproduction, along with the whole set of electro-acoustic transformations, that place us, modern listeners to an invisible voice, under similar conditions.[16]

A final set focuses on the modern transmission of the variant term "acousmate."

M11. Marc Battier, "What the GRM brought to music":
a. This term ["acousmatic"]...has to be extended through the notion of "acousmate," which gave a mystical dimension to the phenomenon of hidden sound. Sound technologies have increasingly reinforced the idea of acousmate as a number of great mystics have given witness, supporting our listening to voices without bodies. Voices without bodies: this addresses itself to the idea that with sound technology one can transport or reproduce sound without its being associated with the material that produced it.[17]
b. Here is what the Dictionary of the Académie française says in its fifth edition of 1798: "ACOUSMATE. Noun singular. Noise of human voices or instruments that one imagines one hears in the air."...[This definition] can be found copied exactly in the notebooks of the young Apollinaire. The poet gave this title of acousmate to two of his poems.[18]

By overlaying and comparing the various statements, it is possible to construct a synthetic key myth.

M′. The key myth:
The term "acousmatic" refers to the disciples of Pythagoras who heard the philosopher lecture from behind a screen, curtain, partition, or veil (M1, M2b, M3,

M4, M5, M6, M7, M8, M9, M10). *The reason they remained on the far side of the veil was to promote a form of concentrated listening (M2a, M8b) or to emphasize the master's message (M1, M3, M6, M7, M9) undistracted by the visual aspects or physical presence of the speaker (M1, M3, M6, M7, M8, M9). In addition to keeping a vow of silence for five years (M1, M5, M6, M7, M10), this exoteric ritual formed part of an initiation into the Pythagorean school where pupils would then see the master (M1, M5, M6). From the experience of the acousmatics, we derive the adjectival sense of the term, meaning a sound that one hears without seeing or being able to identify the originating source (M2a, M2c, M3, M4, M6, M7, M8a, M8b, M10, M11a). The term was transmitted by Diderot in the* Encyclopédie *(M1, M5) and in the pages of* Larousse *(M6, M10). A related term, "acousmate" (M1, M3, M11), was found in the* Dictionnaire *of the Académie française (M11b), as well as* Larousse *(M3). Apollinaire, a lover of rare words, used "acousmate" as the title of two short poems (M1, M11b). These poems tell of voices heard in the air (M1, M11b). The writer Jérôme Peignot was the first to employ "acousmatic" as a term for describing* musique concrète *(M2c, M3). Schaeffer learned about the term from Peignot (M1) and, by attaching it to the phenomenological epoché, developed a concept of acousmatics that formed a significant part of this theory in the* Traité *(M2d). Modern audio technology preserves the ancient acousmatic tradition of the Pythagorean veil (M10) or its mystical variants (M11). Acousmatic music continues the tradition of* musique concrète *Pythagoreanism by veiling sounds, through the use of the loudspeaker, of all causal and contextual associations (M2a, M4, M8b).*

Based on M′, it is possible to throw some of the most unusual and idiosyncratic statements into relief.

1. Some of these statements simply contain mistakes, but do not appear to intentionally misconstrue the facts in the name of some particular purpose or aim. Take, for instance, the mytheme concerning the name of the exoteric disciples of Pythagoras. While the majority of the statements claim that "acousmatics" designates a group of disciples in the Pythagorean school, M3 traces the name back to a single disciple. M8 preserves the group, but claims that they were called the "acousmates," not the "acousmatics."[19] Of course, this claim is not externally supported by the classical sources, nor does it internally agree with other mythemes.[20]

2. Some of the statements omit information in order to promote a particular set of interests. For example, five of the statements address the mytheme concerning the initiatory aspects of Pythagoras' teaching and the five-year vow of silence (M1, M5, M6, M7, M10). Four of those statements mention that this vow was followed by a promotion whereby the initiated were entitled to see the master face to face. One sample (M5) from that set of four explicitly names the class of initiated disciples the "Esoterics." The only sample that mentions the vow without discussing the latter initiation is M7, an important text by Jérôme Peignot. Explicitly stated in M1, but implied in M2c and M3, Peignot was the first to use the adjective "acousmatic" to describe *musique concrète* and he is the figure from whom

Schaeffer was introduced to the term. (I have more to say about Peignot's role in the following chapter.) His omission of the esoteric side of Pythagoreanism is noteworthy, for it portrays the acousmatics as listeners trained in a certain attentive mode—a listening that is unconcerned with the physical source from which the sounds are emitted—rather than as a competing sect in the Pythagorean school.

Peignot's intellectual milieu, his set of particular interests, his institutional role within Schaeffer's organization, his work as a poet, and his influences clarify the meaning of M7. For Peignot, the adjective "acousmatic" was intended to replace the term *musique concrète*. Schaeffer, by originally calling his music "concrete," was trying to position it against the "abstract" music he heard coming from Germany and invading the borders of France after the war. However, concrete music was not intended as a reactionary term; rather, it was chiasmically allied with abstract painting, in that both art forms sought direct encounters with their material conditions—on one side, sound unmediated through the note; on the other, a direct experience with color and line unmediated by the figure.[21] Peignot's insistence on "acousmatic" as the proper descriptive term, rather than *musique concrète*, emphasizes the dictionary definition of the term more than its Pythagorean connotations. The experience uniquely afforded by *musique concrète*, in contrast to other forms of electronic music, was that of a "distance which separates the sounds from their origins."[22] Peignot is not describing the esoteric act of composition; he is interested in offering an *aesthetic* of *musique concrète*, one that is quite content to hold itself to the far side of the Pythagorean veil. Peignot, as a critic and advocate for *musique concrète*, takes his seat with the audience, facing the loudspeakers, not behind the mixing board.

3. Some of the statements augment or embellish mythemes with new details, and thus appear idiosyncratic in comparison to M'. François Bayle, the author of M2, embellishes the standard mytheme that the "acousmatics" heard the master lecture from behind a veil by claiming that Pythagoras lectured in the dark. Of course, this raises a puzzling question. If Pythagoras was already behind a curtain in order to hide his appearance, why did he also require the cover of darkness?[23] Pragmatically, this seems a bit overdetermined. But the puzzle is solved when one realizes that this embellishment is strategically placed with a goal in mind—to set up a mimetic identification between the ancient philosopher and Bayle's own practices as a composer. Bayle popularized the term *"musique acousmatique"* (a term that Schaeffer did not use to describe his compositional work) to describe his particular brand of *musique concrète*, which employs arrays of loudspeakers to create complex patterns of sonic diffusion and projection. This usually requires the creation of a special hall, called an *acousmonium*, in which to perform these pieces. He refers to his compositional practice as "cinema for the ear,"[24] encouraging the propagation of mental images by placing listeners in darkened rooms and exposing them to evocative sounds moving through space. Francis Dhomont (M3) inscribes Bayle into that history, as the inventor of a sound art akin to cinema, in which sounds are "shot and developed in the studio, [and] projected in halls."[25] Thus, by placing Pythagoras' lectures in the dark, Bayle can forge a mimetic identification between the philosopher and himself that authorizes his own cinematic practice of *musique acousmatique*. Pythagoras' darkness is really Bayle's darkness, a penumbra

that hides the loudspeakers to facilitate the promotion of images in the listener.[26] Bayle's fashioning of the Pythagoras legend should be read as a cipher intended to ground his own practices.

Additionally, Bayle deploys what is perhaps the most central mytheme in M'—the tale of the Pythagorean veil. It is present in all but one of the statements (M11). Within the tradition of *musique concrète*, the Pythagorean veil is most often used to organize a set of mimetic identifications, by binding ancient terms to their modern counterparts. The composer occupies the position of Pythagoras, unfolding a musical discourse or projecting a sonic message into the dark while remaining hidden. The audience occupies the position of the *akousmatikoi* (the "hearers," "listeners," or "auditors") who—like Peignot—receive the discourse while remaining outside the veil, listening with concentration to the emissions of the invisible master. The loudspeaker, the mixing console, and the technical tools of the studio occupy the place originally held by the Pythagorean veil. By describing Pythagoras as the inventor of an "original device [*dispositif original*]," Bayle (M2b) grants to the ancient philosopher the aura of an engineer, one that fits well with the extensive use of recording devices (editing stations, signal processors, etc.) in the production of *musique acousmatique*. This claim echoes Schaeffer (M10), who first described the Pythagorean veil as a *dispositif* while making the identification of ancient and modern outright: "In ancient times, the apparatus [*dispositif*] was a curtain; today it is the radio and the methods of reproduction...that place us, modern listeners to an invisible voice, under similar conditions." In addition to recuperating the *technical* aspects of the veil to the practice of *musique concrète*, Bayle also draws out its *aesthetic* consequences. In M2b, he claims that the veil was employed by Pythagoras as a device for developing the technique of "concentration," and in M2a, he calls this a situation of "pure listening." One assumes that these special listening techniques are underscored in order to foreshadow Schaeffer's *écoute réduite*, various kinds of sonic *solfège*, or even the link between the Husserlian technique of phenomenological *epoché* and acousmatic listening (M2d).

The strength of this mimetic configuration depends on the degree to which these identifications bind together the ancient and modern terms into a self-supporting structure. Past and present are stitched together in a pattern that effaces historical, cultural, and technological difference. As I argued in the previous chapter, Schaeffer's thinking, as well as those who follow him closely in their theories of acousmatic experience, promotes an ahistorical view of technology, sound, and listening. As an instrument for the obliteration of time, Levi-Strauss claims that myth "overcomes the contradiction between historical, enacted time and a permanent constant."[27] In the end, these historical contradictions are effaced because they obscure the transmission and arrival of an ancient heritage. An acousmatic horizon, originally disclosed by the ancient technology of the Pythagorean veil, is relived and reanimated by the loudspeaker. Being modern, we have rediscovered that we were always already ancient.

To put it bluntly, the primary role of this tale of the veil is mythic. Although I have borrowed aspects of Levi-Strauss's structural analysis of myth, my particular usage of the term is far more indebted to the work of Jean Luc-Nancy.[28] In Nancy's analysis, presented in *The Inoperative Community*, myth acts as a "founding fiction, or a foundation by fiction" deployed to organize the interests of a community.²⁹ "Concentrated within the idea of myth is perhaps the entire presentation on the part of the West to appropriate its own origin, or to take away its secret, so that it can at last identify itself, absolutely, around its own pronouncement and its own birth," Nancy writes.³⁰ This applies to the mythic discourse on acousmatic sound. For the Schaefferian tradition, rather than squarely address its historical and cultural origins, the Pythagorean veil becomes the origin of the acousmatic horizon— or in Schaeffer's strongly phenomenological terms, its "originary experience." The invocation of Pythagoras is an attempt by the practitioners of *musique concrète* to determine their own origin. It is an act of autopoiesis, or self-foundation. The tale of the Pythagorean veil is the primal scene of electroacoustic music, organizing its self-appropriation, retroactively founding an arché, and projecting a telos. "All myths are primal scenes," says Nancy, "all primal scenes are myths."³¹ A scene like other scenes, the tale of the veil possesses many trappings of theatrical fictions: curtains, offstage voices, a darkened auditorium, and the imposition of silence. And, like all primal scenes, its veracity is as dubious as its grip is powerful.

In addition to operating as a founding fiction, Nancy indicates two other features of myth relevant for the acousmatic discourses under examination. First, myth need not only operate on speculative or fictional material, but functions even in *the accounts one gives of the transmission of that material*. "The scene is equally mythic when it is simply the apparently less speculative, more positive scene of the transmission of myth."³² As we move from the depths of ancient Pythagoreanism and the birth of philosophy to the crisp, clear, and distinct prose of Diderot or the disembodied and ambient poetry of Apollinaire, we cannot say that we are moving from the discourse of myth to the discourse of Enlightenment or modernism. The account of the transmission of myth—who told what to whom, when, and why?—is still the warp and woof of the myth. It authorizes the modern-day acousmatics to give an account of their patrimony, the survival of their knowledge, and to position themselves as the appropriate (if not singularly suited) recipients of the myth. Second, myth operates indifferently on material that may or may not have been invented, since its main concern is the function to which this material is put to use. "We know that although we did not invent the stories (here again, up to a certain point), we did on the other hand invent the function of the myths that these stories recount."³³ When Bayle authorizes his practice by constructing a set of mimetic identifications with Pythagoreanism, his myth operates on material that is both partially discovered and partially invented. The Pythagorean veil was available for appropriation while simultaneously being modified for particular ends.

However, one should not single out the Schaefferians for fabricating stories about Pythagoras on such scant evidence or for such autopoietic purposes. People have been telling stories about Pythagoras for a very long time, often for the purpose of defending positions with little concern for historical accuracy. Indeed, one constant in the reception of Pythagoras is that it necessitates the creation of a stockpile of stories about him. Since we know so little about Pythagoras, he functions as a nearly blank slate upon which to inscribe acts of personal and institutional *Nachträglichkeit*, where the past is rewritten in accord with the demands of the present. Pythagoras is the perfect figure to anachronistically authorize some latter day privileged claim or position.

By stepping back from these legendary tales and fabrications, I hope to inquire into the word "acousmatic" from a position outside the Schaefferian tradition. With the awareness that one cannot outwit myth by pretending to shatter it with the force of history—for documents and archives never speak for themselves—I will, at the very least, try to "interrupt" it, as Nancy suggests. The form of that interruption will involve the comparison and investigation of our mythemes against a series of ancient sources. (Note: As I develop the comparison, I will mark ancient sources with the prefix "S" to differentiate them from our initial mythic statements that are marked by "M.")

I want to give the reader fair warning that this interruption will involve quite a lot of detailed investigation into the extant sources. But that is always the case when working diligently with a corpus of texts as old and fragmentary as those in the ancient world. For readers less concerned with the historical origins of the term "acousmatic," simply leafing through this and the following chapter might be good enough to satisfy their curiosity without taxing their patience. But for those who are interested in the transmission of this term, the contexts in which it was used, and the meanings it has accrued, I hope the patient investigations presented will be edifying. I have tried to provide all the evidence required, and nothing more than what is required, to answer the following questions: Who were the *akousmatikoi*? What was their relation to the *mathematikoi*? How was the division of the Pythagorean school understood? What were the roles of seeing and hearing for the Pythagoreans? When and where does the Pythagorean veil emerge, and for what ends? How was the word "acousmatic" transmitted, and what was its reception? What is the relationship between *acousmatic* and *acousmate*? When, where, and why did the latter term emerge?

In posing these questions, I am tracking more than the word "acousmatic" and the legend of the acousmatic veil; I am also investigating three related terms: first, the Greek word *akousmata*, the "things heard" or oral saying of Pythagoras; second, the word *akousmatikoi*, the class of Pythagorean disciples whose name derives from their status as "auditors" or "listeners"; and finally, the unusual French word *acousmate*, meaning the "sound of voices or instruments heard in the air." In this chapter and the next, I divide this constellation of terms into two large phases: first, a phase that investigates the classical literature for information on the *akousmatikoi* and Pythagorean *akousmata*, tracing usage and transmission up to the revival of Pythagoreanism in the Renaissance; second, a phase that begins with the baptism of the word *acousmate* in the French Enlightenment, and traces its reception by Apollinaire, Peignot, and eventually Schaeffer. Although the impact of this interruption may diminish the efficacy of using acousmatic sound in its current mythic role, it may also have the effect of clarifying the articulation of the processes of autopoiesis and appropriation operative in acousmatic myth. At the very least, we will know more about how the Pythagorean veil and its related mythemes function as a founding fiction and as the fiction of a foundation.

AKOUSMATIKOI AND *MATHEMATIKOI*

As demonstrated in M′, a basic set of mythemes promotes the view that the acousmatics were a group of Pythagorean disciples positioned on one side of a screen, veil, or curtain, unable to see the master's face but able to hear his lectures. Iamblichus (c. 245–325 C.E.) is the most influential source of evidence concerning the acousmatics (*akousmatikoi*) and the Pythagorean veil—the uncited source for Schaeffer's account (M10), as well as Diderot's entry on the *l'acousmatiques* in *L'encyclopédie* (M5).[34] The Iamblichan account divides the Pythagorean school into two classes of disciples separated by a veil: The *mathematikoi*, seated inside the veil close to Pythagoras, were not only able to see the master lecturing but were entitled to witness demonstrations of his theories; the *akousmatikoi*, seated outside the veil, were only entitled to hear the master's propositions and were not given the privilege of seeing the demonstrations. The Greek term *mathematikoi* is often translated as "the students," while the term *akousmatikoi*, which literally translates as "those who hear" or "the auditors," derives from the belief that they heard the sayings (the *akousmata*) of Pythagoras from outside the veil.

The mytheme concerning the rigorous silence of the disciples (M1, M5, M6, M7, M10) also derives from the Iamblichan account. Before entering into the Pythagorean school, Iamblichus describes the extensive examination undergone by hopeful students. Pythagoras, after examining their relations with parents and kin,

(S1) watched them for untimely laughter, and silence and chatting beyond what was proper. Also, he looked at the nature of their desires, the acquaintances with whom they had dealings and their company with these. Most of all, he looked at the leisure occupations in which they spent the day, and what things gave them joy and pain. He observed, moreover, their physique, manner of walking and their whole bodily movement. Studying the features by which their nature is made known, he took the visible things as signs of the invisible character traits in their souls.[35]

Following this initial inquiry, Iamblichus claims that potential disciples underwent a three-year probationary period to see if they were disposed to a "true love of learning" (*alethines philomatheias*).[36] The probationary period was then followed by an initiatory period during which the new pupils, in order to test their capacity for self-control, were compelled to observe a vow of silence for five years. "The subjugation of the tongue," writes Iamblichus, "is the most difficult of all victories."[37]

During the period of silence, *akousmatikoi* participated in Pythagoras' discourse, (S2) "through hearing alone, being outside the veil and never seeing him."[38] But, after passing the period of initiation, (S3) "the candidates themselves, then, if they appeared worthy of sharing in his teachings, having been judged by their way of life and other virtuousness, after the five year silence, became 'esoterics' and heard Pythagoras within the veil, and also saw him."[39] Consistently, Iamblichus describes the position of the disciples as *exo sindonos* (outside the veil) or *entos sindonos* (inside the veil). The word *sindon* (σινδων) means "a fine cloth, usually linen," but also "anything made of such cloth," such as a shroud, winding sheet, or a napkin.[40] Because the word *sindon* can refer to both a curtain and a veil, standard accounts use both terms, often interchangeably.[41] Whether veil or curtain, the *sindon* that hangs between the two camps within the Pythagorean school does not function primarily as a means for demarcating spatial locations or separating vision from hearing. Rather, the veil is emblematic of two different kinds of pupils, representing two different orientations within Pythagoreanism and, ultimately, representative of competing views of Pythagoras within the ancient world. The real difference between the *akousmatikoi* and *mathematikoi* is not simply, or even primarily, determined by the physical position of the students inside or outside the veil.[42] Rather, it is a difference of kind and orientation—a difference between Pythagoras as teacher of scientific wisdom and Pythagoras as "shaman," to use Walter Burkert's handy phrase.

Most scholars do not dispute that there were distinct types of Pythagorean disciples, tracing the distinction as far back as Aristotle.[43] Rather, the challenging questions emerge when one inquires into the central features that distinguish the *akousmatikoi* from the *mathematikoi*. Which group represents the oldest, most original forms of Pythagoreanism? Who were the authentic disciples?

The *akousmatikoi*, or exoteric disciples, are typically described as religious Pythagoreans. **(S4)** "The philosophy of the *Acousmatics* [*akousmatikon philosophía*]," according to Iamblichus, "consists of oral instructions without demonstration and without argument: e.g., 'In this way one must act.'"[44] Since the acousmatics do not have access to the proofs and demonstrations of the master, they have become the inheritors of a motley assortment of doctrines and sayings (*akousmata*) that lack explanation.[45] Following the master's precepts and proscriptions, they scrupulously observe a series of taboos and rites concerning bathing, diet, and other matters of everyday life and worship.[46] In many respects, the *akousmatikoi* are the disciples who treat Pythagoreanism as a "way of life," or what is often referred to as the Pythagorean *bios*. One unusual source of information concerning the *akousmatikoi* comes from the middle comedies (middle–late fourth century B.C.E.), such as those by Alexis or Aristophon, in which grubby and dimwitted Pythagoreans are satirically depicted with a variety of unsavory personal characteristics: as barefooted, oddly dressed, clothed in ragged and dirty garments; as covered in filth due to their taboo on bathing; as adherents to the doctrine of vegetarianism; as believers in metempsychosis; as prohibited from eating beans and drinking wine; and as drawing undue attention to themselves by their conspicuous habit of keeping silent. Jokes played at the expense of these *akousmatikoi* are plentiful and derisory, designed to expose the hypocrisy of their ascetic lifestyle. For example, Alexis targets the Pythagorean taboos on eating meat. After it is mentioned that Pythagoreans are prohibited from eating anything animate, someone objects that Epicharides, who is a Pythagorean, eats dogs. To this, the response follows: "yes, but he kills them first and so they are not still animate."[47]

In contrast, the *mathematikoi*, or esoteric disciples, are characterized as scientific Pythagoreans, often identified as the genuine disciples of the group. Unlike the *akousmatikoi*, who do not have access to the proofs and demonstrations of Pythagoras, the *mathematikoi* possess knowledge in the fields of learning grouped under the rubric of *mathemata*—such as arithmetic, geometry, music, and astronomy.[48] In the history of philosophy, especially through the filter of the Academic tradition, the great majority of writers have emphasized the "scientific" aspects of Pythagoras' teaching, from the Pythagorean theorem to his discovery of the basic harmonic proportions of music, to his rational cosmology and founding of the word "philosophy." The *mathematic* Pythagoras has also been transmitted in the history of music theory as a thinker who discovered principles capable of associating cosmological motions and musical proportions.[49] However, since this view is shaped by the Platonic and

Academic tendency to view Pythagoras as a natural scientist, rational philosopher, and mathematician, the claim for its authenticity must be closely scrutinized.

WHO ARE THE GENUINE PYTHAGOREANS?

Within Pythagoreanism, a debate rages over which group, the acousmatics or the mathematics, are the genuine disciples. One way to track this debate is to focus on the figure of Hippasus of Metapontum (end of sixth cent. B.C.E.–early fifth cent. B.C.E.). Hippasus is one of the earliest Pythagoreans discussed in the ancient literature, appearing in Aristotle's writings on the Pythagoreans.[50] Aristoxenus (370–300 B.C.E.) claims that Hippasus undertook musical experiments with bronze discs of various sizes and thicknesses.[51] In contrast with the experiments attributed to Pythagoras, some modern scholars claim that Hippasus' experiments actually produced scientifically verifiable results concerning the numerical proportions involved in musical concords.[52] In addition, he is often described as discovering the harmonic mean.[53] In light of these descriptions, one could see in Hippasus the epitome of a mathematic Pythagorean. Yet, in the story of Hippasus' death, the tension between the religious and scientific faces of Pythagoreanism is legible. According to Iamblichus, Hippasus committed the "impiety" of "having disclosed and given a diagram for the first time of the sphere from the twelve pentagons [i.e., a pentagonal dodecahedron]," and was thus put to death by drowning.[54] Ostensibly, the secrecy and silence that are hallmarks of acousmatic Pythagoreanism also held for the mathemata, which were not to be carelessly disclosed to the uninitiated.[55] Hippasus challenged acousmatic secrecy with his disclosure of mathematic knowledge.

A controversy between the two camps arose over Hippasus' disclosure. On one side, the acousmatics disavowed the significance of Hippasus' impiety by arguing that his mathematical work was his own original invention and could not be traced back to the figure of Pythagoras. For them, Hippasus' mathematical inventions initiated a new and inauthentic line within Pythagoreanism—a mathematical strain—that departed from the religious, social, and communal practices to which they were dedicated. Thus, they argued, Hippasus' mathematic Pythagoreanism was not genuine. On the other side, the mathematics claimed that the work of Hippasus could not be admitted as original and derived from the figure of Pythagoras himself. If they were to admit Hippasus' contributions as original, they would undermine their claim for the priority of mathematical and scientific knowledge in the Pythagorean school. For *mathematikoi* to be the genuine disciples, Hippasus must have been little more than a plagiarist.[56]

The dispute is not easy to resolve. If we take our evidence from Iamblichus, himself a prominent Neoplatonist, one would expect the Academic interpretation, which supports the *mathematikoi*, to hold the day. Concerning the dispute, Iamblichus writes, (S5) "there were two kinds of philosophy, for there were two kinds of those pursuing it: some were acousmatics and others were mathematics. Of these, the mathematics are agreed to be Pythagoreans by the others, but the mathematics do not agree that the acousmatics are Pythagoreans."[57] The argument is predicated on the gift of recognition: If the *akousmatikoi* recognize the *mathematikoi* as Pythagoreans, but not vice-versa, then the *mathematikoi* must be the genuine disciples. Iamblichus supports this argument by claiming that Pythagoras himself identified the *mathematikoi* as his "true followers" and "decreed" that the *akousmatikoi* "show themselves as emulators" of the *mathematikoi*.[58] Yet, Iamblichus' strategy is quite transparent: Put the argument into the mouth of the master in order to make it so. Although an authorial proclamation would help to settle the question, naturally, no evidence for this Pythagorean decree exists.

However, Iamblichus contradicts his own argument a few paragraphs later. Citing the evidence of "a certain Hippomedon...a Pythagorean of the acousmatics," Iamblichus offers a different tale about the origins of the acousmatics and mathematics. Iamblichus claims that Pythagoras originally gave demonstrations and explanations for all of his precepts; but because these explanations were passed down through a series of intermediaries, they were eventually omitted or lost, while the bare precepts remained. Yet, these ancient precepts, even without explanation, preserve wisdom that can be originally attributed to Pythagoras. Because of the ancient patrimony of the *akousmata*, **(S6)** "they who are concerned with the mathematical doctrines of the Pythagoreans (the mathematics), agree that these (the acousmatics) are Pythagoreans, but they claim even more strongly, that what they themselves say is true."[59] Here the argument from recognition is reversed: The *mathematikoi* recognize the *akousmatikoi* as genuine Pythagoreans, even if they have lost the demonstrations and reasons for Pythagoras' precepts. By granting this concession, Iamblichus' weakens his argument for the *mathematikoi* as the genuine disciples. The overemphatic claim that the precepts of the *mathematikoi* are nevertheless true is small recompense for abandoning their stake on the direct historical lineage. Tacked onto the end of this passage, Iamblichus reasserts his mathematic position by countering the acousmatic claim concerning the originality of Hippasus' inventions. He states that, although Hippasus became publicly famous for his disclosures, when it comes to mathematics, "all the discoveries were of that man...Pythagoras."[60]

This strange contradiction in Iamblichus' text has not escaped notice. Walter Burkert, in his magisterial *Lore and Science in Ancient Pythagoreanism*, places great emphasis on the passages in question: §81 and §§87–89 (S5 and S6).[61] Burkert's book offers the most comprehensive and detailed survey of the literature on Pythagoreanism, in order to assess to what degree Pythagoras' teaching was mathematical and scientific, or shamanistic and religious. The origins of the acousmatics and mathematics within the Pythagorean school is a central issue for Burkert, because his project is intended to offer a precise characterization of the earliest practices of Pythagoras and Pythagoreanism. Burkert tries to resolve the contradiction by citing an additional passage from Iamblichus' *De communi mathematica scientia*:

> **(S7)** Of these, the acousmatics are recognized by the others as Pythagoreans, but they do not recognize the mathematics, saying that their philosophic activity stems not from Pythagoras but from Hippasus....But those of the Pythagoreans whose concern is with the *mathemata* recognize that the others are Pythagoreans, and say that they themselves are even more so, and that what they say is true.[62]

This passage repeats Iamblichus' second reading (S6) by granting recognition to the *akousmatikoi* as genuine Pythagoreans. Burkert argues that both S6 and S7 are "correctly reproduced" from Iamblichus' ancient source—likely Aristotle. For Burkert, the Platonized reading of Pythagoras has left us with a distorted view, turning the historical figure of Pythagoras from a shaman and miracle-worker into a proto-scientist. Aristotle's references to Pythagoreans present a greatly contrasting picture. His conjecture that Aristotle is Iamblichus' source is supported by the fact that S6 and S7 are congruent with other evidence concerning the early Pythagorean school in the extant literature by Aristotle.[63]

Aristotle's account suggests that both groups subscribed to the *akousmata*—as the argument from recognition implies—but that the two differed in the manner in which they implemented Pythagoreanism in practice. Here is Aristotle, as copied by Iamblichus:

> **(S8)** Pythagoras came from Ionia, more precisely from Samos, at the time of the tyranny of Polycrates, when Italy was at its height, and the first men of the city-states became his associates. The older of these [men] he addressed in a simple style, since they, who had little leisure on account of their being occupied in political affairs, had trouble when he conversed with them in terms of learning (*mathemata*) and demonstrations (*apodeixeis*). He thought that they would fare no worse if they knew what to do, even if they lacked the reason for it, just as people under medical care fare no worse when they do not additionally hear the reason why they are to do each thing in their treatment. The younger of these [men], however, who had the ability to endure the education, he conversed with in terms of demonstrations and learning. So, then, these men [i.e., the *mathematikoi*] are descended from the latter group, as are the others [i.e., the *akousmatikoi*] from the former group.[64]

Glossing this passage, Philip Sidney Horky has argued that the distinguishing difference between the acousmatics and the mathematics concerns the "type of knowledge" employed. "The acousmatic Pythagoreans only have knowledge of 'the fact' of 'what one is to do,' but the mathematical Pythagoreans, whose knowledge is advanced, understand the 'reason why they are to do' what they should do."[65] One can trace the distinction between the "fact" (*to oti*) and the "reason why" (*ti dei prattein*, literally, "what one is to do") in Aristotle's other writings—textual evidence that helps to establish the authenticity of this passage.[66] In Aristotle's view, the fundamental difference between the acousmatics and the mathematics is not a difference in beliefs; rather, the difference depends on the latter's use of demonstration to provide arguments for their ideas. "While acousmatic Pythagoreans apparently simply accepted the facts as they were, mathematical Pythagoreans engaged in investigations that employed the principles of mathematics in order to makes sense of the world they experienced."[67]

But for Aristotle, the explanations of the *mathematikoi* were flawed. In the *Metaphysics*, Aristotle critiques mathematical Pythagoreans for relying too heavily on homology as a mode of explanation. The *mathematikoi* "were the first to take up mathematics" and began to believe that "its principles were the principles of all things."[68] Because numbers were considered the principles *of* principles, the *mathematikoi* would find homologies between numbers and natural phenomena, such as musical harmonics and astronomy. These homologies provide poor grounds for analyzing natural phenomena and were often extended far beyond whatever usefulness they may have possessed. When their homologies failed to account for the facts,

Aristotle charges the *mathematikoi* with making additions to preserve the coherence of their theory. A telling example comes from the application of number theory to astronomy: "As the number ten is thought to be perfect and to comprise the whole nature of numbers, they say that the bodies which move through the heavens are ten, but as the visible bodies are only nine, to meet this they invent a tenth—the 'counter-earth.'"[69]

How much faith should we place in these Aristotelian sources? Although the authority of Aristotle is great, and the uniqueness of his account helps to differentiate it from the Platonists—who were much more partisan in their high estimation of Pythagoras—the reception, as always, must be measured. Aristotle has an argument to make with the Pythagoreans and writes about them from a philosophical, not historical, perspective. Although Burkert's foundational work argues for the priority of the acousmatics as the genuine disciples of Pythagoras—whom he characterizes as a religious figure or, in his terms, a shaman—he summarizes the basic question concerning the division of the Pythagorean school in a passage that bears repeating:

> Modern controversies over Pythagoras and Pythagoreanism are basically nothing more than the continuation of the ancient quarrel between *acousmatics* and *mathematics*. Is there nothing more in the doctrine of Pythagoras than what is indicated by the *akousmata*, with which the Pythagoras legend and the theory of metempsychosis are of course closely connected? Or was there from the beginning, behind these religious and mythical features, whose existence cannot be denied by the modern scholar any more than it could by the *mathematics*, a new, scientific approach to philosophy, mathematics and the study of the world's nature?[70]

We should note that Schaeffer's reception of Pythagoras, which selectively emphasizes some aspects of the legend at the expense of others, prolongs the very ancient quarrel of the acousmatics and the mathematics. (Perhaps it is *only* this prolongation of the ancient debate, of putting Pythagoras at the origin of one's practice in order to authorize it, that makes Schaeffer a Pythagorean.) One cannot be but struck by the very unusual image of Pythagoras in Schaeffer's *Traité*, one that is quite distinct from the typical mathematic reception of Pythagoras in the history of music theory. Schaeffer's Pythagoreanism (if one can indeed call it that) is neither bound to the monochord, to the natural science of harmonics, string lengths, tuning, the mathematics of musical proportions, nor to the inaudibility and omnipresence of the music of the spheres. Viewing the situation with a very broad lens, Schaeffer's Pythagoreanism is acousmatic in the sense that it is primarily focused on listening; it veers away from mathematic explanation or scientific demonstrations of phenomena. It is a metaphysics of sound objects, not numbers. Schaeffer, at the very end of the *Traité*, criticizes his fellow composers and musical researchers, "fanatics of a digital catechism," for their "mathematical bigotry."[71] Schaeffer's reception of Pythagoras as teacher of techniques of listening—an unusual view in the history of music theory—has, surprisingly, gone almost entirely unnoticed.[72] Yet, if emphasis on listening is what aligns Schaeffer with the acousmatics, then we must press on and investigate how the mythemes on acousmatic listening compare with what is attested in the ancient sources. What do the ancient sources say about acousmatic listening? Is there any basis for the claim (M2a, M8b) that the acousmatics developed a technique of concentrated listening? What was heard on the far side of the veil, and how was it auditioned?

THE ACOUSMATA

Unlike the modern-day acousmatics of Peignot or Schaeffer, the Pythagorean *akousmatikoi* heard nothing from the far side of the veil resembling bells, trains, or creaking doors, or even the music of the spheres; rather, the acousmatics attended to the *akousmata*, the "things heard," the maxims or "oral sayings" of the master. Aristotle's lost treatise, *On the Pythagoreans*, contained a list of Pythagorean *akousmata* that was widely transcribed and thus preserved in the ancient sources. Today, it affords the reader a motley assortment of curious maxims:

—what are the Isles of the Blessed? The sun and the moon;
—an earthquake is a mass meeting of the dead;
—a rainbow is the reflected splendor of the sun;
—one must put the right shoe on first;
—do not walk on roads travelled by the public, or wash oneself in bathing houses;
—do not join in putting a burden down, but join in taking it up;
—the most just thing is to sacrifice; the strongest, insight; the most beautiful, harmony;
—do not speak in the dark;
—do not have children by a woman who wears gold jewelry;
—do not sacrifice a white cock, for it is a suppliant and sacred to the moon;
—a bronze ring, when struck, releases the voice of a daemon;
—abstain from beans; do not break bread; do not pick up food that falls from the table, for it belongs to the Heroes; put salt on the table as a symbol of righteousness;
—spit on your nail and hair trimmings;
—do not look into a mirror with help of artificial light;
—do not stir the fire with a knife;
—do not urinate facing the sun.

Iamblichus, following Aristotle, divides the *akousmata* into three different groups: those concerning what a thing is, those concerning what is the best in any category, and those concerning what it is necessary to do in various situations.[73] Although Burkert describes this threefold division as "artificial and inconsistently followed," it provides some orientation in trying to parse out different functions served by the *akousmata*.[74] Starting with the final group, one might describe these *akousmata* as delineating rules about the performance of sacrificial rituals, how to show honor to the gods, dietary restrictions, and prescriptions for everyday behavior. The mixture of maxims shows the intercalation of sacrificial and ritual practices, basic medical and biological concepts, and a "religious awe before the elementary forces of nature."[75] The second group, concerning what is the best in any given category, approaches various topics of concern for the Pythagorean school, from medicine and religion to number theory and philosophy. "What is the most just? To sacrifice. What is the wisest? Number; and in the second place is that which gives names to things. What is the wisest of things among us? Medicine. What is the loveliest? Harmony. What is the strongest? Insight."[76] As Christoph Riedweg says, "in their brevity these sayings are simultaneously enigmatic and suggestive," offering up bits of wisdom in a form that lacks clear explanation. The implication is that such bits of knowledge about the best in any category will act as directives for leading an ethical life.[77] Yet, these riddles, though not wholly opaque, are presented in a form that lacks supporting reasons and demonstrations; thus, they leave each inquiry underdetermined by preserving an explanatory gap between question and answer.

The remaining group, *akousmata* concerning what a thing is, is perhaps the most mysterious of all. Porphyry (235–305 C.E.), a Neoplatonist contemporary of Iamblichus and author of a life of Pythagoras, reproduces *akousmata* of this variety: The constellations are "the tears of Kronos," Ursa Major and Minor are "the hands of Rhea," and the planets are "Persephone's dogs." Riedweg describes these *akousmata* as "early examples" of allegory; they "decode the true, real meaning in a (figurative) mythical mode of expression."[78] The answers provided are often nearly inexplicable. If there was indeed a vow of secrecy within the Pythagorean school, then one might speculate that such answers or explanations were internal secrets of the school—perhaps relying on tacit knowledge to understand what kind of interpretive or allegorical framework was being employed. Such procedures were not foreign to other ancient ceremonies or initiations into the mystery cults. As Riedweg notes, these forms of initiation often had a didactic purpose, whereby initiates were "introduced to the relevant cultic myth and its correct interpretation."[79] Such a procedure would correspond well to the division of the Pythagorean school into exoteric and esoteric groups. In this way, the Pythagorean school may have been continuing interpretive practices rooted in older forms of archaic Greek lore and oracular practice. Burkert identifies the *akousmata* with *griphos*, or riddles, which were used in the "promulgation of oracles," in order to argue that "the *akousmata* are, rather than simple, commonsense wisdom in abstruse form, ancient magical-ritual commandments."[80]

Whether the Pythagoreans themselves understood the *akousmata* as literal, figural, ritual, or practical is an open question. What is perhaps more important for my purposes is that the later traditions of Neoplatonists, early Christians, and Gnostics certainly did take such maxims as allegorical, especially around the time when Iamblichus' *Vita* was composed. Burkert writes, "The prevailing view in antiquity was that what was desired [in interpreting the *akousmata*] was not compliance to the letter but comprehension of the deeper meaning."[81] Since the *akousmata* lacked definitive explanations, transmitted as a list of sayings in Aristotle's lost text, philosophers who wished to indulge in speculation about Pythagoras and Pythagoreanism could try their hand at either offering explanations and allegorical interpretations of the sayings or accounting for their superficial appearance.[82] Porphyry claims that the philosophy of the Pythagoreans died out because it was "enigmatical" and was written in "Doric," which was an "obscure dialect."[83] Because these teachings could not be fully understood, they were misapprehended and suspected as spurious by later generations. Moreover, he argues that the best parts of the Pythagorean wisdom were appropriated by Plato, Aristotle, and their various followers, who then maliciously characterized whatever was leftover as "Pythagorean" in order to cast contempt on Pythagoreanism.[84] Androcydes, an earlier source than Porphyry (and a Pythagorean of whom little is known), also regards the *akousmata* as *ainigmata*, which "clothe a lofty wisdom in unintelligible language."[85] About this later reception of the *akousmata*, Burkert writes, "allegorical interpretation, here as elsewhere, was the necessary means of adapting ancient lore to new ways of thinking, and thus preserving its authority."[86]

VEILED UTTERANCES

The Pythagoreans, according to Iamblichus, "protected their talks with one another and their treatises" by the use of "symbols," casting the *akousmata* in a cryptic form.[87] Such encryption remained in accordance with their vows of silence and method of instruction, which forbade the transmission of Pythagorean wisdom to the uninitiated. Thus, Iamblichus argues, the *akousmata* are only properly disclosed through a proper method of exegesis:

> **(S9)** If someone, after singling out the actual symbols, does not explicate and comprehend them with an interpretation free from mockery, the things said will appear laughable and trivial to ordinary persons, full of nonsense and rambling. When, however, these utterances are explicated in accord with the manner of these symbols, they become splendid and sacred instead of obscure to the many... and they reveal marvelous thought, and produce divine inspiration in those scholars who have grasped their meaning.[88]

However, this method of encryption was not equally available to all the Pythagorean disciples. As is typical in Iamblichus' text, there is a great division between the acousmatic and mathematic understandings of the *akousmata*. The acousmatics justify the obscurity of the *akousmata* by claiming that, originally, **(S10)** "[Pythagoras] declared the reasons and gave demonstrations of all these precepts [i.e., the *akousmata*], but because they were handed down through many intermediaries, who became progressively lazier, the reason was omitted, while the bare precepts remained."[89] But the mathematics have a different story to tell. Citing the Aristotelian account presented above (S8), Iamblichus argues that the obscurity of the *akousmata* derives from Pythagoras' practice of withholding demonstrations from those unprepared to fully understand them. The difference between the *akousmatikoi* and the *mathematikoi* depends on the kind of philosophical study one makes and the aptitude one has for "scientific lessons." The *mathematikoi*, knowing the demonstrations, hold the key to the proper allegorical exegesis of the *akousmata*.

Immediately following this account, the veil appears. This is significant indeed, for Iamblichus moves from the question of *veiled* utterances to the meaning of the veil itself. The veil differentiates "two types of philosophical study." **(S11)** "Those who heard Pythagoras either within or without the veil, those who heard him accompanied by seeing, or without seeing him, and who are divided into the esoteric and exoteric groups,"[90] are all ways of describing the fundamental distinction between acousmatic and mathematic Pythagoreans—that is, those who know how to properly interpret the master, and those who do not.[91] Yet, that distinction is not developed in terms of the difference between seeing and hearing, or between the spatial locations of the various disciples. After committing to the view that the *akousmata* are coded, as Iamblichus does, what benefit could there possibly be in seeing or not seeing the master?[92] If the key to unlocking the *akousmata* depends on the degree of initiation of the receiver, knowledge of the demonstrations, and the gift of allegorical exegesis, then the visual aspects of the speaker would be simply irrelevant. The Pythagorean veil is in excess of the hermeneutic situation; how the *akousmata* are decoded is indifferent to the issue of whether one does or does not see the speaker. Yet, in the Schaefferian myth of the Pythagorean school (M′), which borrows generously from Iamblichus' text, the veil is taken as a physical divider or screen, a *dispositif* or device that serves a host of purposes—none of which are specifically hermeneutical. It spatially divides the Pythagorean school and separates the auditory effect from its physical source; it distinguishes the eye from the ear in order to reduce the visual contributions to audition; it creates new conditions of listening that rely solely on the ear and not on the visually overdetermined context.

Indeed, the more we look into the ancient sources for the Pythagorean veil, the more we find its literal and physical status questionable. In Book V of the *Stromateis*, Clement of Alexandria (c. 150–c. 215 C.E.) contrasts the *akousmatikoi* and *mathematikoi* and then introduces the veil only after treating the question concerning the proper interpretation of the *akousmata*.[93] For Clement, the *akousmatikoi* and *mathematikoi* are separated not by spatial location but according to their level of "genuine attachment to philosophy,"—the *akousmatikoi* aligned with the curious multitude, and the *mathematikoi* aligned with the genuine students of philosophy.[94] The veil, although present, hangs nowhere in Clement's account. Simply put, Clement's veil is allegorical.

In Book V, §58, Clement cites examples from Greek philosophy where allegories were used to conceal esoteric wisdom:

> **(S12)** It was not only the Pythagoreans and Plato then, that concealed many things; but the Epicureans too say that they have things that may not be uttered, and do not allow all to peruse those writings. The Stoics also say that by the first Zeno things were written which they do not readily allow disciples to read, without their first giving proof whether or not they are genuine philosophers. And the disciples of Aristotle say that some of their treatises are esoteric, and others common and exoteric. Further, those who instituted the mysteries, being philosophers, buried their doctrines in myths, so as not to be obvious to all. Did they [the ancient philosophers] then, by veiling (*katakrupsantes*) human opinions, prevent the ignorant from handling them; and was it not more beneficial for the holy and blessed contemplation of realities to be concealed (*epekruptonto*)?[95]

Both the words *katakrupsantes* and *epekruptonto* share a common root, the verb *krupto* (κρυπτω)—the root from which we derive the word "cryptography"— which means "to hide, cover, cloak; to cover in the earth, bury; hide, conceal, keep

Myth and the Origin of the Pythagorean Veil

secret."[96] The veiling, hiding, or coding of the *akousmata* preserves the meaning of the discourse from the uninitiated or ignorant. For Clement, the wisdom of Greek philosophy was transmitted within a rhetorical economy of covering and uncovering, revealing and concealing—a strategy of encoding messages to ensure their appropriate understanding by the initiated. The common trope that the *akousmata* require exegesis is reworked into a generalized tradition of a cryptographic hermeneutics.

Clement's key sentence about the Pythagorean veil appears immediately afterward. Pythagoras and Plato concealed many ideas, things that

> (S13) are to be expounded allegorically, not absolutely in all their expressions, but in those which express the general sense. And these we shall find indicated by symbols *under the veil of allegory* (ηυπο παρακαλυμματι τη αλληγορια). Also the association of Pythagoras, and the twofold intercourse with the associates which designates the majority, the *akousmatikoi*, and the others that have a genuine attachment to philosophy, the *mathematikoi*, hinted that something was spoken to the multitude, and something concealed from them.[97]

The word used to describe the veil in this passage, *parakalummati*, means "anything hung up; a covering, cloak, curtain."[98] It could be referring to an actual veil, were it not for its connection to allegory; rather than sounding forth from behind a veil, the *akousmata* are themselves veiled—they are presented *under the veil of allegory*. Unlike the physical veil that hangs in Iamblichus' account, Clement's veil is woven from figural language. Moreover, the veil not only functions as an allegorical figure, it becomes the figure of allegory. As the figure of figurality—the icon of the hermeneutical power of meaning to be concealed—the veil of allegory figures the power of language to be simultaneously communicative and opaque, encoded for the initiated, but banal for the multitude.

A comparison of two passages in Clement and Iamblichus (S13 and S11) interrupts the transmission of the Pythagorean veil in the history of philosophy and should give us pause. On the one hand, considering that Iamblichus tells his story of the separating veil immediately after describing Pythagoras' use of allegorical *akousmata*, is it not possible that we, as readers, are being encouraged to read this tale of the veil as being itself figural? What reason prevents us from considering that Iamblichus himself is not acting like Pythagoras, presenting the auditor with a riddle that, unless suitably explicated, remains "laughable and trivial to ordinary persons, full of nonsense and rambling?"[99] Perhaps Iamblichus would find laughable our dogged literality and lack of interpretive skill, envisioning the master hidden behind a screen rather than understanding the point of the figure. Perhaps our inability to read these subtleties in Iamblichus' text maintains our exoteric position. (We might say the same about Schaeffer and the Schaefferians who, being no less literal than Diderot [M5], unhesitatingly accepted the Pythagorean veil as real rather than figural.) On the other hand, considering that Clement introduces the juxtaposition of the *akousmatikoi* and *mathematikoi* at the same moment he mentions the veil of allegory, is it not possible that he tropes upon previous accounts of the literal veil? Could it be that his account of allegorical veiling knowingly replicates and refashions an older convention of a literal dividing veil, by transforming it into a figure in homage to Pythagoras' own excessive use of figurality?

If there is an ancient Greek source for the Pythagorean veil, we are lacking evidence in the historical record to prove it. I can find no extant trace of it earlier than Clement's *Stromateis*. Moreover, given Clement's encyclopedic reading and constant referencing of other sources, it seems reasonable to assume that he might have mentioned such an account. The only potential candidate I can find is Timaeus, the historian of Southern Italy (350–260 B.C.E.).[100] Since Pythagoreanism had a long tradition in that part of the ancient world, Timaeus is a significant and authoritative source of information.[101] Although Timaeus' historical works are no longer extant, Diogenes invokes him as an authority in a passage that touches on the topos of seeing and hearing Pythagoras.

(S14) According to Timaeus, [Pythagoras] was the first to say "Friends have all things in common" and "Friendship is equality"; indeed, his disciples did put all their possessions into one common stock. For five whole years they had to keep silence, merely listening to his discourses without seeing him, until they passed an examination, and thenceforwards they were admitted to his house and allowed to see him.[102]

This passage is similar to Iamblichus (S2), with the veil substituting for the house. However, in a second passage that is uncited, Diogenes claims that Pythagoras was so greatly admired that **(S15)** "not less than six hundred persons went to his evening lectures: and those who were privileged to see him wrote to their friends congratulating themselves on a great piece of good fortune."[103] Here the penumbra of evening and the size of the gathering preclude the auditors from seeing Pythagoras. This new passage complicates things tremendously. Was Pythagoras obscured by the darkness, or did he lecture from inside his house? R. D. Hicks, in the Loeb edition of Diogenes' *Lives*, skates over the question by adding a critical footnote to the end of S14, claiming that Pythagoras was heard but not seen since he lectured at night, and referring the reader to S15.[104] But these two accounts cannot be so quickly reconciled. Is Pythagoras obscured by darkness or by some physical device? To posit both would be overdetermined.[105]

Philological evidence, based on replications of vocabulary and grammatical construction in extant texts, shows that Timaeus was indeed a source for both Diogenes and Iamblichus.[106] Textual analysis can be used to establish Timaeus as the source concerning common property in the Pythagorean school, the period of five years of silence, and initiation after a test or trial. Yet, it does not establish him as a source for the Pythagorean veil. On that point, if there is an ancient source, we are left with a lacuna.[107] At the very least, though there is some evidence that there may have been an older tradition concerned with whether Pythagoras was visible or obscured, there is no evidence for the veil as a central part of this tradition.

THE METHOD OF CONCEALMENT

Lacking an ancient source, it appears safe to say that the veil is something of a latecomer to the Pythagorean legend, a product of late antiquity appearing in the literature no earlier than the second or third century C.E. Moreover, I would conjecture that the topos of the Pythagorean veil, based on the available evidence, begins with Clement as a figural veil, operating as part of his interpretive strategy of hiding and revealing known as the "method of concealment."[108] According to this method, some piece of knowledge or divine wisdom, rather than being spoken literally and disclosed to all, is concealed in figural language. The coded utterance can transmit its content to those prepared to receive the message, while avoiding misinterpretation by those unprepared. The method of concealment allows for the same utterance to be available to esoteric and exoteric listeners, while being properly understood only by the former. Henny Fiska Hägg offers three reasons why Clement employs the method of concealment: First, Clement holds to the position that not all students can be taught in the same manner, so that the teacher must adapt his or her message to the appropriate situation; second, language is itself inadequate for expressing divine truth in a literal manner, so symbolic or figural language is required to convey its sense; third, Clement believes that the authors of scripture and Greek philosophical works employed this method and is thus authorized to follow after their model.[109] Additionally, Clement's method of concealment is motivated by a suspicion toward writing—one that is consistent with his emphasis on vocality and logos (discussed below)—and has many of the features of classical logocentrism.[110] According to logocentric premises, oral transmission is always superior to writing because the teacher is in the position of directly assessing how his discourse is being understood. The teacher can distinguish "the one who is capable of hearing from the rest," by keeping an eye on their "words and ways, their character and life, their impulses and attitudes, their looks, their voice," and so forth.[111] Writing, reduced to being an aide-memoire, is merely a necessary evil.[112] Having none of the safeguards that privilege oral speech, written texts are open to misinterpretation. Citing Plato's *Second Epistle*, Clement claims that "once a thing is written there is no way of keeping it from the public" where it can "make no response to a questioner beyond what it written." The written text is destitute, possessing "no voice," relying on support from its author or some external defender following in the footsteps of the author to defend its claims.[113] Given the ease with which the written text is open to misinterpretation and misuse, the method of concealment is a strategy for transmitting knowledge to only those qualified to interpret the teaching.

Clement finds authorization for this method of concealment in texts that predate Christianity—in Egyptian, Hebrew, and Greek writings:

(S16) In accordance with the method of concealment, the truly sacred Word, truly divine and most necessary for us, deposited in the shrine of truth, was by the Egyptians indicated by what were called among them adyta, and by the Hebrews by the veil. Only the consecrated—that is, those devoted to God, circumcised in the desire of the passions for the sake of love to that which is alone divine—were allowed access to them. For Plato also thought it not lawful for "the impure to touch the pure." Thence the prophecies and oracles are spoken in enigmas, and the mysteries are not exhibited incontinently to all and sundry, but only after certain purifications and previous instructions.[114]

Clement's method of concealment was a central textual strategy in the rise of Christian philosophy in late antiquity. Literary theorist Frank Kermode has explored forms of early Christian concealment in his work on literature and secrecy.[115] According to Kermode, Jesus' parables are designed as riddles or allegories for the initiated. "When Jesus was asked to explain the purpose of his parables he described them as stories told to them without—to outsiders—with the express purpose of concealing a mystery that was to be understood only by insiders."[116] A thematic passage from the Gospel of Mark supports Kermode's claim. Jesus says to the apostles, "To you has been given the secret of the kingdom of God, but for those outside everything is in parables; so that they may indeed see but not perceive, and may indeed hear but not understand."[117] For Kermode, such parables are at the very origin of hermeneutics—a discipline named after the god Hermes, the patron of "heralds and what heralds pronounce, their *kerygma*. He also has to do with oracles, including a dubious sort known as *kledon*, which at the moment of its announcement may seem trivial or irrelevant, the secret sense declaring itself only after long delay, and in circumstances not originally foreseeable."[118]

The method of concealment also possesses another side—revelation. Just as Clement describes the need to encode the transmission of divine knowledge, he also addresses its manner of unveiling. The voice of the divine logos, which manifests itself most completely in Christ, can also be demonstrated in the harmony that exists between Christian, Greek, Hebrew, and Egyptian texts—a harmony that is predicated on the idea that a single divine logos speaks in all and utters a single truth.[119] When these ancient authors spoke, the resulting voice was not their own, but that of the divine logos. In essence, these ancient authors were involved in an elaborate act of ventriloquism. According to David Dawson, Clement's "voice-based hermeneutic" allowed him to produce acts of "revisionary reading" that were intended to intervene in the cultural and religious situation of early Alexandria, with its complex and fragmented overlapping of Christianity, Gnosticism, and paganism. Through this strategy, the "authorial specificity of [Clement's] precursors is irrelevant to the fact that when subjected to his revisionary reading, they express the same underlying voice or meaning."[120]

The figure of the veil appears regularly in Clement's *Stromateis*, especially in Book V where he discusses Pythagoras and the use of allegory as a technique within the method of concealment. To understand the pervasiveness of this figure in Clement's discourse, I present a few cases where the concealment and revelation of the divine logos are read back into Greek and Hebrew texts and associated with figures of the veil. Paradigmatically, Clement identifies the figure of the veil with the method of concealment: "All then, in a word, who have spoken of divine things, both Barbarians and Greeks, have veiled the first principles of things, and delivered the truth in enigmas, and symbols, and allegories, and metaphors, and such like tropes."[121] The ancient Greek poets—such as Orpheus, Linus, Musaeus, Homer, and Hesiod—are also understood as veiling their discourse in order to convey the divine logos only to the initiated: "The persuasive style of poetry is for them a veil for the many."[122] The Pythagorean *akousmata*, which possessed a venerable tradition of allegorical interpretation even before Clement's time, are appropriated by Clement to this same model.

While the veil functions tropologically as the generalized figure for allegories, metaphors, symbols, and other forms of figural language, Clement also discusses literal veils. Describing the tabernacle that houses the Holy of Holies, Clement draws upon biblical sources where literal veils are present to create a consecrated space that distinguishes the proper realm of the priests from the multitude.

> (S17) Now concealment is evinced in the reference of the seven circuits around the temple, which are made mention of among the Hebrews.... In the midst of the covering and veil, where the priests were allowed to enter, was situated the altar of incense, the symbol of the earth placed in the middle of this universe; and from it came the fumes of incense. And that place intermediate between the inner veil, where the high priest alone, on prescribed days, was permitted to enter, and the external court which surrounded it—free to all the Hebrews—was, they say, the middlemost point of heaven and earth.... The covering, then, the barrier of popular unbelief, was stretched in front of the five pillars, keeping back those in the surrounding space.[123]

Yet, in both cases, literal and figural, the *function* of the veil remains the same. Namely, the veil separates the initiated from the uninitiated, the exoteric from the esoteric. But the emphasis differs when the veil is treated as literal or figural. When the veil is figural, Clement highlights the linguistic act of concealing truths in obscure language; when the veil is literal, it becomes the "barrier of popular unbelief." The emphasis shifts to the difference between the authentic and inauthentic recipients of the encoded message. The method of linguistic encryption is soft-pedaled in favor of addressing the techniques of reception that distinguish the various grades of disciples.

Iamblichus' deployment of the literal veil emphasizes the act of reception. Such an emphasis jibes well with the particular exigencies of Iamblichus' own political and religious situation. For Iamblichus, who comes from a Neoplatonist tradition that had already canonized Pythagoras as an important precursor to Plato and had emphasized (even to the point of fabricating) the authenticity of the mathematics as disseminated through Plato's works and the Academic tradition, the stakes of Pythagoras' reception in late antiquity were great. Against the Christian appropriation of Pythagoras as a precursor of Jesus, the Neopythagorean emphasis on mathematic esotericism allowed them to differentiate between an authentic and inauthentic reception of Pythagoras' legacy. For the pagan Neoplatonists, like Iamblichus, it was important to emphasize the *singularity* of Pythagoras as a thinker who transmitted philosophy and the *mathemata* from Egypt to Plato and on through the Academy, rather than assimilate Pythagoras' voice to ventriloquism of the divine logos.

One way of legitimating Neoplatonism against the Christians was to focus on the role of Pythagoras in the transmission of the *mathemata*. In Clement, the *mathemata* are noticeably absent; Pythagorean harmonia, mathematics, music, and natural science play no role in the transmission of the divine logos. While Clement produces interpretations of Pythagorean *akousmata* (alongside those of Hebrew and Egyptian symbols), there is little emphasis on the symbolism of numbers, on the role of geometrical knowledge, or on the primary principles of the one and the many. Yet, these remained central and distinctive aspects of the Pythagorean tradition for Iamblichus. In *De Mysteriis*, writing under the pseudonym "Abamon" (an Egyptian priest), he articulates the first principles of "Egyptian" cosmology—principles that bear a striking and obviously intentional resemblance to Pythagorean doctrines associated with the *mathemata*:

> **(S18)** And thus it is that the doctrine of the Egyptians on first principles, starting from the highest level and proceeding to the lowest, begins the One, and proceeds the Many, the Many being in turn governed by the One, and at all levels the indeterminate nature being dominated by a certain definite measure and by the supreme causal principle which unifies all things.[124]

The musical aspects of the Pythagorean *mathemata* are also not neglected. Again, in *De Mysteriis*, Iamblichus outlines the role of "musical theurgy" in terms that are clearly Pythagorean and Platonic.[125]

> **(S19)** Before it gave itself to the body, the soul heard the divine harmony. And accordingly even when it entered the body, such tunes as it hears which especially preserve the divine trace of harmony, to these it clings fondly and is reminded by them of the divine harmony; it is also borne along with and closely allied to this harmony, and shares as much as can be shared of it.[126]

The particular emphasis on the Pythagorean *mathemata*—such as the connection between metempsychosis and anamnesis, the audibility of divine harmony to the soul, and its echo in earthly life and the principles of the one and the many—carve out Iamblichus' brand of Pythagoreanism in sharp contrast to Clement's appropriation of the master. Rather than subjugate Pythagoras to the unity of the divine logos, Iamblichus attempts to place him as a *singular* figure within a tradition—an initiate and teacher of ancient hermetic wisdom—of which he himself is a part. Traces of the agon with Christianity appear in the texts of Neoplatonic authors like Iamblichus and Porphyry.[127] By emphasizing Pythagoras' wonder-working and coded knowledge, the ancient philosopher begins to resemble a pagan saint in order to compete with the miraculous workings of Jesus and the martyrs.

Historian Peter Brown argues that the period of late antiquity saw the beginnings of the Christian Church and a style of religious experience marked by the rise of the "friends of God," individuals who could claim dominance over "earthly" forces by possessing a special, direct relationship with heaven. The rise of Christian martyrdom and texts like *The Acts of the Martyrs* placed emphasis on the "friends of God" and the new form of religious power that was implied in their experiences. According to Brown, "Over against the secular hierarchy of an increasingly 'pyramidal' society there stood, in clear outline, a spiritual hierarchy of 'friends of God,' the source and legitimacy of their power in this world held to rest unambiguously on a heavenly origin."[128] The shadowy origins and shamanistic practices of Pythagoras made him an attractive candidate for revisionary readings, where he could occupy the position of a pagan friend of God. Pythagoras became a figure ripe for spiritualization, capable of challenging the new crop of Christian martyrs and saints. "Late in the [third] century, in the circles of Plotinus, Porphyry, and in the early fourth century, with Iamblichus, the image of the 'divine man' takes on firmer outlines among pagan philosophers. The appearance within one generation of two major lives of Pythagoras is the symptom of this change."[129] It is perhaps no coincidence that the miracles of Pythagoras, as they appear in the pagan literature, resemble those of Jesus, with Iamblichus' *On the Pythagorean Life* often assuming "characteristics of a pagan alternative to the Gospels."[130]

DISSOLUTION AND FUSION IN RENAISSANCE PYTHAGOREANISM

The eventual Christianizing of Europe would resolve the battle between Christian and pagan over the status of Pythagoras. However, the pagan mysteries, which were closely associated with the figure of Pythagoras, did not disappear; in fact, they underwent a great revival in the Renaissance. Broadly speaking, for Renaissance thinkers, Pythagoras had become assimilated into the Christian tradition. Thus, one might say that Clement won the argument, although victory was Pyrrhic. Not only is he a far less important interlocutor for Pythagoreanism than Iamblichus—whose works were translated by Ficino and widely circulated—but the basic Renaissance reception of the Pythagoras legend followed along the lines of Iamblichus' account.[131] Rather than consider Pythagoras to be speaking—or rather spoken by—the divine logos, he became significant as a node in the transmission of ancient knowledge. For instance, scholars like Pletho (c. 1355–1452), an important figure in the rise of Renaissance Platonism, argued that Pythagoras transmitted Zoroaster's learning and knowledge of magic to Plato. Pletho's student, Cardinal Bessarion (1403–1472), emphasized that Plato learned his method of secrecy and the importance of exercising memory and discipline from Pythagoras. The role of secrecy was to keep divine doctrines away from the hands of the masses. This method was eventually adopted by the Christians and incorporated into their doctrine. Thus, Pythagoras, in the eyes of Bessarion, is a significant figure in the transformation from ancient philosophy into Christianity, not only as a precursor but also as a source of Christian inspiration.[132] In a more surprising move, Ficino (1433–1499) (in his introduction to *Pimander*) traces a line of venerable ancient theologians from Moses, to Hermes Trismegistus, to Orpheus, to Pythagoras—at one point even claiming that Pythagoras was Jewish.[133] Johannes Reuchlin (1455–1522) gave Pythagoras a prominent place in his argument that the Kabbalah was a prime source for Christianity. In his letter to Pope Leo X, Reuchlin takes up the idea that Pythagoras was an inheritor of Mosaic wisdom, and that it was he who rescued the mystical Hebraic texts and incorporated them into his doctrines. The Kabbalah, which Reuchlin understands as an ancient oral tradition of secret and esoteric doctrines, presented in the guise of symbols, numbers, and enigmas, was first presented to Moses and then revealed to the Greeks by Pythagoras. According to Reuchlin, "Pythagoras drew *his* stream of learning from the boundless sea of Kabbalah [and] led his stream into Greek pastures from which we, last in the line, can irrigate our studies."[134] Thus, Pythagoras is imagined to be *the* central node in the transmission of the Kabbalah to the Renaissance.[135]

Although the centrality of Pythagoras to Renaissance thought is unquestionable, for our purposes, we must note that interest in the details of the Pythagorean school—in the *akousmatikoi* and the *mathematikoi*, in the specific nature of the *akousmata*, and in the topos of the Pythagorean veil—diminishes by the time of the Renaissance. Rather, these features of Pythagoras' legend were assimilated into a generalized reception of Pythagoras as a node in the transmission of esotericism, Orphic mysteries, Egyptian knowledge, Jewish Kabbalah, and such. In the Renaissance Platonists, one can see the fusion of Clementine and Iamblichan strains. Pico (1463–1494) evinces this fusion when describing Pythagoras' use of allegory:

(S20) Pythagoras had the Orphic theology as the model after which he molded and formed his own philosophy. In fact, they say that the words of Pythagoras are called holy only because they flowed from the teachings of Orpheus: thence as from their primal source flowed the secret doctrine of numbers, and whatever Greek philosophy had that was great and sublime. But, as was the practice of ancient theologians, Orpheus covered the mysteries of his doctrines with the wrappings of fables, and disguised them with a poetic garment, so that whoever reads his hymns may believe there is nothing underneath but tales and the purest nonsense.[136]

I can find no sentence better than that final statement to demonstrate the combination of the two strains, as if Pico has intercalated sentences from Clement and Iamblichus. We see Clement's trope (S13) of the *akousmata* being covered in the veil of allegory—in this case, doctrines being wrapped in a poetic garment—concatenated with Iamblichus' claim (S9) that, to the uninitiated, the *akousmata* appear "full of nonsense and rambling." The Renaissance may have given birth to the revival of Greek knowledge—and the prolongation of pagan mysteries—but it also spelled the dissolution of the specificity of the *akousmata* and the divisions in the Pythagorean school as a living tradition. Although the *akousmatikoi* persevere in volumes on the history of ancient philosophy, we will have to wait a few centuries to discover the rebirth of the term *akousmata* in an unexpected place and in a foreign tongue—without a trace of Pythagoreanism.

3

The Baptism of the Acousmate

When Pierre Schaeffer premiered his *Cinq études de bruits* on Tuesday, October 5th, 1948, no listener would have associated the sounds filling the airwaves with the ancient word "acousmatic." The link would take nearly 20 years to forge. In 1954, Schaeffer left the studio and was put in charge of France's overseas radio network, developing radio stations, programming, and a media theory appropriate for the French colonies in Africa. The *Groupe de recherche de musique concrète* (GRMC) was placed in the administrative hands of Philippe Arthuys, with Pierre Henry as the artistic director. When Schaeffer returned to take back the reins of the GRMC in 1958, he was disappointed in what the organization had become. In particular, he was critical of the GRMC's emphasis on individual works over collective research. "*Musique concrète* has not, in effect, made great progress," Schaeffer lamented, "[it] has not produced a great work, nor yet confirmed a theory."[1] In his letter to Albert Richard, which opens a special volume of *La revue musicale*, Schaeffer bluntly stated his dissatisfaction:

> I dreamt of a return to collective involvement with procedures and systems, like those of the Conservatory, atonalism or cybernetics. At last, in place of concerts and festivals where snobbism is the law…I dreamt of an honest approach to the phenomenon of listening, of experimentation on diverse publics and an ethics of listening in which the musician would rediscover…his rules and his standards. Nothing like that has happened.[2]

Schaeffer decided to dissolve the GRMC and reform it as the *Groupe de Recherche Musicale* (GRM), which would place a larger premium on musical research by attempting to consolidate and systematize the results of the first decade of compositional and technical work.[3] During the change of the institutional name, the moniker *musique concrète* was also dropped. If the focus was no longer on concrete music but rather on musical research, what name could meet the task of describing these new objectives? Schaeffer preferred the term "musique expérimentale."[4]

In 1960, Jérôme Peignot, a poet, critic, commentator, and collaborator of the GRMC, offered his recommendation in an issue of *Esprit* dedicated to new music:

> **(M7)** To try and finish in good time with the expression "musique concrète," why not use the word "acousmatic," taken from the Greek word akousma, which

means "the object of hearing." In French, the word "acousmatique" already describes those disciples of Pythagoras who, during five years, only heard his lessons hidden behind a curtain, without seeing him, and keeping a rigid silence. Pythagoras was of the view that a simple look at his face could distract his pupils from the teachings that he was giving them. If one gives the word an adjectival form, acousmatic, it would indicate a sound that one hears without being able to identify its origin.[5]

Originally, Peignot's suggestion was stillborn. But after the publication of Schaeffer's *Traité*, with its exploration of the connection between the acousmatic situation and the phenomenological *epoché*, the term began to stick. François Bayle would eventually champion it during his tenure as the director of the GRM in order to differentiate his style of tape music from other forms of electroacoustic practice. Dhomont and Chion, and then (from Chion) Dolar and Abbate (M1, M4, M6, M9) would also prolong Peignot's suggestion by repeating his embellishment to the Pythagoras legend—the claim that Pythagoras used the veil so that the disciples would not be distracted by the look on his face.

However, it was not the first time Peignot had used the term "acousmatique" to describe the strange sounds heard when listening to the works of the GRMC. Five years prior, he had employed the term in a 15-minute radio broadcast, in collaboration with GRMC composer Philippe Arthuys:

> What words could designate this distance that separates sounds from their origin.... Acousmatic sound means (in the dictionary) a sound that one hears without revealing [*déceler*] the causes. Ah, good! Here is the very definition of the sound object [*l'objet sonore*], this element at the base of musique concrète, the most general music that there is....[6]

Here, Peignot deploys the term without its typical train of Pythagorean justifications. Rather, the word "acousmatic" emphasizes the distance between the sonic source and its audible effect. This idea of acousmatic sound—a sound that is disembodied, autonomous, and separated from its sources—will have a long and varied reception, both within the Schaefferian tradition and beyond.[7] For the moment, I want to hold that reception at bay, in order to pursue a historical question: How did Peignot discover this term?

Surprisingly enough, the key myth (M′) is not explicit on this point. Just before the turn of the century, Apollinaire wrote two poems entitled "Acousmate."[8] Under the heading "*acousmatique*, history of the word," François Bayle reproduces a stanza from each of these poems in the lexicon of his *Musique Acousmatique* (the source for M2). Since Peignot was himself a poet, the implication is that he discovered the term from reading Apollinaire. Marc Battier also cites the "Acousmate" poems in order to suggest a line of transmission from Apollinaire to Schaeffer by way of Peignot. Although neither Battier nor Bayle unequivocally makes this claim, the connection is quite plausible. Apollinaire's ideas would not have been foreign to Peignot, who grew up in an environment saturated in French modernism. His father, Charles Peignot, an important figure in the development of typography in the 20th century, owned the type foundry Deberny et Peignot. Charles associated with intellectuals and artists like Cocteau, Gide, and Le Corbusier, and was a great supporter of typographical experimentation. He founded two influential journals in the history of graphic design: *Divertissements typographiques* and *Arts et métiers graphiques*.[9] Jérôme, in his own *typoèmes*, pursued typographical experimentation along lines opened by Apollinaire's *Calligrammes*.

Yet, Apollinaire's "Acousmate" poems are quite immature works. They display neither the daring typographical avant-gardism of the *Calligrammes* nor the radical montage of the conversation poems. They are not undergirded by a commitment to cubism, and they do not demonstrate Apollinaire's aesthetic of Simultanism. Instead, they reveal the poet as a young man whose imagination was set ablaze in the rarified air of hermeticism. The "Acousmate" poems, like much of Apollinaire's writing in the period before he came to Paris, cultivate their subject matter from "the ruins of the past."[10] In notebooks from the period, Apollinaire often transcribed definitions of rare and unusual words from the dictionary, like "acousmate," "dendrophones," and "Argyraspides"—in 1899, the word "acousmate" was quite a *rara avis* indeed. This homemade glossary of hothouse words seemed to serve Apollinaire well, for he used them in poems from the period, as well as mining the quarry in poems from much later dates.[11]

Bayle and Battier are likely correct to imply that Peignot's discovery of the word was due to his familiarity with Apollinaire; however, there still remains a glaring question: Aren't "acousmate" and "acousmatique" different words? The statements that concern the word "acousmate" (M1, M3, M11) seem to treat the pair as if they were simply synonymous. Although I am not disputing the possibility that Peignot may have discovered the word "acousmate" from reading Apollinaire, there should be some accounting for their difference. What is their relation?

ACOUSMATE VERSUS ACOUSMATIQUE

(S21) ACOUSMATE, noun. The noise of human voices or instruments that one imagines hearing in the air [*ACOUSMATE, s. m. Bruit de voix humaines ou d'instruments qu'on s'imagine entendre dans l'air.*][12]

This definition from the *Dictionnaire* of the Académie Française is transcribed in Apollinaire's notebooks. Like the definition of "acousmatique" found in Larousse (and cited by Schaeffer), an "acousmate" could be understood as describing an audible sound produced by invisible sources. The potential invisibility of an acousmate is amplified by the definition's modal characterization—it is a sound that one *imagines* hearing. If the source is imagined, as in an auditory hallucination, one might simply assume its invisibility. But by definition, an acousmate does not require the criterion of invisibility. One could plausibly imagine visualizing the source of the sound while hearing it in the air. Furthermore, there is perhaps something supernatural about an acousmate, like hearing the voices of angels, or of extraordinary natural events, like the sound of thunder. If an acousmate does possess this supernatural or extraordinary character, it may contrast with acousmatic sounds, which (again, taken at the letter) are often said to be ubiquitous in the era of mechanically reproduced sound.

It is easy to imagine that Peignot, upon finding the word acousmate in Apollinaire, picked up his copy of Larousse and found, right next to it, the word acousmatique. Although many modern (and abridged) editions of Larousse do not contain both words, older editions place "acousmate" and "acousmatique" in successive entries.

(S22) ACOUSMATE, masculine noun (a-kouss-ma-te—from Greek *akousma*, that which one hears). Imaginary sound, a sound where one does not see the cause, the author [ACOUSMATE s. m. (a-kouss-ma-te—du gr. *Akousma*, ce qu'on entend). Bruit imaginaire, bruit dont on ne voit pas la cause, l'auteur.]

ACOUSMATIQUE, adjective (a-kouss-ma-ti-ke—root, *acousmate*) Pertaining to a noise that one hears without seeing the instruments, persons, or real causes behind it. [ACOUSMATIQUE adj. (a-kouss-ma-ti-ke—rad. *Acousmate*). Se dit d'un bruit que l'on entend sans voir les instruments, les personnes, les causes réelles dont il provient.]

—noun. Name given to the disciples of Pythagoras, who, for five years, would listen to his lectures hidden from behind a curtain, without seeing him, and observing the most rigorous silence [—Subst. Nom donné aux disciples de Pythagore, qui, pendant l'espace de cinq années, écoutaient ses leçons cachés derrière un rideau, sans le voir, et un observant le silence le plus rigoureux].[13]

If Peignot slid from acousmate to acousmatique, Larousse facilitated it. The Académie makes no mention of the *visual status of the object* imagined to be heard in their definition. Larousse, on the other hand, adds a clause to acousmate concerning the visual status of the sound's source, and thus effaces what might be a crucial sensory difference between an acousmate and an acousmatic sound. Additionally, Larousse provides an etymology that makes it appear as if the adjectival form of acousmatique is derived from acousmate. (Shortly, we will see why this is incorrect.) By altering the Académie's definition and tying together their roots, Larousse makes it appear that the adjective acousmatique and the noun acousmate are really two forms of the same word, alternately describing and referring to an audible sound lacking a visual source.

In constituting the key myth, Bayle, Battier, and Chion (M1, M3, M11) all tacitly accept Larousse's identification of acousmate and acousmatique. In his brief history of the term acousmatique, Bayle cites a few lines of Apollinaire's poems, preceded by this terse clause: "Note in the *Poèmes retrouvés* of Guillaume Apollinaire, 1913, this 'rediscovery' [*re-trouvaille*]."[14] Punning on the word *retrouver*, the implication is that the real find in Apollinaire's collection is simply the discovery of the word acousmate, and thus, the rediscovery of an ancient acousmatic tradition. Apollinaire becomes a crucial node in the transmission of this ancient horizon by prolonging the term acousmatique through its synonym acousmate. Thanks to Apollinaire, Peignot can pass the term onto Schaeffer. Yet, if acousmate is not synonymous with acousmatique, then what exactly did Apollinaire rediscover? Why not focus the attention on Peignot? Isn't he the one who makes the more significant *retrouvaille* by giving Schaeffer the word acousmatique and initiating its mythic icon, the Pythagorean veil? If we are truly being presented with a historical account, then Apollinaire's role can be little more than accidental, or at most, incidental.

BATTIER'S ARGUMENT

The musicologist Marc Battier, in a recent essay on the GRM, makes a case for keeping Apollinaire in the story. He argues that the central experience of phonography has less to do with the mechanical reproduction of sounds than with the separation of sonic sources from their effects. The splitting of sources from effects takes place at two levels in recorded sound: First, the source is separated from its effect and the effect is captured as a "physical inscription"; then, separation of source and effect is evident at playback, when the listener is given the responsibility for "reconstituting the sound image which traces the sound of [the] origin."[15] It is in these two tendencies, inscription and reconstitution, that "one finds the sources of the creation of phonographic sound."[16] Battier does not conceive of phonography as *initiating* the separation of auditory sources from effects; rather, the experience of such separation *predates* the invention of recording. He contests the technologically determined association of acousmatic sound with mechanically reproduced sound or "schizophonia," because the separation of sonic source from effect can already be detected in the word acousmate. This is why Apollinaire is such an important figure for Battier; the self-named *poète phonographiste* of the 1910s gives voice in his "Acousmate" poems to pre-phonographic experiences of "voices without bodies" and "sounds without their causal source."[17]

Battier considers three moments in Apollinaire's career that can be read as anticipations of Schaeffer's theories:

1. In *Le Roi-Lune*, Apollinaire imagines the Moon-King sitting at a keyboard where, through the use of microphones, he transports the sounds of the world into his chamber. Running his fingers over the keyboard, the king plays a "symphony, made by the world."[18] Battier cites the phrase to evoke Schaeffer's early works of *musique concrète* and his dreams of a "symphony of noises."[19]
2. Battier describes a trip Apollinaire made in December of 1913 to the Archives of the Voice, where he recorded himself reciting three poems. Upon hearing these recordings, Apollinaire experienced the uncanny effect of hearing his voice played back to him. "After the recording, they played my poems back to me on the apparatus, and I did not recognize my voice in the slightest."[20] Battier describes this moment in strongly Schaefferian terms. He calls Apollinaire's uncanny reaction an experience of "blind listening," perhaps overdetermining the invisible character of that experience.[21]
3. Battier cites a passage from André Salmon, who, upon hearing recordings of Apollinaire, noted that they made "very profound and delicate [aural] perceptions" of the voice available, perceptions that are stifled when the voice is heard in its usual context, emitted from the mouth of the poet. Salmon writes, "Thus at the second hearing we heard ourselves... for the first time." Battier intends the reader to understand this phrase as anticipating Schaeffer's claim that the acousmatic situation creates new conditions for listening—that we are often deceived about what we hear because of the context. By alienating and

separating the voice from its contextual and causal situation—by blinding us to the source of the voice—the voice returns to the ear anew.

When the poet René Ghil went into the studio, Apollinaire experienced a similarly chilling and uncanny effect upon hearing Ghil's disembodied voice, describing it as "aerial music."[22] Battier places great emphasis on this little phrase, for "aerial music" links the pre-technological world of Apollinaire's early "Acousmate" poems—where shepherds listen to angelic voices or a melancholic poet hears the quiet voice of the absent—directly to the alienating disembodiment of modern phonography. Identifying the experience of an acousmate (the sound of voices or instruments that one imagines to hear in the air) with that of phonography, Battier argues that the essence of both experiences inhabits the horizon defined by the separation of source and effect. The experience of recording prolongs (and reinforces) the pre-technological experience of an acousmate. Battier writes,

> (M11 complete) This term [acousmatic], henceforth used in reference to the musical work of the GRM, has to be extended through the notion of "acousmate," which gave a mystical dimension to the phenomenon of hidden sound. Sound technologies have increasingly reinforced the idea of acousmate as a number of great mystics have given witness, supporting our listening to voices without bodies. Voices without bodies: this addresses itself to the idea that with sound technology one can transport or reproduce sound without its being associated with the material that produced it. Historically, the idea of "acousmate" is linked to mysticism.[23]

In the past, shepherds may have heard angelic voices, but for us, "this is the role now played by phonography, to make voices without bodies or sounds without their causal source heard."[24] Apollinaire's poetry prefigures a technological experience yet to come.[25]

In addition to the argument that recording and the experience of an acousmate both inhabit a horizon of auditory effects split from their sources, Battier develops another argument concerning music and mysticism. By emphasizing the mystical connotations of the word acousmate, connotations evoked in Apollinaire's early poems, Battier makes a claim about the centrality of the term to musical experience generally. While canvassing the history of the term, Battier exposes a tantalizing passage on St. Cecilia, the patron saint of music. The passage appears in a volume from 1807, in which various definitions from the Académie's *Dictionnaire* are selected, critiqued, and commented upon. Battier cites the following:

> (S23) Biographers have written that St. Cecilia, ready for her martyrdom, heard within herself the songs of angels from which derives her title as the patron saint of music. If this historical point is correct, St. Cecilia was in a state of acousmate, or of enchantment, for these two words in the language of learned metaphysicians are essentially synonymous. Both designate a mental condition, which few physiologists know how to distinguish. The condition is rarely morbid, sometimes endemic; but those who suffer from it, when they are not saints, have often imputed it to witchcraft.[26]

What could be more mystical—and more acousmatic—than the moment when St. Cecilia hears the angelic chorus while turning her eyes from the earthly musicians? By connecting St. Cecilia to the term acousmate, Battier establishes a myth of musical listening that is predicated on an essential trait, the separation of sources and effects—in this case, the splitting of audible music from invisible choristers. This horizon of auditory severance, in which Battier had already inscribed the experience of recorded sound, is now extended back to its founding moment; St. Cecilia functions as the icon of that tradition. Paralleling Schaeffer's use of an "originary experience" to situate the loudspeaker in the horizon of the Pythagorean veil, Battier can now locate his own originary experience of acousmatic listening inside the iconic ear of St. Cecilia.

ACOUSMATE AND THE *ACADÉMIE FRANÇAISE*

Of course, making claims about St. Cecilia is a venerable strategy in the history of music aesthetics, as Wackenroder, Kleist, Schopenhauer, or Nietzsche could attest.[27] As the apotheosis of the musical listener, claims about Cecilia resonate across the whole of music, for the terms of her mystical listening become an ideal standard to which the rest of us mere mortals aspire. It would indeed be symbolically powerful if a tradition supported the notion that Cecilia heard an acousmate. But that is not quite what Battier argues; rather, Cecilia does not *hear* an acousmate but is *in a state of* acousmate. There is a subtle swerve in the meaning of acousmate from an object—the *sound* of voices and instruments heard (or imagined to be heard) in the air—to a mental state, a particularly auditory form of mystical experience. Great weight is placed on the idea that one can be in a "state of acousmate, or of enchantment," to cite the author of S23. Battier's usage of the term acousmate, most often appearing without a definite or indefinite article, demonstrates a shift in the term away from being an object *heard* toward a state of enchantment within hearing. (This occurs in both instances of the term in M11 and in Battier's habitual phrase "the idea [or notion] of 'acousmate.'"[28])

Upon closer examination, Battier's source (S23) cannot sustain the weight he places upon it. The passage comes from a volume entitled *Remarques morales, philosophiques, et grammaticales, sur le dictionnaire de l'académie françoise*, which Battier incorrectly attributes to Antoine-Augustin Renouard.[29] The author is not Renouard but rather Gabriel Feydel, who published the work under the pseudonym P* P* P* with Renouard's press in 1807. The volume is organized alphabetically, and it systematically reproduces entries from the Académie's *Dictionnaire* in order to challenge and question particular definitions. In the case that concerns us, Feydel detects a logical error in the Académie's handling of two words dealing with mysterious mental afflictions, *acousmate* and *incantation*. I reproduce the entire passage:

(**S23 complete**) Acousmate. Masculine noun. Sound of human voices or instruments that one imagines to hear in the air.

Incantation. Fem. noun. Name that one gives to the absurd ceremonies of swindlers who pose as magicians.

Remark. Biographers have written that St. Cecilia, ready to receive her martyrdom, heard inside herself the song of angels; whence she has been given the title, patron saint of musicians. If this story is true, St. Cecilia was then in a state of Acousmate or Incantation; for these two nouns, in the language of learned metaphysicians, are essentially synonyms. Both designate a mental affection that few physiologists know how to distinguish; affections seldom morbid, sometimes endemic, but for those who suffer from it, when they are not saints, often ascribed to the power to witchcraft.

When Sganarelle, the lumberjack posing as a doctor, threatens to give a peasant a fever, he boasts of a power that makes the peasant apprehensive, and of which the academicians happily do not believe. But the academicians, whose duty is at no point to believe that a man can give a fever to another by looking at him askance, have nevertheless committed an inexcusable fallacy, if they contradict the existence of the fever. The Académie française has implicitly done this, by the manner in which they have written their entry on incantation.[30]

In Feydel's first paragraph, concerning St. Cecilia, he argues that acousmate and incantation are essentially synonymous—they both refer to mental states that cannot be effectively distinguished. If Feydel can establish this point, then he can accuse the Académie of offering a definition of acousmate that does not question its veracity as an intentional state while simultaneously offering a definition of incantation that does. In other words, if the terms are synonymous, how can they possess different valuations? To illustrate the fallacy, Feydel selects an example from Molière's *Le médecin malgre lui*. Sganarelle, posing as a doctor, lacks the ability to give a fever to a peasant, because a fever is not the kind of thing that can be transferred by a crooked glance. But just because Sganarelle lacks such powers, it does not follow that one should deny the existence of fevers altogether. In Feydel's analogy, Sganarelle occupies the same place as a magician; and, to complete the analogy, just because the Académie is skeptical of the power of a magician to invoke a state of incantation doesn't mean that they should deny the reality of such states.

That same year, Feydel's publication evoked a bitingly witty response by André Morellet, a member of the Académie and *philosophe*. Morellet's response is worth quoting in full:

(S24) The critic claims that these two terms acousmate and incantation are essentially synonymous; and that one did not recognize this claimed synonymy. The Greek term "acousma" signifies the thing, the noise that one hears. Incantation employed by fake sorcerers, can have many diverse and different ways of producing sounds for those upon which they practice their art.

After this grammatical observation the critic speaks to us of St. Cecilia, *who hears inside herself the songs of angels*, and passes from that, one knows not how, to speak truthfully: *when Sganarelle threatens to give a peasant a fever, he boasts of power of which the academicians happily do not believe; but they have nevertheless committed an inexcusable fallacy if they contradict the existence of the fever, which is implicitly done by the manner in which they have written their entry on incantation.*

The Academy has not spoken of Sganarelle or of any fever given by magicians. One does not know [how]...after this article, the Academy could appear to deny the existence of a fever.[31]

As Morellet implies, Feydel is a bit feverish in his critique of the Académie. First, the term acousmate is related to the Greek noun *acousma*, "the thing heard," and does not describe the act of hearing. It refers to a sound heard in the air, not the mental state of the listener hearing that sound. Fedyel elides this difference and assumes a synonymy that was never claimed by the Académie. And Morellet is correct; the peculiarity of Feydel's thinking is brought into relief when one looks at the historical usage of the term acousmate, for I have been able to locate no other passage that invokes being in a "state of acousmate" other than Feydel's. Second—and similarly—the association of St. Cecilia with the term acousmate is idiosyncratic and unsupported by historical usage as well. Again, there is no other locus than Feydel for the attribution.[32] These findings contest Battier's symbolic identification of St. Cecilia with a tradition of sounds severed from their causes and grouped under the term acousmate.

In fact, an acousmate has little to do with music, St. Cecilia, or mystical unions; rather, it is much more closely associated with extraordinary visual spectacles like eclipses, meteor showers, and the northern lights. It is supposed to transpose some of the supernatural effect of such spectacles into the auditory register. If it were not an oxymoron to say so, perhaps the term acousmate could be defined as an "auditory phenomenon."

PHILOSOPHY, PHYSICS, AND THE BAPTISM OF ACOUSMATE

Since acousmate is such a rare word in the French language, it is possible to trace many of its appearances—even its baptism. It first appears in the pages of the *Mercure de France*, in December of 1730. The word is coined in an article by M. Treüillot de Ptoncour, the Curé d'Ansacq, that concerns an extraordinary, supernatural auditory event. Ansacq is a small village in a hilly, wooded region of France, about 40 miles due north of Paris. According to the Curé, on the evening of January 27th and 28th, 1730, an "extraordinary noise like human voices" was "heard in the air by several people in the parish of Ansacq."[33]

> (S25) Saturday, Jan. 28 of the present year, the noise pervaded the parish of Ansacq, near Clermont en Beauvoisis, that the preceding night several Individuals of both sexes, having heard in the Air a prodigious multitude like human voices of different sounds, sizes and brightness, of all ages, of all sexes, speaking and crying all at once, without the Individuals being able to distinguish what the voices said; that among this vocal confusion, one recognized and distinguished an infinite number who emitted lugubrious and lamentable cries, like distressed people, others cries of joy and peals of laughter, like persons amusing themselves; several added that they clearly distinguished among these human voices, allegedly, the sound of different instruments.[34]

In the article, written in the form of a letter, the Curé displays his initial skepticism toward the event, arguing that he was "quite pyrrhonian" toward tales of spirits and witches' sabbaths, for such tales circulate and terrify "coarse and ignorant minds, like those of most country-folk."[35] After speaking with ear-witnesses to the extraordinary event, and especially with townspeople described as men of "honor and probity, quite enlightened and incredulous" (2809), the Curé begins to take the events more seriously. After a second set of strange sounds on the evening of May 9th and 10th, he decides to initiate a series of depositions in order to inquire into the nature of the event. The major portion of the article is comprised of transcriptions of these depositions followed by a series of reflections upon them—and this is where the word acousmate is introduced. Throughout, the Curé maintains a position of neutrality, offering his inquiry to the readers of the *Mercure* as a curiosity and amusement:

> **(S26)** What happened, whether Spirits, Goblins, Sorcerers, Magicians, Meteors, conflicts of vapors, or battles of elements, I leave to the curious to choose or to find other causes: it suffices me to assure [the reader] that the witnesses of this alleged wonder, appear to me to be in good faith, and that they have been interrogated several times, and having been given the chance to contradict themselves by forgetting in their second, third and fourth depositions, that which they confessed in the first, they nevertheless support such wonders and they have never varied in the least circumstance.[36]

Whether the Curé is indeed as neutral as he claims is another question—for the strategy may be to feign neutrality in order to report evidence of some kind of miraculous event, one that the Curé knows well will not be easily explained. But, given the amount of detail in the article—which includes depositions, reflections, and even topographical descriptions of the town environs—there seems little reason to cast suspicion on his motives.

Details about the mysterious sound emerge from the villagers' testimony. The first set of depositions come from Charles and François Descoulleurs, two middle-aged laborers heading back to Ansacq late on the evening of January 27th from the neighboring village of Senlis.[37] Charles details the progression of the sound, attesting that while traveling along the north wall of the town, he "was suddenly interrupted by a terrible voice, which appeared in the area of five paces...."[38] Another voice resembling the first could then be heard at the other end of the village. Following upon that dialogue, a confusion of voices broke out in the space between the first two—voices of elders, children, men, and women, all speaking in an unintelligible jargon and accompanied by the sound of instruments. Although some voices appeared to come from quite high in the air, "about twenty or thirty feet," others were emitted from about the height of an ordinary man or even from the ground. The whole event ended in "peals of delicate laughter, as if there had been three or four hundred people who began to laugh with all their force."[39] The volume of the noise was so loud that the brothers report struggling to converse with each other. On the opposite side of the village, Louis Duchemin, a glove seller, and Patrice Toüilly, a bricklayer, were traveling from Ansacq to Senlis and report hearing the same event as the Descoulleurs brothers. After stopping briefly to listen more closely, they hastily continued onward to Senlis, determined to keep their distance from the noise.

Villagers safely inside the town walls also reported hearing the noise. Sir Claude Descoulleurs, the retired guard of the town's gate, reported hearing the events of both January 27th and May 9th. The sounds were loud enough to rouse him from a slumber. By the time the aged Sir Claude dressed himself and headed out to his courtyard, he reports that the "Aerial Troupe" was already quite distant. When asked about the sound, Sir Claude offers one of the more colorful descriptions, comparing it to a "Fair" or "demonstration," where "two or three thousand persons form a kind of chaos or confusion of human voices, of women, of elders, of young people, and children." He draws a comparison to the port of Hales in Paris on the day of a great march, the halls of a palace before a great audience, or the Fair of St. Germain in the evening, "when it fills an infinite world."

> (S27) Doesn't one hear in all these places…a dreadful racket, in which one understands nothing in general, while each one in particular speaks clearly and makes himself distinctly understood? Add to that the sound of Violins, Basses, Oboes, Trumpets, Flutes, Drums and all the other instruments that one plays in the House of Spectacles, and which blend themselves into this confusion of voices, and you would have a good idea of the noises that I heard.[40]

Alexis Allou, a churchman, reports being awakened by his wife and heading downstairs to see about the noise. After opening the door, he hears "an innumerable multitude of persons, some pressing out cries of bitter words, others cries of joy," accompanied by the sounds of instruments. The sound travels along the road, passing by his house, before heading toward the town church. After being seized with a shiver and overcome with fear, he heads back upstairs to comfort his wife. Nicolas Portier, a laborer, reports a similar experience of being roused by the noise. It was so loud and terrifying that his dogs were also disturbed and threw themselves at the door in terror. Other depositions from townsfolk report experiences of being rudely awakened, noting the movement of the sound along the road toward the church, and describe a similar composition of voices and instruments.

In the series of reflections that flank the depositions, the Curé balances his trepidation about publishing such a report—and the skepticism with which the general public may receive it—with the potential interest it holds for the scientific community. Perhaps a touch self-serving, the Curé imagines that this extraordinary sound might capture the attention of the populace and the scientific community in the same way that an astronomical appearance had done a few years earlier. In 1726, the aurora borealis could be seen in a spectacular display all across Northern Europe. The display was so remarkable that it caused a panic among some crowds in France, triggering the government to commission an investigation by Jean-Jacques Dortous de Mairan into its causes. Mairan's book became the most widely known treatise on the subject of aurorae in the first half of the 18th century.[41] Even in England, where there had been remarkable displays of the aurora in the previous decade, the phenomenon merited an article in the *Transactions of the Royal Society*, describing the prismatic color of the lights and their formations into "Coronas," "Canopies," "Arches," and "Streams."[42] While the popular imagination saw portentous and ominous oracles in the displays of the aurora, they gave burgeoning natural scientists an opportunity to hypothesize explanations and dispel folk anxieties. The Curé had phenomena of this variety in mind when publishing his account in the *Mercure*. For just as he characterizes himself as an enlightened man of good faith, distrusting the superstitious and overactive imagination of the townsfolk, he encourages the *Mercure*'s readers to explain the extraordinary sound through the application of skillful reasoning. To that end, he even attaches a "topographic description" of Ansacq to aid "those who believe themselves to be able to explain this by natural causes and to exercise their Physics and their Philosophy."[43]

In presenting so much material about the acousmate d'Ansacq, the Curé was participating in a typical practice in early 18th-century natural science, whereby scientific data and accounts were widely shared.[44] Just as the circulation of eyewitness reports of the aurorae aided Mairan's work in establishing a history of such phenomena and discerning recurrent cycles that would facilitate prediction of future aurorae, the Curé was disseminating information to those who might be interested in investigating extraordinary sonic events. Similarly, the Curé coined terms to aid in the identification and specificity of such events. According to Patricia Fara, "Natural philosophers developed a new vocabulary for providing detailed accounts of aurorae, thus consolidating their claims to intellectual possession of a phenomenon governed by the laws of nature."[45] The baptism of the word acousmate fulfilled this function.

> (S28) Everyone knows that Phenomenon [*Phénomène*] is a Greek word, which has been Gallicized as much as any other, because one cannot find in our language full of terms a signification energetic enough, to express by itself objects that appear extraordinarily in the air. Our language does not furnish us with many expressions to designate the extraordinary noises that exist, nor those, which might be heard [*où qui pourraient se faire entendre*]. But as the latter [i.e., extraordinary noises] are less common than the former [i.e., Phenomena], no one has been shrewd enough thus far to Gallicize a Greek word to express it.
>
> Doesn't the event in question authorize me to do it myself, and to appeal in the same way to what the Ancients called Phenomena [*Phénomènes*], the extraordinary objects that appear in the air; can't I, for the same reason, designate the surprising and prodigious noise heard with the word, *Akousmène*, or to speak more properly Greek in French, *Akousmate*? The first (i.e., *Phénomène*) signifies a thing that appears extraordinarily; the second would signify a thing that makes itself heard extraordinarily.[46]

The word "akousmate" is intended as an auditory parallel to the visual phenomenon. Etymologically, phenomenon is derived from the Greek verb *phanein*, which means to cause, to appear, or to show. Although philosophical use of the term generalizes it to cover any kind of sensory perception, traditionally, the word has a visual basis and cannot be easily applied to auditory events.[47] In the Curé's case, there was an established tradition of using the term to speak about natural appearances, especially astronomical events. Eudoxus had used the word in the title of his treatise on astronomy, and his usage continued in the natural scientific journals of the 18th century.[48] Moreover, the Curé's justification for his coinage clarifies one odd feature of the word. The "-ate" ending is uncommon in French, which helps to make akousmate such a rare word in the language. But it is apparent that the Curé knows his Greek grammar and is simply transcribing the plural Greek noun "akousmata," the things heard, into French. Naturally, the vowel sound at the end of the Greek becomes silent when transcribed. So, akousmate is really just a revival of our old Greek friend, the word akousmata. Or is it? In fact, the word suddenly reappears without a trace of the Pythagoreanism that had been attached to it at the time of Clement or Iamblichus. The Curé's acousmata, although perhaps portentous or supernatural, are definitely not coded utterances or allegories. They encrypt no esoteric wisdom, participate in no ritual initiations, and conceal no Orphic mysteries. The only thing mysterious about them now concerns their cause. When reborn in the 18th century, acousmata are unexplained effects.

After coining the term, the Curé appeals to the interest of scientists by addressing the specifics of each sense modality, seeing and hearing. The nature of each modality differentiates the kind of extraordinary events that will be perceived. The appeals of the Curé challenge the nobility and privilege of sight in the name of equal treatment:

> (S29) In effect, if the extraordinary aerial phenomena which appeared five years ago greatly excited great minds and has given material for several assemblies of the Members of the Academy of Sciences, why shouldn't an event which falls under another sense, which is no less real, no less essential to man than the sense of sight, merit as much attention and curiosity from the same scholars?...This principle established, I claim that if the Phénomènes are pricking the curiosity of the scholars, the Akousmates, should do no less....[49]

The first objection to be countered concerns the wide availability of aurorae to a whole hemisphere of viewers, in comparison to the miniscule reception of the acousmate by a few auditors in Ansacq. One might argue that the greater number of witnesses to the aurora gives more credibility to the event. No one can doubt the veracity of its appearance in the sky with so many witnesses to verify it. The Curé rebuts that since "the Sphere of activity of vision is much more extended than that of hearing, Phenomena and Akousmates do not require the same number of witnesses to yield authenticity."[50] Because vision and hearing differ in their modes of operation and spheres of activity, different criteria must be applied to the eye and the ear. Next, the Curé turns the argument around, claiming that despite the smaller sphere of activity of the ear, it is a more reliable witness because "the sense of vision is ordinarily more subject to error and illusion than the sense of hearing. To prove it, just recall all the extravagances which were reeled off on the occasion of prior Phenomena, especially that of October 1726. How many saw appear in the sky a fiery Dome, some a dove...others an angel, altar, or dragon...all things which had reality only in their active imaginations, or in their fatigued and dazzled eyes?"[51] The greater number of people involved in seeing aurorae also exposes great discrepancies in their testimony—and vivid cases of imaginative seeing. Of course, this argument is unconvincing, because the figures that one imagines seeing in aurorae do not affect the evidence for the existence of aurorae in the first place. (For example, one can see all kinds of things in a cloud without doubting the existence of the cloud.) If the depositions can be believed, there is a great correspondence among the acousmate's ear-witnesses concerning the kinds of sounds heard: a great multitude of voices of all ages speaking in an unrecognizable tongue, peals of laughter, and the sounds of instruments. Unlike aurorae, where everyone sees something different in the lights, the uniformity of the acousmate's descriptions must count for something. The Curé implies as much when he anticipates, and questions, the possible rebuttal that the "dreadful noise" could have been caused by "several howling wolves," "cries of geese," or "wild ducks."[52] How could such an unusual sound be so precisely described by all involved?[53]

One further aspect of the Curé's reading is noteworthy. There is an implicit claim that the senses are separate and distinct from one another, and that the standards of evidence change in accordance with the mode in question. Even before the heyday of *Les Philosophes*' sensualist epistemology, the logic of the senses at work in the Curé's argument was shaped by contemporary philosophy and physics. The necessity to find a word that will parallel in the auditory domain extraordinary visual events registers the separation and development of the senses as sites of research and inquiry. The coinage of the term acousmate evinces a logic of separate but equal—audition *should* be of interest to scientists, and of as much interest as visual events, while requiring its own vocabulary, tools for investigation, and canons of evidence.

The separation and coordination of the senses had become a pressing philosophical issue by the time of Locke's epistemology. According to Locke, our ideas of external bodies come in two kinds. Secondary qualities of objects—such as color, taste, scent, or sound—"have admittance only through one sense," while primary qualities—such as space, extension, figure, rest, and motion—"convey themselves into the mind *by more senses than one*."[54] Sensations of primary or secondary qualities revealed something significant about the relationship between sensations of the world and the world itself. According to Locke:

> The ideas of primary qualities of bodies are resemblances of them, and their patterns do really exist in the bodies themselves; but the ideas, produced in us by these secondary qualities, have no resemblance of them at all. There is nothing like our ideas [of secondary qualities] existing in the bodies themselves...what is sweet, blue, or warm in idea, is but the certain bulk, figure, and motion of the insensible parts in the bodies themselves.[55]

Sensations of taste, color, or heat were known to vary with the perceiver and the health of their senses, even while the object remained unchanged. As any skeptic would tell you, the taste of an apple might vary if the eater were sick, or the sensation of an object's heat might change relative to the temperature of one's own hands. Thus, it was unwise to think that our *sensations* of secondary qualities resembled the *bodies* themselves. Not so for primary qualities. Sensations of figure, extension, and motion, regardless of the health of the perceiver's sense organs, resembled the figure, extension, and motion of a body's smallest parts. Moreover, one had to appeal to these primary qualities of objects in order to give an account about the cause of sensations of secondary qualities. In the passage above, Locke attributes the cause of our sensation of "what is sweet, blue, or warm" to the "bulk, figure, and motion" of its smallest parts, that is, to primary qualities of the object.

William Molyneux, an attentive reader of the first edition of Locke's *Essay*, noticed some "rather strange implications" of Locke's epistemology and posed them in a letter to him in 1693.[56] The question he posed was the following: Would a man born blind, able to distinguish between a sphere and a cube by touch, upon regaining his sight be able to immediately identify the shapes of the objects that he formerly knew by touch alone? Molyneux argued, based on Locke's principles, that the man would not be able to immediately identify the shapes through vision—a conclusion that was later confirmed by developments in the medical practice of curing blindness through the removal of cataracts.[57] Although Locke, in the second edition of the

Essay, would agree with Molyneux's reasoning, it was a blow to his epistemological project. One of Locke's goals was to fill the gap between sensation and the world by assuring us that our ideas of primary qualities (precisely those at stake in Molyneux's query) resembled the actual inherence of those qualities in real objects. Molyneux's problem challenged the primacy of primary qualities by showing them to be no more secure than secondary qualities. One might eventually learn through experience to correlate the feel of a shape with its look, but there would be no necessary connection between them.[58] In Jonathan Rée's words, "They would be as separate from each other as the feel of apples and their smell, or the smell of a fire and its sound. Indeed, none of us, however perfect our sensory equipment, could even dream of matching the deliverances of one sense with those of another."[59]

The blow to Locke's project left the philosophical epistemology of the senses no better off than it had been in Descartes's physiological and natural scientific writings from the early 1630s. For instance, in the *Treatise on Light*, Descartes contrasts the sensation of light with its mechanical causes and argues that there is no necessary resemblance between them:

> In putting forwards an account of light, the first thing I want to draw to your attention is that it is possible for there to be a difference between the sensation that we have of it, that is, the idea that we form of it in our imagination through the intermediary of our eyes, and what it is in the objects that produces the sensation in us, that is, what it is in the flame or in the Sun that we term "light."[60]

Glossing this, Descartes compares the relationship between the ideas produced by sensation and the mechanical causes of such sensations to that of a word and the thing it signifies. The relationship of the signifier to the signified is taken as the model for sensation generally.[61] Descartes spins out the same theme when describing the act of hearing:

> Do you think that, when we attend solely to the sound of words without attending to their signification, the idea of that sound which is formed in our thought is at all like the object that is the cause of it? A man opens his mouth, moves his tongue, and breathes out: I see nothing in all these actions which is in any way similar to the idea of the sound that they cause us to imagine.[62]

The lesson drawn by Descartes's readers was not a retrenchment to skepticism, but a reminder of the efficacy of natural scientific explanation. Since purely philosophical reasoning could not establish a necessary relationship between a sensation and its object, the mechanistic and natural scientific explanation of the cause of a sensation was the best explanation to be had. Readers like Fontenelle, a great promoter of Descartes's natural scientific writings, were extraordinarily influential in France during the period in which the Curé published his article.[63] Two of Fontenelle's philosophical works, the *Histoire des oracles* (1686) and the *Origine des fables* (1724), are noteworthy for their attempt to explain religious myths and national legends through the use of natural science.[64] Fontenelle exposed the supernatural as a "purely man-made phenomenon, resulting from...the gullibility of the populace, and the all-too-human craving for *le merveilleux*."[65] In contrast to this craving for the supernatural, the clear and distinct knowledge offered by natural science could explain such events. In many respects, the *acousmate d'Ansacq* would be a perfect object for investigation in the style of Fontenelle. It afforded the opportunity to extend scientific knowledge about the sense of hearing and the nature of sounds and to investigate the causes of a marvelous occurrence.

THE DEBATES IN THE *MERCURE*

For next 15 months, the acousmate was hotly debated in the pages of the *Mercure*, appearing no less than nine times between the original account in December 1730 and March of 1732.[66] The first response was a profession of faith from an anonymous Burgundian: "I believe willingly that there is no physical explanation to discover in all of this, and I think that, like Saint Paul assured, the air is full of demons...."[67] The cause is supernatural and should be taken as proof of the hubris of natural scientists and savants: "Perhaps God permits the reality of the event at Ansacq to oblige philosophers to admit that there exist aerial spirits...."[68] But the profession of faith is tempered by a more rational question: If this is indeed evidence of demons or aerial spirits, shouldn't there have been precedents before the events at Ansacq? The Burgundian encourages historians to look into the matter, and relates a few personal anecdotes of past experience with similar acoustical oddities. But such professions of faith are the minority in the *Mercure*; most of the accounts accept the Curé's wager and try to account for the acousmate d'Ansacq by looking for its potential natural scientific causes or cases of human deception.

In March 1731, another anonymous writer attributes the cause of the acousmate to a human source—a ventriloquist. Challenging the Curé's claim that hearing is much less easily deceived than vision, the author of the letter argues that

> **(S30)** One is fooled quite easily in judging where a sound comes from when the eyes do not help in judging from where it comes, that there are people called *Ventriloquists*, who by squeezing their throat and making certain muscular contractions in the lower abdomen, articulate a sound... [such that] one believes they hear a very distant voice.[69]

As Steven Connor demonstrates in his magisterial history of ventriloquism, *Dumbstruck*, the 18th-century practice of ventriloquism was quite different from our present-day practices of wooden dummies voicing the loquacious projections of the id. The end of the 18th century saw the legitimizing of ventriloquism as an urban entertainment, whereas it had previously been associated with itinerant fairs, country sideshows, oracular utterances, and even the demonic. Connor describes how ventriloquists like Fitz James, Charles Mathes, and Vattemare Alexandre popularized the genre by offering performances that involved the recreation of various natural and domestic scenes, with illusions of sounds approaching and receding or coming from various parts of the theater. In 1803, William Nicholson, in the pages of the *Edinburgh Journal of Science*, described a performance by Fitz James. The routine involved the supposed extraction of a tooth.

> [Fitz James] went behind a folding screen in one corner of the room, when he counterfeited the knocking at a door. One person called from within, and was answered by a different person from without... the latter was in pain and desirous

of having a tooth extracted.... The imitation of the natural and modulated voices of the [doctor], encouraging, soothing, and talking with his patient, the confusion, terror and apprehension of the sufferer, the inarticulate noise produced by the chairs and apparatus, upon the whole, constituted a mass of sounds which produced a strange but comic effect. Loose observers would not have hesitated to assert that they heard more than one voice at a time....[70]

Ventriloquial scenes, in addition to reproducing domestic situations, might also involve the recreation of natural phenomena. An anonymous author from the *Mercure* gives an example:

(S31) I have known a young man who, behind a screen, imitated in an amazing manner a choir of voices and instruments... with a skillet, tongs and his voice, he would imitate, to the point of being fooled, the noise of a house in a tumult, the cry of the husband, of the wife, of children, dogs, of country-folk, of a watchman and of a commissioner.... Under the name Loeillet, he would go into the houses where he was summoned, and without leaving the chamber where he was, hidden behind a screen or a curtain, he would make a noise so varied and so amazing, imitating a concert, a scuffle, peals of laughter, or other noises of this kind, that one could not imagine that he was alone. Having a late dinner meeting with a celebrated actor of the Royal Academy of Music, it is said that he [Loeillet] made the guard appear more than once, who tranquilly found both of them sitting face to face at the table, and who, hearing the same noise again, retraced his steps a moment later... and quite tranquilly found them just where he had left them.[71]

Thus, the author hypothesizes that a good ventriloquist, poised inside the town wall, could have easily produced the confused noise of the *acousmate d'Ansacq*. By traveling along the wall, he could create the illusion of the sound moving across the town, in front of houses, and toward the church. Moreover, the author argues that one need not possess the talent of a Loeillet to have scared the townsfolk, for if such an auditory illusion can be produced for those who *know* that it is a trick, like Loeillet's audience, how much easier it must be to fool those roused in the middle of the night or frightened on a dark country road. Unlike the ancient association of ventriloquism with oracular utterance and demonic possession, Connor argues that during the 18th and 19th centuries, the practice transitioned into a secularized entertainment: "It became a form of trickery bound up with human susceptibilities and powers rather than supernatural or spiritual phenomena."[72] The account in the *Mercure* registers this transition, whereby the debunking of demons or aerial spirits depends on an attribution to the trickery of a ventriloquial source.

Although we are assured that a single person can produce a polyphonic effect, the claim is undermined by the author's repeated attempts to explain how such polyphony is possible. The author doth protest too much, trying to fend off the worry that no ventriloquist can produce the sheer quantity of polyphony described in the depositions. To counter, he draws a parallel between visual and auditory illusions:

(S32) If one rapidly moves a firebrand, or an ember, in a circle, the eye apperceives only an uninterrupted ring of fire; the reason is sensible; the charcoal successively travels through all the points of the circle; the impression that makes on vision when it was at its height persists even while it is at the bottom. The rapidity with which it is moved, makes the eye see it all at once in all the points of the circumference, and apperceives only a continuous circle of fire. Different sounds that would rapidly succeed one after another would, for the same reason, seem to be heard at the same time. One person alone, if they could render all these different and passing sounds successively, passing very quickly from one to another, would thus imitate the confused and continuous noise of different voices, and consequently those who would hear it without seeing the author of the noise, would believe to simultaneously hear all the different voices that were imitated.[73]

The faulty nature of this parallelism ultimately weakens the explanation. The persistence of vision does not hold for the ear, at least in the days before the digital sample. If the depositions are correct in their description of the acousmate, the ventriloquial explanation cannot hold.

The author of the next response, one Monsieur de la R., isolates precisely this problem with the production of polyphony:

(S33) One person alone can be in a room, behind a screen, making a considerable and varied noise of sounds that imitate the voices of men, of women, of children and of different animals; if one mixes into it skillets, tongs, etc., the confusion of all these things might be able to imitate the din of a house full of sounds. But this could not represent in the same instant several mixed voices.... I have heard the racket of Loeillet; he has amused me greatly, but while as fast as his play was, one could distinguish clearly that it came from only one person. The skillet, the tongs, the chairs, that he employed, form at each moment only the sound of each of those things, and produce in the room and its surroundings, only the value of each sound heard distinctly one after another.[74]

After debunking the ventriloquial theory, M. de la R. offers a tentative explanation based on the experience of his friend, Monsieur P***, a physicist. In the middle of the night in early October 1730, M. P*** claims to have been awoken by an acousmate similar to that at Ansacq, filled with the sound of voices and instruments. Following the sound to various locations, but never finding the source itself, M. P***, returned to the front steps of his house, only to be struck by a new acousmate, which sounded like "many whistles of different tones," first filling the air with sound and then "dying out as it receded like a wave."[75] The transformation of voices into whistling tones gave the physicist a clue about the source of the sound—air masses. Moreover, the location of these sounds was in a wine-growing region in the south of France at the base of a mountain where it meets the plains, topographically similar to the environs surrounding Ansacq. The source of the acousmate is attributed to "billows [of air] striking in different ways [the landscape's] uneven surfaces, now flat, now convex, and now concave."[76]

In line with the hypothesis about air billows, the rest of the explanations in the *Mercure* are scientific in character—no more ventriloquists or supernatural causes. Aside from other conjectures and amusements concerning the acousmate, the debate in the pages of the *Mercure* devolves into a discussion between two figures,

M. Laloüat de Soulaines, an advocate to the French government, and M. Capperon, the ex-headmaster at Saint Maxent, who had a small reputation for work in natural science. Laloüat bases his explanation on personal experience, having experienced acousmates in the village of Sézanne, and on his observations of cloud formation and various phenomena concerning the movement of "columns of air."[77] Without going into detail, Laloüat attributes the source of the sound to corpuscles of air that, when placed under various kinds of atmospheric conditions, burst and produce all varieties of motion; upon striking the ear, the corpuscles can sound like any number of things—voices, instruments, laughter, and so forth. Capperon claims that the sounds are caused by the action of "cold fermentation" occurring in the bowels of the earth; again, under extraordinary atmospheric and geological conditions, the motions caused by fermentation can strike the ear and sound like a variety of sources. Throughout the latter half of 1731 and early 1732, Laloüat and Capperon rebut each other, each trying to demonstrate how the other's theory is incapable of explaining away the mystery of the *acousmate d'Ansacq*.

Needless to say, the mystery remains unsolved; no definitive cause is attributed to the acousmate. Like so many things that capture the fickle attention of the public, the interest in the acousmate quickly faded. The term is not to be found in issues of *Mercure* in the years immediately following, other than a humorous reference to the event in 1738.[78] I have been able to find only one other attempt to solve the mystery of the acousmate d'Ansacq. On November 18th, 1901, a paper was read by one M. Thiot before the *Société académique d'archéologie, sciences et arts* from Oise, the *département* where Ansacq is located. Thiot, revisiting the acousmate, attributes the source of the sound to a strong "polysyllabic echo" caused by the high, flat facade of the town's château.[79]

EXPANSION OF THE SEMANTIC FIELD

What happens to the word after the controversy in the *Mercure* dies away? It is not easy to track the vicissitudes of a word as it moves from context to context, accruing and sloughing off significations. But when the word is rare, like the word acousmate, tracking is greatly facilitated. In dictionaries of the period, one detects a *fracturing* of the word's semantic field along two vectors. The first vector preserves the original sense of the term, as an extraordinary but literal object of hearing. One finds this sense of an acousmate preserved in a dictionary of French words "whose meaning is unfamiliar" authored by the Abbé Prévost. In the edition from 1755, the Abbé enfolds the original, baptismal sense of the word into his definition:

(S34) Acousmate, s.m. A term newly formed from the Greek, to describe a phenomenon that makes heard a great noise in the air, comparable, one says, to that of several human voices and diverse instruments. The Mercure of 1730 & 1731, provide a description of an event of this nature, occurring near Clermont en Beauvoisis.[80]

Other dictionaries from the period tend to replicate the Abbé's definition, preserving the original sense of the term.[81] We see the second vector in the *Dictionnaire* of the Académie—the same volume that formed the centerpiece of the debate between

Feydel and Morellet. Recall that the Académie defined acousmate as the "Noise of human voices or instruments that one imagines to hear in the air." Rather than be defined as a real sound, ontologically akin to phenomena like the aurora borealis, the Académie demotes the acousmate to a hallucinated or imagined sound.[82]

These two vectors of signification penetrate the various disciplinary contexts in which the word is received, exchanged, and replicated. In particular, three contexts of reception—scientific, psychological, and literary—shape the signification of the word. In the scientific context, the original signification of the Curé is preserved. An acousmate is an extraordinary sonic event that demands a causal explanation, and such an explanation helps to debunk the supernatural or miraculous characterization of its reception. In a long chapter from *Traité historique et critique de l'opinion* concerning the production of sound and the motion of air, Gilbert Charles Le Gendre, the Marquis de Saint-Aubin-sur-Loire, refers to the *acousmate d'Ansacq* as one example in a long line of similarly extraordinary auditory events, combing the pages of Pliny, Livy, Clement, and various historians for instances. The acousmate is no longer a terrifying event, but a case study in the extraordinary workings of nature.[83]

In the psychological literature of the 19th century, acousmate is used as a technical term to describe the object heard in an auditory hallucination. The *Nouveau dictionnaire de médecine*, from 1821, defines an acousmate as "a noise that one believes to hear in the air, and which is purely imagined. Uncommon."[84] As a medical term, the word finds its way into English, appearing in 1881 as "an imaginary sound."[85] The word becomes associated with its root and generates a series of "acousma-" words, which are used to describe various sorts of auditory hallucinations. For example, the 1881 volume contains acousma, "a species of depraved hearing in which sounds are imagined as if they were really heard." A 1907 volume makes acousma synonymous with acousmate as "an auditory hallucination or imaginary sound," and introduces a slew of new "acousma-" words: acousmatagnosis, the failure to recognize sounds due to mental disorder, mind-deafness; acousmatamnesia, the failure of the memory to call up the images of sounds.[86] These words are all associated with the pathology of listening rather than, say, the physiology of the ear or the physics of sound.

In the literary context, acousmate takes on the signification of an uncanny, supernatural event—a creepy effect without a cause—that can invoke a feeling of terror, dread, or a state of awe. In 1805, the word appears in a frightening tale by Le Gorse entitled "*Le sabbat des esprits*," with all the requisite mise-en-scène: an uninhabited castle, rumors of inexplicable events, ominous inscriptions, sudden noises, trembling limbs, and trapdoors. In a scene inside the throne room, a sudden sound occurs: "One could almost call it an acousmate, for the excellent actors in this magic symphony were skillfully veiled by the sumptuousness of ornamental decoration, that one might believe it comes from the air."[87] By 1833, the word appears as part of a scene of music in the uncanny "*Les Pressentiments*" by Petrus Borel. When Ténobie, the protagonist, begins a *marche funèbre* by Beethoven, this Hoffmanesque passage follows:

(S35) O Beethoven! Who does not feel touched by this music so mystical, so grave, so melancholic, so majestic, so mortifying! Do you not hear in the distance a procession slowly advancing [with the] sound of the sepulchral voice, their hymns of death.... Do you not hear from the [cathedral] apse a melodious acousmate, or rumbling in the crypt a furious masterpiece?[88]

The association of acousmates with the uncanny or supernatural, preserved in the literary context, is also preserved in studies of folklore and religion, which explored the mythological association of acousmates with witches' sabbaths and the dance of the dead.[89]

MYTH INTERRUPTED

Within the literary reception of the term acousmate, one can perhaps make sense of Feydel's insistence on an acousmate being a mental condition akin to incantation, rather than an object of hearing, one that is associated with St. Cecilia. If so, this contextualization of acousmate might allow us to nuance Battier's reading of Feydel's passage. Battier is not mistaken to say that the term acousmate is related to "voices without bodies" and "sounds without their causal source." But a relation is not an essence—it is a moment. Feydel's response to the Académie could have been understood as part of a non-essentializing account of acousmatic sound that interprets moments of "sounds without their causal sources" and "voices without bodies" in historically and culturally specific situations, not as part of a mythic "originary experience."

In fact, I would argue that historical accounts of acousmatic sound have been led astray by remaining overly fixated on the words acousmate and acousmatique. Fixation on the words has forced together moments that do not form a single tradition; the need to maintain this tradition, to provide a "foundation by fiction," perpetuates the mistaken assumption that acousmatique and acousmate are synonyms when they are, at best, homonyms. Rather than rely on the presence of the words acousmate or acousmatique to provide historical orientation, what would it be like to tell a history of acousmatic sound that treated the words as two names (among others) that refer to various sets of cultural practices concerning the relationship of seeing and hearing? Perhaps we could reconceptualize acousmatic listening as a collection of techniques for manipulating the senses; we could be guided by moments from a long history of sensation when de-visualized listening is *privileged for culturally specific and historically situated ends* (and the invention of *musique concrète* would be one of those instances); we could write a history that eschews all claims about the essence of acousmatic sound (so, no originary experiences associated with the Pythagorean veil or St. Cecilia) in favor of a history of listening that remains faithful to historical agents' attempts to make sense of acousmatic sound in whatever terms were available to them—whether philosophical, metaphysical, mystical, technological, scientific, religious, or aesthetic.

Imagine a scenario where the Curé coined a different term to describe his extraordinary auditory phenomenon—some other odd word to entice Apollinaire's hermetic fantasy. Perhaps Peignot would have used *it* to describe his experience of *musique concrète* and, thus, we would never have had Schaeffer's restitution of the Pythagorean veil. Yet, the experience of de-visualized listening would still remain. It is the cultural and historical specificity of those experiences that should matter in a history of acousmatic sound. If Schaeffer owes something to Apollinaire (or Diderot, or Pythagoras), the debt must be spelled out in terms stronger than those of assumed etymological filiations. The important question is this: How does a practice of acousmatic listening get localized, configured, and deployed in specific situations to perform some particular kind of cultural work? Acousmatic is just a name; it could have been called, among other monikers, de-visualized listening, blind audition, or sound unseen.

PART THREE

Conditions

4

Acousmatic Phantasmagoria and the Problem of *Technê*

Lovers of strange words owe a debt of gratitude to Karl Marx. Without his evocative use of the term "phantasmagoria" to describe the power of the commodity, fetishized under capitalism, the word would likely have become as obsolete as the visual technology from whence it came.

Wildly popular at the turn of the 19th century, phantasmagoric performances used a special type of magic lantern, placed behind a screen, to project extraordinarily realistic images of approaching skeletons, apparitions, and specters, to frightening effect. In contrast to many contemporary pieces of visual technology, art historian Jonathan Crary notes something unique about phantasmagoria. Thaumatropes, phenakistiscopes, dioramas, and stereoscopes functioned by placing the viewer into the device, or by placing the device in the viewer's hands. Spectators would manually manipulate the device, place cards into it, press it against their eyes, and conform their bodies to the conditions required to view the illusion. Take, for instance, the stereograph, which places two images into a device that, when looked through, produces the effect of seeing an image in three dimensions. At the same time they produced their illusions, these optical devices exposed to the viewer their "operational structure" and the "form of [bodily] subjection that they entail."[1] They underscored for the spectator that the illusions they saw were mechanical productions. In contrast, phantasmagoria produced images without the spectator's awareness of the technical means involved. Stereographs and similar devices were eventually displaced and made obsolete by technologies that better obscured their means of production. In Crary's words, such devices eventually disappeared because they were "insufficiently 'phantasmagoric.'"[2]

Returning to Marx, one could not say that he failed to sufficiently appreciate the phantasmagoric character of the commodity. In his famous analysis of commodity fetishism, he invokes the term to describe the commodity's strange "metaphysical subtleties," which cannot be derived from its use value or its exchange value, but only from the form of the commodity itself.[3] Marx is keen to show how the commodity takes on a special form of appearance (*Erscheinungsform*) that obscures the labor involved in its production. In Marx's analysis, the commodity, which is fundamentally a relation between people, assumes "the phantasmagoric form of a relation between things."[4] Like religious idols—"autonomous figures endowed with a life of their own" while being, in reality, manmade creations—the commodity appears to its worshippers to be imbued with an intrinsic power that obscures the social relations effectuating its creation.[5] It effaces its traces of production like a phantasmagoric illusion, donning the appearance of a natural, even supernatural, thing.

Theodor Adorno, echoing Marx, employed the term when describing Wagner's attempt to create a work of art as seamless and timeless as the products of nature, where the blending of the various elements that compose the work effaces all traces of its manufacture. In Adorno's assessment, phantasmagoria is the keynote of Wagner's *oeuvre*: "the occultation of production by means of the outward appearance of the product...is the formal law governing the works of Richard Wagner."[6] It is epitomized in Wagner's *Venusberg* music, scored to produce a spatialized effect of tremendous volume at a great distance—"the image of loudness from afar."[7] The spatialized effect is supplemented with harmonic progressions that perpetually move while paradoxically lacking any sense of direction. In their harmonic stasis, time seems to stand still. Wagner's musical phantasmagoria spatializes time in order to depict an ahistorical or mythic temporality where, in the words given to Gurnemanz, "time becomes space."

Throughout the late 1920s and '30s, identifying and interpreting moments of phantasmagoria were preoccupations of Adorno and his associate Walter Benjamin. While Adorno isolated moments of phantasmagoric occultation in Wagner's work, Benjamin was compiling material for his unfinished *Arcades* project. In Benjamin's research on 19th-century Paris, the phantasmagoric effects of commercial objects and entertainments are tracked through a patient analysis of French material culture. From the glass-enclosed arcades lined with products on display; to daguerreotypes and panoramas, visual technologies that supplemented nature by imitating it, capturing it, and making it available for consumption and circulation; to the interior decor of the bourgeois home and Baudelaire's Parisian *flânerie*, phantasmagoria was Benjamin's magic world, where the dazzling dreams of commodity culture could be frozen by the allegorical gaze of the critic and interpreted from the perspective of its catastrophic results.

In most Marxist discourse, the word phantasmagoria appears to be little more than a colorful synonym for another magic word, reification. The two words share a strange affiliation. On the one hand, phantasmagoria and reification can be considered synonymous. Both terms refer to the economic process whereby the commodity, a product of laborers who have become alienated from the fruits of their labor, becomes endowed with an autonomous existence, thus masking a social relation in the form of something objective, natural, or timeless. On the other hand, phantasmagoria rubs against reification's sense when something abstract, like a concept or an idea, is treated as if it were something concrete or real. One can only describe the strange transformation of phantasmagoria from a concrete piece of visual technology into an abstract concept, denoting the occultation of production, as the inverse of reification.

If we can indeed continue to employ the word phantasmagoria, I employ it to indicate the occultation of production. This is the sense of the term that I intended when, in chapter 1, I objected that Pierre Schaeffer perpetuated phantasmagorical thinking about technology in his theory of the sound object. I argued that Schaeffer phantasmagorically masks the technical specificity and labor involved in the production of the sound object in order to present an autonomous realm of sonic effects without causes. Building upon the previous two chapters—where I have tried to interrupt the mythic narrative that is typically presented in accounts of acousmatic sound—I would like to suggest that the practice of acousmatic sound is not tied to the Pythagorean legend or the baptism of the term acousmate, but to a tradition of musical phantasmagoria. This tradition is sutured to the birth of Romanticism, the aesthetics of absolute music, and the intercalation of the production of music, the commodity, and technology. The history of absolute music has been told elsewhere, and the world does not need another account of its vicissitudes.[8] Yet, the relationship between musical phantasmagoria and the acousmatic separation of seeing from hearing has not been adequately thematized.

In order to remedy that situation, this chapter and the following "interlude" take up two tasks. Since musical phantasmagoria begins where technical means of production are occluded, in this chapter, I begin by clarifying the relationship between the acousmatic situation, musical aesthetics, and music's technical conditions. This investigation focuses primarily on the 19th century, at the moment when aestheticians begin to privilege acts of hearing music without seeing its performers to a radically new degree. I will also establish some basic philosophical conditions concerning the production of acousmatic phantasmagoria. After establishing these basic conditions, in the interlude that follows, I return to Schaeffer and *musique concrète* and pose a seemingly simple question: Must all *musique concrète* be phantasmagoric?

PHANTASMAGORIA AND BODILY TECHNIQUES

A good place to begin the investigation into acousmatic phantasmagoria is with the philosopher Arthur Schopenhauer. He writes, "Music as such knows only the tones or notes, not the causes that produce them."[9] The phantasmagoric occultation of production is blatant. What is Schopenhauer's reason?

Broadly speaking, Schopenhauer's aesthetics could be described as quasi-Platonic.[10] Like Plato, Schopenhauer believed that the most objective contact a subject could make with reality was by means of an idea, the "persisting form of a whole species of things."[11] But Schopenhauer disagreed with the low valuation that Plato gave artworks. Plato criticized art for being a form of imitation. Since art was a copy of reality, and reality was itself a copy of ideas, art was doubly removed from ideas. Thus, in the Platonic view, ideas were not best contemplated in artworks. In contrast, Schopenhauer saw art as the best means for attaining access to the idea. In aesthetic contemplation, the particular object beheld is taken as an expression of an idea, as a representative of its kind. At the same time, following Kant, the aesthetic beholder becomes wholly disinterested and abandons all traces of his or her own willfulness. In aesthetic experience, both the beholder and the artwork beheld are transformed. In Christopher Janaway's elegant formulation, "the will-less contemplation of the beauty of a particular thing transforms the individual [i.e., the beholder] into a pure subject of knowing and is 'at one stroke' also the apprehension of an Idea which gives us our most objective contact with reality."[12]

To understand Schopenhauer's phantasmagoric statement, the subject and object of aesthetic contemplation should be treated separately. Schopenhauer's treatment of objects takes its inspiration from Kant, in that the external world of objects and things appears under a priori forms of representation, such as space, time, and causality. The world is conceptualized and individuated as particular objects bound together in an inexorable web of causal relations. In aesthetic contemplation, however, these forms of representation are loosened. Rather than appearing as particular spatiotemporal objects bound by causal relations, the objects depicted in painting, sculpture, and poetry express the ideas. In their particular content, the idea shines through. Music, however, holds a special place at the pinnacle of Schopenhauer's aesthetics. Since music does not depict objects, it does not express ideas; rather, due to its contentlessness, music bypasses the expression of ideas altogether and directly expresses the will. What is truly real—the will—is grasped as a whole in music's wordless and conceptless objectification. Only the relations of tone to tone are capable of expressing the endless longing and striving of the will, while music's instrumental causes have no place in this order. Causality is sloughed off in the transformation effected by aesthetic contemplation.

Aesthetic contemplation also transforms the subject. Typically, Schopenhauer describes the subject as divided between knowledge of oneself (which is known through one's feelings, will, desires, and interests) and knowledge of the exterior world (which is known through perception). Schopenhauer's subject is founded on an economy that operates as a zero-sum game: The suppression of the will results in the heightening of perceptual powers and vice-versa. "The more conscious we are of the object," writes Schopenhauer, "the less conscious we are of the subject; on the other hand, the more this occupies consciousness, the weaker and less perfect is our perception of the external world."[13] Aesthetic contemplation operates within this economy. Schopenhauer lauds "the capacity to remain in a state of pure perception, to lose oneself in perception, to remove from the service of the will the knowledge which originally existed only for this service... the ability to leave entirely out of sight our own interest, our willing, and our aim... in order to remain *pure knowing subject*, the clear eye of the world."[14] Aesthetic contemplation, by being wholly absorbed in perception, holds all traces of willfulness and interest at bay. Thus, Schopenhauer often describes aesthetic contemplation as a state of "pure will-less knowing."

However, a strange problem arises. Since the state of pure will-less knowing is a denial of the will, it cannot itself be willed. Rather, it must come into being through processes and acts that draw the subject away from the will toward a state of pure perception. Typically, the genius is the figure endowed with the capacity to enter directly into the state of pure will-less knowing. But Schopenhauer, surprisingly enough, suggests processes and techniques for individuals who are not endowed with genius to suppress the will, increase the intensity of their perceptual experience, and thereby facilitate the state of aesthetic contemplation. Schopenhauer was fluent with contemporary scientific and medical writings, and often appealed to the physiological bases of conscious experience. In his view, physiological conditions could be used to still the perpetual throbbing of the will and help a subject enter into a state of aesthetic contemplation. He writes, "The state required for pure objectivity of perception has in part permanent conditions in the perfection of the brain and of the physiological quality generally favorable to its activity; in part temporary conditions, in so far as this state is favored by everything that increases the attention and enhances the susceptibility of the cerebral nervous system, yet without the excitation of any passion."[15] To help an individual get into the right state, Schopenhauer suggests various bodily techniques to increase attentiveness, such as "a peaceful night's sleep, a cold bath, and everything that furnishes brain-activity with an unforced ascendency by a calming down of the blood circulation and of the passionate nature." Such techniques aid in "making the object more and more detached from the subject," and producing a state of "pure objectivity" and "pure will-less knowledge," in which objects stand before it with "enhanced clearness and distinctness."[16]

Interesting and unusual as these prescriptions are, it is important to underscore their place in Schopenhauer's overall argument. Bodily techniques operate as preparatory acts or rituals that allow for the attainment of aesthetic truths—the expression of the idea, the objectification of the will, etc. Such techniques are always *simply preparatory*, as one can see in Schopenhauer's phantasmagoric aesthetics. Once the state of pure will-less knowing has been attained, reflection on bodily techniques grows silent. They have done their work, dissolving into the idea, without remainder.

Schopenhauer's strategy of phantasmagorically deploying and then dismissing bodily techniques after attaining the pure state of aesthetic contemplation has a venerable history. Wackenroder employs bodily techniques for musical phantasmagoria in the final tale of his *Outpourings*, concerning the fictional musician Joseph Berglinger. Describing Joseph's comportment at a concert, Wackenroder writes: "Whenever Joseph [Berglinger] was at a big concert, he seated himself in a corner, without looking at the brilliant gathering of auditors, and listened with the very same reverence as if he were in church—just as quietly and motionlessly and *with his eyes fixed upon the ground* before him. Not the slightest tone escaped him and, at the end, he was very weak and fatigued from the intense attentiveness."[17] By means of a bodily technique to invoke the acousmatic situation—in this case, averting one's eyes from the musical performance—Joseph directs his attention away from the site of musical production and onto the tones. Schopenhauer's claim that heightened attention toward the object makes the subject less self-aware finds a precursor in Wackenroder. Joseph, at the height of aesthetic contemplation, undergoes an experience where it is "as if [his soul] were detached from his body and were flitting about it more freely, or as if his body had become part of his soul."[18] Alongside the use of bodily techniques to focus attention, musical phantasmagoria follows, although one can note a divergence in its particular form. Whereas for Schopenhauer, the tones of music were so general as to demonstrate "no resemblance between its production and the world as representation,"[19] Wackenroder's form of phantasmagoria relies on the supplementation of the tones with additional visual images. If Schopenhauer promotes a kind of proto-reduced listening, where all signification has been evacuated from the tones, Wackenroder is less strict. Joseph does not hear the objectification of the will itself in music; rather, he experiences a variety of accessory images and feelings while listening attentively. During "gay and charming" passages, he seems to be "watching a lively throng of boys and girls dancing" in a meadow; other times, the tones are "so clear and penetrating that...they seemed to him to be *words*"; and elsewhere, the music causes a strange and wonderful sensation of "gaiety and sadness."[20] Although all these experiences are made possible through the bodily technique of the averted glance, none acknowledge the actual production of the music being heard.

For argument's sake, Schopenhauer can be bookended between Wackenroder, at the turn of the 19th century, and Wagner at its end. Just as Wackenroder described bodily techniques that promoted the acousmatic situation and musical phantasmagoria, the architecture of Bayreuth could be interpreted as an attempt to institutionalize the phantasmagorical power of the averted glance. As is well known, one of the notable architectural features of Bayreuth was the "invisible orchestra"—an unusual design for an orchestral pit tucked under the proscenium stage. The orchestra, placed behind a partition, sits in a pit containing a series of terraced steps that rake downward at the same angle as the floor of the auditorium, thereby making them invisible to the audience. George Bernard Shaw, captivated by the effect, explicitly notes how Bayreuth's architecture is designed to alter and channel the attentiveness of the audience. "Unlike the old opera houses, which are constructed so that the audience may present a splendid pageant to the delighted managers," Shaw writes, "[Bayreuth] was designed to secure an uninterrupted view of the stage, and an undisturbed hearing of the music, to the audience."[21]

The hidden orchestra at Bayreuth was preserved from the early designs that Gottfried Semper made for Wagner in the 1860s, when Munich was considered as a site for the *Festspielhaus*. Although is uncertain whether the idea first originated with Semper or Wagner, in either case, the acousmatic experience of the hidden orchestra loomed large in Wagner's aesthetic writings and practical essays concerning his plans for a festival theater. A letter to Wagner's fried Felix Dräseke illustrates the point. As a young man living in Paris, Wagner describes attending a rehearsal of Beethoven's Ninth Symphony at the Conservatoire. Arriving late, he waited in a room separated from the main hall by a partition or half wall; entranced by the sound coming over the divider, Wagner writes that music, when freed of the visual aspects of its mechanical production, "came to the ear in a compact and ethereal sort of unity."[22] There is no drastic change in the acoustical signal of the orchestra whether it is seen or not seen—other than the filtering or dampening that may occur due to the partition. Yet, according to Wagner, the whole *effect* of the music is transformed when the ugly mechanism of production is phantasmagorically veiled.[23] Later in his life, Wagner would claim this acousmatic experience as the motivating factor for the architecture of Bayreuth. "To explain the plan of the festival-theater now in course of erection at Bayreuth," he writes, "I believe I cannot do better than to begin with the need I felt the first, that of rendering invisible the mechanical source of its music, to wit the orchestra...."[24] At Bayreuth, where the orchestra is obscured from view, Berglinger's averted glance is transformed from a deliberate bodily technique to an architectural a priori.

Wagner was not the first to imagine strict architectural controls over the eye and the ear, even if he was the one to bring them to fruition. Precursors for Bayreuth abound in a variety of proposals, aesthetic treatises, and imaginative essays that span the period between Wackenroder and Wagner. Camille Saint-Saëns mentions two French examples that Wagner would not have likely known.[25] The first is from the composer Grétry's *Mémoires* of 1797. Grétry proposes plans for a theater appropriate for dramatic performances with music, where the orchestra would be placed "out of sight [*voilé*] so that neither the musicians nor the lights of the music desks can be seen by the audience."[26] According to Grétry, "This would create a magical effect," as no one would expect the orchestra to be there. The second is from Choron's *Manuel de musique* from 1838. He argues that "the attitudes and the movements required for instrumental execution are among those most contrary to [theatrical] illusion," and

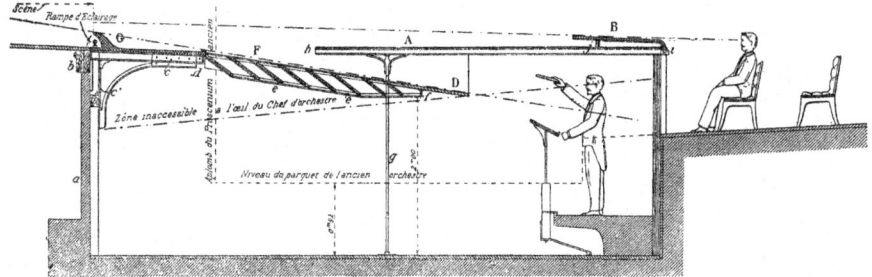

Figure 4.1 Sketch for a design of a covered orchestra by Jean Girette. Reproduced from *Die Musik* 5/8 (1905/06). Courtesy of the Yale University Library.

that "the presence of the orchestra, playing in full view... is every bit as disturbing as would be the sight of the back-stage machinery and the stage-hands working away on it." To overcome this disturbance, Choron proposes a theater where the orchestra no longer sits in front of the proscenium—which takes up considerable space—but is positioned in a room or cage built into the frame of the theater's structure. This cage should be wood-lined in oak or fir (for acoustic reasons) and hidden from view by a masonry wall punctured with openings to allow the sound to reflect and fill the hall.[27] For a German precursor, one could turn to Ignaz Ferdinand Arnold's *Der angehende Musikdirektor* of 1806: "In order to increase the pleasure of music, the musician should be invisible, either covered with a screen [*Taffetnen Schirme*], or a music hall should be built so that one does not notice the musicians."[28]

In the wake of Bayreuth grew an entire concert reform movement, which sought to import Wagner's architectural inventions into other performance spaces. New concert halls were designed to cover not only the orchestra but the stage as well, permitting performances of choral music where the musicians and singers would be veiled (figure 4.1). Older spaces were rigged with coverings, screen, scrims, and veils, or decorated with flowers, palms, and laurels strategically placed to obscure the performers (figure 4.2). Colored lights could be used to add atmosphere, as well as scents. In addition to architectural changes, the reformers sought to unify concert programs, expressing the tendency toward unity and integration that had begun with Beethoven and was carried on through Wagner's *Gesamtkunstwerk*. The aim of these reforms was primarily conservative and "spiritual," an attempt to aesthetically shape the listening public, and in turn the social body, by creating the conditions for performances that could properly channel music's transcendent content.

In 1899, Wilhelm Mauke, a composer and music critic active in the circle of concert reformers, published an article in the *Frankfurter Zeitung* suggesting architectural, decorative, and even olfactory reforms to the concert hall, designed to improve the performance of lieder. His ideas prolonged and embellished those of Grétry, Choron, Arnold, and, most importantly, Wagner.

> The auditorium, whose seats are arranged in the shape of an amphitheater, affords only enough light to permit the audience to read the text of the lieder.... [The singer's] voice touches our hearts by being heard through a sea-green web of liana plants. An aroma of Heliotrope passes through the hall when sensuous sultry

Figure 4.2 Sketch for a floral music screen. Reproduced from Paul Marsop, "Vom Musiksaal der Zukunft," *Die Musik* 3/21 (1903/04), 170. Courtesy of the Yale University Library.

love songs are sung. Serious lieder are heard with incense that comes from rows of columns that are embraced by the holy groves or cypress. The hymns of summer night rock one to sleep in the midst of large umbellated buds, violet-colored clouds, stars that glitter gently—everything is in mystical darkness. Passionate cries of erotic songs speak to the imagination and to intimate emotions of the audience, which is thrilled with perplexity and pain.[29]

Perhaps the most extended statement of the concert reform movement came from the writer and educator Paul Marsop. In an ambitious series of articles published in *Die Musik*, Marsop advocated for "*der Musiksaal der Zukunft*," with all of its Wagnerian overtones. Marsop documents his transformation of the *Stephaniensaal* in Graz for a music festival organized by the *Allgemeiner Deutscher Musikverein* in 1905. The auditorium became a "veritable garden," replete with dense thickets and shrubs surrounding the conductor's podium. The performers were hidden behind a screen of vegetation backed with an olive green cloth. The conductor on the opening night of the festival, June 1, was none other than Gustav Mahler. According to Marsop, everyone at the festival was delighted with the decorations and their effect on the performances—the audience, the musicians, and even the conductors, except Mahler. Wanting to remain visible to the audience and to have a full view of his orchestra and soloists, Mahler forbade the lights of the concert hall to be extinguished, against Marsop's wishes.[30] Mahler's own views about Marsop were somewhat more extreme: "What an ass!" he told Richard Wickenhauser. Rather than worry about the concert hall's effect on the audience's sensorium, Mahler felt that it was the job of the music to produce its own mode of attentive blindness: "The music must be so beautiful that the audience can no longer see anything, that it is blinded! It all depends on the performance, not the concert hall!"[31]

But perhaps Mahler and Marsop were not so very far apart. At the June 1st concert, Mahler performed 13 of his orchestral songs, which included the *Kindertotenlieder*.

Ernst Decsey, a music critic in Graz, attended the rehearsal earlier in the day and described Mahler's attitude and comportment while conducting "*Ich bin der Welt abhanden gekommen.*" Mahler, with his eyes closed and head bowed, directed the orchestra while in a state of complete absorption, as if "transported into the world he had created, and oblivious to his surroundings."[32] Mahler's bodily technique of closed-eye conducting, in line with Schopenhauer and Wackenroder, allowed him to experience the full effect of musical phantasmagoria, even while leading the orchestra. Marsop's architectural techniques to block the eye and control the ear were aimed at the same goal, the fully absorptive experience of music's spiritual content. Aside from Mahler's desire to make a spectacle of his absorption, both moments are united in their production of musical phantasmagoria.

Before Graz, screens and scrims had been installed in other concert halls. In *Die Musik*, Marsop documents experiments made by Philipp Wolfrum for a music festival in Heidelberg from 1903. The festival was held in honor of the dedication of a new city hall, which featured an auditorium with a sunken orchestra, similar to the one at Bayreuth. In a photograph taken on the floor of the *Stadthalle*, a scrim covers the choir and a pit hides the orchestra from view (figure 4.3). In a second photograph, taken from another angle, the photographer peers down into the orchestral pit—breaking the illusion by revealing the means of production (figure 4.4). Marsop also documents a screen installed by Louis Glass and Georg Hoeberg for an evening of chamber music in the smaller hall at Copenhagen's *Konzertpalais*. The screen, decorated with paintings of classical figures, hides the instrumentalists (figure 4.5). The covering was installed in front of the stage, retrofitting a traditional performance space into a primitive acousmonium. Unlike the Heidelberg hall, where the architecture of the space, with its sunken pit, encourages the splitting of the sensorium, Copenhagen's improvised solution is far more primitive. In both cases, however, the masking of musical production depends on screening or blocking visual access to the physical bodies performing the music—the *Musikapparates*, to borrow Marsop's dehumanizing term.[33] Occasionally, he includes a photograph from the apparatuses' point of view (figure 4.6).

Carl Dahlhaus, addressing the 19th-century fascination with the "invisible orchestra," claims that Wagner's practice at Bayreuth (and other concert halls like those of Heidelberg and Copenhagen) reproduces the "the prevailing doctrine of nineteenth-century music—the idea of 'absolute music,' divorced from purpose and causes...."[34] This doctrine gave rise to "the demand for an 'invisible orchestra' concealing the mundane origins of transcendental music."[35] Commenting on the Copenhagen hall, Dahlhaus writes, "when the screen hiding the musicians is covered with paintings... the end of a purely abstract conception of music is thwarted by the means."[36] Duly noted, but even with painted screens, the conjunction of musical phantasmagoria and the divided sensorium lives on. In comparison with Heidelberg, where arabesques on the screen and partition emphasize formal construction and hieratic designs, Copenhagen's screen, decorated with classical figures and instruments, occludes one set of performing bodies with images of another.

All of these techniques, whether bodily or architectural, are intended to split the sensorium—to separate the ear from the eye—and intensify the act of listening. It matters less whether this act of listening becomes wholly abstract, as demanded by Dahlhaus's characterization of absolute music and embodied in the Heidelberg hall,

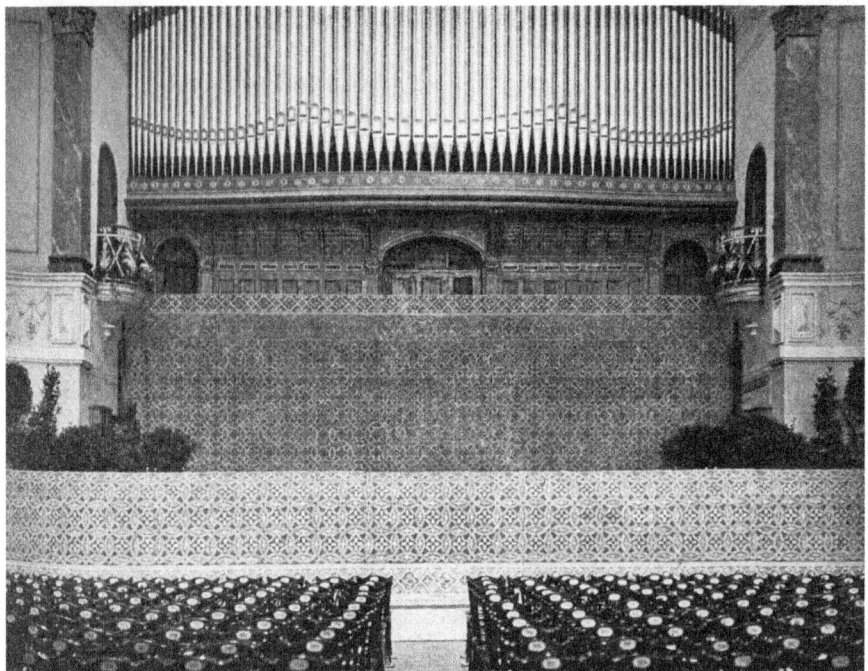

Figure 4.3 Hidden orchestra and choir in the Heidelberg *Stadthalle*, 1903. Reproduced from *Die Musik* 3/4 (1903/04). Courtesy of the Yale University Library.

Figure 4.4 Photograph from above the hidden orchestra, Heidelberg *Stadthalle*, 1903. Reproduced from *Die Musik* 3/4 (1903/04). Courtesy of the Yale University Library.

Figure 4.5 Decorated music screen from the small hall in the Copenhagen *Konzertpalais*. Reproduced from *Die Musik* 5/16 (1905/06). Courtesy of the Yale University Library.

Figure 4.6 A view from the orchestra platform, Dessau *Hoftheater-Orchester*. Reproduced from *Die Musik* 5/16 (1905/06). Courtesy of the Yale University Library.

or supplemented with images, such as Wackenroder's descriptions of Berglinger's musical experiences or the screen placed in front of the stage in Copenhagen. More importantly, in both cases, acousmatic situations are being exploited for a common end: to hide "the mundane origin of transcendental music." Phantasmagoria is the means of production of musical transcendence. By "transcendence," I simply mean the positing of any sphere—whether it be religious, secular, philosophical, ethical, aesthetic, or otherwise—that exists outside the bounds of the mundane world, and that is manifested in this world only at special or singular moments. Transcendence depends on separation, on the articulation of differences in kind. With its strict separation of the eye and the ear, an especially potent form of this phantasmagoria employs the acousmatic situation to occlude the mechanism of musical production for the sake of musical transcendence. The more the body is hidden, the less the eye sees, and the more grandiose are the claims about music's power.

INVISIBLE ORCHESTRAS AND ANGELIC CHOIRS

The kinship between phantasmagoria, the acousmatic situation, and musical transcendence—exploited by Schaeffer and Wagner alike—was not the invention of the 19th century. Before Wagner and Wackenroder, the experience of music was often understood as offering earthly listeners a prefiguration of the heavenly angelic choir. In 1619, Paolo Morigia, a Jesuit scholar, described Milanese religious music this way:

> Every Saturday evening at the hour of Compline one sings the Salve Regina...thus, at the appointed hour therein one finds music, the organist and the sacerdotes...the music begins and the organ responds, and then the organ and the music [sound] together, with such sweetness and such beautiful harmonies, which, because they seem an angelic choir, generate in the hearts of the listeners a whole-hearted composure and a holy devotion to the Mother of God.[37]

Federigo Borromeo, the archbishop of Milan, employed the trope of the angelic voice in his discourses on music. In his Assumption Day sermon, Borromeo began with the topos of the angelic song, offering it as a model to be imitated in musical performances by nuns.[38] The model went hand in hand with a prohibition on vanity during the nuns' concerts, which, for Church officials, had begun to veer uncomfortably close to the kinds of spectacular musical performances taking place outside the cloister.

In the period following the Council of Trent, when the practice of *clausura* was instituted, it was declared that, without exception, nuns were to be confined within the walls of the convent and kept away from the eyes of the congregation.[39] Many of the convents were walled in, with only grilles to allow for the passage of sound while obstructing visual access. Martin Gregory, an English Catholic scholar, described the experience of *clausura* during his travels to Italy:

> [The Tridentine Reforms are] so esteemed not only in Rome but through all Italy that thou shalt never see Nonne out of her Cloister, and being in the Churche thou shalt only hear their voices singing their service most melodiously, and the Father himself, that is, their Ghostly father heareth their confession through a

grate in a wall, where only voice and no sight goeth between: and I have seen the blessed Cardinal Borromaeo say Masse in their Chappel at Millan before them, when I could not possibly see any of them... and in Bononie [Bologna] and Rome having been many times at their service in the Chappels and hearing the goodly singing, never did I yet see one of them.[40]

Sound, which penetrates and pierces enclosures, became an important mechanism that reminded the world beyond the convent of the cloistered nuns inside it. Although the voice of the nuns can resemble the voice of the angels even without any kind of visual reduction, *clausura* can be understood as a technology that, despite its obviously repressive aspects, produced acousmatic situations in order to make the sensuous audition of the angelic voice all the more transcendent. The architecture often supported the resemblance. For instance, the interior of the Convent of *Santi Domenico e Sisto* in Rome possessed an extraordinarily high altar with grated windows above to the left and right, and, high up near the vaults, a series of grated openings that circled the church. The voices emanating from these high grates were juxtaposed against the frescoed ceilings depicting images of the heavenly host. The architectural space reinforced the fantasy: The listeners were encouraged to identify the vocalic body, heard in the nuns' voices, with the celestial figures floating above their heads.[41] Like the painted screen of the Copenhagen *Konzertpalais*, one set of bodies replaced another.

But the practice of *clausura* could never completely secure the phantasmagoric transcendence of the angelic voices. The nuns' voices could just as easily be associated with an angelic source as with the mundane body from where it emerged. For Rousseau, the dialectics of the angelic voice fascinated and maddened him on his trip to Venice in 1743. He writes:

> Every Sunday, in the church... motets are sung during vespers, for full choir and orchestra, composed and conducted by the greatest masters in Italy and sung in the grilled galleries by these girls, the oldest of whom is under twenty. I cannot conceive of anything so pleasurable or so moving as that music.... Never did Carrio or I miss those vespers in the Mendicanti, and we were not the only ones. The church was full of music-lovers; even singers from the opera came here to have a real lesson in tasteful singing from these excellent models. What distressed me were the accursed grilles, which only let the sound through but concealed those angels of beauty—for the singing was worthy of angels—from my sight.[42]

Rousseau's erotic drive to peer behind the grilles and behold the actual bodies fantasized by the nuns' voices leads to a cruel and misogynist joke. After begging, Rousseau is taken to meet the young women.

> As we entered the room where sat these beauties I had so desired, I felt such an amorous trembling as I had never known. M. Le Blond introduced me to one of these famous singers after another, whose names and voices were all I knew of them. "Come, Sophie."...She was hideous. "Come, Cattina."...She had only one eye. "Come, Bettina."...She was disfigured by small pox.... Two or three, however, seemed passable to me; *they* only sang in the chorus.[43]

The curiosity to peer behind the screen looms large in any attempt to produce musical transcendence by veiling the source. Thus, to deal with its possible failure in practice, the acousmatic situation is often treated as an aesthetic duty: that, lacking bodily techniques and architectural spaces, the traces of musical performance *should be* erased. In the period between the end of *clausura* and the invisible orchestra of Bayreuth (the period that covers the birth of Romanticism and the rise of absolute music), the grilles of the convent were converted into deontological demands that the musicians should be heard and not seen. Some, by the middle of the 19th century, treated the demand as nearly self-evident. Robert Zimmerman, in his *Allgemein Aesthetik als Formwissenschaft*, writes, "The sonorous element in music...[is] the ultimate consideration. The visual element of the performance does not belong to the work's essence.... It is for this reason that orchestral musicians rightly appear in the simplest clothes; it would be best if they were not visible at all."[44]

The anonymous author of a short fictional piece entitled "Our Concerts," which appeared in the *Musikalische Eilpost* in 1826, articulates a stronger version of this position. The narrator, a music lover, eagerly anticipates a concert featuring Beethoven's famous C-Minor Symphony, only to be disappointed by the chatter of the audience. In the piece that follows Beethoven's, the narrator employs a tactic to preserve his enjoyment: "In order to escape from all disturbance, I closed my eyes tightly, and then all was once again at peace within myself, and I was the lord of my own mood."[45] But this strategy of closing or averting the eyes—which worked so well for Joseph Berglinger—is foiled by the "whispering and tittering" of the audience in the hall, a reaction to the hideous pantomimes and expressions of the singer onstage. "She waved with her arms and bent her whole body so frantically back and forth, as though she had been set upon by a swarm of bees.... The flutist, who with bespectacled eyes gazed spookily from behind the old music stand out at the audience like Banquo's ghost at *Macbeth*, made such an adverse impression that a pregnant woman withdrew at once, afraid of things going wrong."[46] If the problems with the concert first began with the audience's insolent noisemaking, the author increased the ethical demand for invisibility by satirizing the gestures and infelicitous appearance of the performers.

Later, in a pub, the narrator meets with a cellist, a "passionate musical amateur," and the two commiserate about the horrors of the concert hall.[47] The cellist relates to the narrator that he has not been to a concert in 20 years, ever since he took a fateful boat trip on the Danube. Recalling the event, he describes how, one sunny afternoon, a pair of young virtuoso horn players performed impromptu duets on the deck of the boat. Although the musicians performed impeccably, the cellist could not "obtain any enjoyment" from the music, despite his best intentions.

As the cellist articulates the reasons why he could not enjoy the music, he ventriloquizes a musical aesthetic in miniature for the readers of the *Musikalische Eilpost* that brings together the acousmatic situation, phantasmagoria, transcendence, and bodily techniques. The cellist posits that the transcendent power of music is anything but foolproof. "Does music simply need to sound forth, no matter when, no matter where and under what circumstances, for the soul...to cast everything else away and aside and listen to the notes?"[48] No—the soul must be in a "special mood for it." All "distracting and disruptive influences must be eliminated," and one could use bodily techniques to diminish distraction and put the soul in a state of readiness.

Acousmatic Phantasmagoria and the Problem of *Technê*

Like Schopenhauer's baths, the body must be prepared for the musical transcendence—a condition that underscores the very tenuousness of such transcendence. While the boat bobbed on the water, the eye was distracted with the play of light upon the river and the sights of the villages on the shore. The soul was unprepared to enjoy music's power. With the fall of dusk, however, a transformation happens. The approaching darkness of the evening brings on a contemplative mood, when one can "forget the outer world and all its pretenses" in favor of the cherished world of "innermost dreams." The penumbra of evening diminishes the sense of vision and alters the perceptual ratio of eye to ear. Then, suddenly,

> ...horn tones arose from the wooded mountain nearby. Heavens, what a world of feeling was awakened by those sounds! Ah! They pressed upon the deepest deepness of the heart and awakened the entire slumbering past, so that it sprang forth, as though through magical pictures called up by a magic lantern, many-colored, brightly shining, and then gloomy once again—sweet and melancholic images.[49]

All the motifs are drawn together. In the acousmatic situation produced by the darkness of the evening, we hear the distant call of the horn, whose source is not clearly discerned. With the soul properly prepared, the special feeling of musical transcendence washes over the listener. That experience is compared, quite explicitly, with the effect of the magic lantern—literally, phantasmagoria. The author's choice of instrument is deliberate. Drawing a contrast between the young musicians' horn duet, heard in the clear light of day, and the distant, nocturnal horn, acousmatically auditioned under the cover of night, the author prepares the reader for his conclusion. By keeping the instrument the same but changing the conditions of reception in which it is heard, the author hones the argument on the conditions of reception in the production of musical transcendence:

> What, now, can explain the completely different effect that the same instrument had upon the same hearts? Was it not the bright day, with its many-faceted, rich images, which took the power away from my friends' tones; and was it not likewise the night, which veiled all things from the eyes, and distant, unseen nature of the horn player in the forest, which so facilitated this?[50]

Since the conditions of reception are so important in obtaining the music's proper effect, the cellist proposes some reforms for the concert hall that reproduce the acousmatic conditions encountered on the Danube:

> Imagine a hall in which, first of all, the orchestra with its people and instruments is hidden from the audience's view by a light curtain, this would put a whole crowd of destructive demons in chains, not to mention how much more atmospheric music becomes when it resounds unseen. Imagine further that instead of the many burning candles there is a single hanging light, which gives forth only as much subdued illumination as wretched decency demands.... Would not the dim light, full of foreboding, compose the souls of those who entered, purifying away the dross of everyday life and setting them into that mood which alone is appropriate for the enjoyment of art? Would not the springlike sounds, coming

as though from another world, lift these poor earthly worms, swimming in the sludge of the everyday world, for a moment at least into the bright, heavenly regions of a more beautiful world?[51]

In these reforms, each feature of the experience on the Danube is transposed onto the design of the concert hall. The night, "which veiled all things from the eyes," becomes the "light curtain." The shimmering of the stars becomes the subdued illumination of the single hanging light. The mesmerizing effect of the distant forest horn calls would be preserved by the sounds coming through the curtain, "as though from another world." The nocturnal mood in which one can "forget the outer world and all its pretenses" would be preserved by the music's purifying power to wipe away the "dross of everyday life." It is not a large step from this imaginary concert hall to Bayreuth, Schaeffer's loudspeaker, Bayle's darkened acousmonium, or Peignot's *akousmatikoi* listening on the far side the Pythagorean veil. But we must remember that the cellist's experience on the Danube is no repetition of the originary experience of the Pythagorean veil. Rather, the power of the Pythagorean veil to elicit from us images of nocturnal transcendence, otherworldliness, distance, and sublimity comes not from its "originary" status, but from its inscription into a series of modern, historical, aesthetic practices: the erasure of the musical performer, the effective power of blindness to transform sound by changing modes of listening, and the use of bodily techniques in the production of transcendence and musical phantasmagoria.

One of the most extraordinary accounts of the transcendental power of musical phantasmagoria comes from another anonymous article, this time written by a "student," and entitled "Another Evidence of the Wide-Spread Influence of Theosophical Ideals," in *Century Path*, a theosophical journal published in California in the first decade of the 20th century. In it, the author describes how the theosophist Katherine Tingley inaugurated an "absolutely novel" custom of hiding the musicians behind a screen in musical and dramatic performances. The point of the practice was sheer transcendence:

> Undisturbed—one comes very near writing "undismayed"—by the personality of the performers, with naught appealing to the eye and sense save that beauty which Nature holds forever secure from the possible and wretched rivalry of counterfeit, with naught save the impersonal and beautiful between the soul of the hearer and the soul of the music itself, the listener is lifted into a new world of feeling and aspiration.[52]

The widespread influence mentioned in the title refers to the recent fashion of having performances of music with hidden orchestras, as if the theosophists had invented the practice. Aspersions are cast on latecomers, like the New York Symphony Orchestra, which had recently announced an upcoming concert with covered orchestra in their 1908 season—following suit from the German concert reform movement. A paragraph later, influence becomes confluence, and the author takes the sudden trend in covered performances "to prove the existence, so to speak, of Marconi messages on the mind and soul planes." The example of Marconi's wireless telegraphy, where dematerialized messages are transmitted in the ether, was taken as an indication of the existence of vibrations and planes of consciousness that transcended the materialism and egotism of the individual. The "veil of Isis," as Madame Blavatsky called it, was ready to be lifted and was prefigured in phantasmagoric musical performances. A transcendent cosmic consciousness was behind it all, manifesting itself in the scientific discoveries of X-rays and wireless transmissions, pushing humanity toward universal brotherhood. Musical phantasmagoria, where soul communicated directly with soul through the impersonal power of sound, was just another piece of evidence for the coming ascent of man's consciousness—his entry into a new, transcendental plane.

PHYSIS AND TECHNÊ

Not all claims about music's transcendental power are the same. Unlike the confidence of the anonymous theosophist in the power of musical phantasmagoria to produce musical transcendence, the author of "Our Concerts" also acknowledges its tenuousness. Never is the power of music simply surefire. Unlike the Orphic or Apollonian myths of music, the very sound of music does not simply soothe the savage beast, lift the veil of Isis, or transport the listener to a heavenly sphere. Rather, the subject must employ various forms of *technê*, from bodily attitudes and management of the sensorium to the creation of artificial architectural spaces, in order to facilitate music's power. Music's nature, its *physis*, requires the supplementation of *technê* to make it effective.

By juxtaposing *physis* and *technê*, I am invoking the classical distinction between art (*technê*) and nature (*physis*). Philippe Lacoue-Labarthe, in his masterful essay "Diderot: Paradox and Mimesis," offers a quick sketch of the relationship of *physis* and *technê* that draws upon Aristotle's treatment in the *Physics*.[53] As Lacoue-Labarthe demonstrates, Aristotle offers two different accounts. First, at 194a, "art imitates nature" [*he tekhné mimeitai ten phusin*] in the sense that *technê*, which follows after the products of *physis*, develops by copying from the works of nature. *Technê* would take what *physis* has already provided as a model and duplicate it. This is what Lacoue-Labarthe calls the "restricted form" of the relationship, where *technê* "is the reproduction, the copy, the reduplication of what is given (already worked, effected, presented by nature)."[54] Second, Aristotle offers a competing account: "On the one hand, *technê* carries to its end [accomplishes, perfects, *epitelei*] what *physis* is incapable of effecting [*apergasasthai*]; on the other hand, it imitates" (199a). In the second account, which Lacoue-Labarthe refers to as the "general form" of the relationship, *technê* "reproduces nothing given... [but] *supplements* a certain deficiency in nature, its incapacity to do everything, organize everything, make everything its work—*produce* everything."[55]

Aristotle's competing views about *physis* and *technê* cannot be easily reconciled. How can *technê* both imitate nature, and thus duplicate the model that nature provides, while simultaneously perfecting or accomplishing what nature cannot achieve? Where would *technê* have learned its skills? If nature is its teacher, how could it learn to fulfill nature's ends better than nature itself? How can *technê* be both subordinate to and master of *physis*? Lacoue-Labarthe notes that Aristotle's strange double stance on *physis* and *technê* follows the logic of the supplement in Derrida's sense: "that the outside be inside, that the other and the lack come to add themselves as a plus that replaces a minus, that what adds itself to something takes the place of a default in the thing, that the default, as the outside of the inside, should be already within the inside, etc."[56] If *technê* comes to the aid of *physis* and brings *physis* to completion, then *physis* cannot be conceived as a simple plenitude or potentiality without lack. The inability of *physis* to realize its ends without the aid of *technê* reveals that the relationship cannot be one of simple subordination. In fact, when describing the supplement in *Of Grammatology*, Derrida isolates the term *technê* and concatenates it among other supplements that organize the discourse of Western philosophy: "The supplement adds itself, it is a surplus, a plenitude enriching another plenitude.... It cumulates and accumulates presence. It is thus that art, *technê*, image, representation, convention, etc. come as supplements to nature and are rich with this entire culminating function."[57] The equivocation within the Aristotelian definition reveals the logic of the supplement at the heart of Aristotle's thinking on *technê*.

The supplementary relationship of *physis* and *technê* underlies the production of musical transcendence in the 19th century. If the nature of music is transcendent, if it comes "as though from another world," if it "purifies the dross of everyday life" and lifts the soul from the confines of its earthly prison, why does such transcendence also require the use of *technê* to achieve its ends? The bodily techniques for directing attention and the architectural constraints of screened musical performance—from Wackenroder and Schopenhauer, to the *Musikalische Eilpost* and Wagner—can be understood neither as subordinate to music's nature nor imitations of it. The tenuousness of musical transcendence and its dependence upon the use of such forms of *technê* underscore the original supplementarity of musical *physis* and *technê*. The two never appear separately. The constant presence of *technê* as a remedy for the tenuousness of musical transcendence should force a further question: If music's nature is lacking, if it requires technical supplementation to achieve its ends, should we not wonder why, in claims of music's transcendent nature, *technê* has been so quickly dismissed? What is the challenge of thinking about musical *technê*?

The question is acute in Wagner's writings, especially in passages where he attempts to distinguish the transcendent nature of music from its technical effects. In *Opera and Drama*, his desire to produce the artwork of the future springs, in part, from his contempt for cheap transcendence that wholly relies on *technê* for its striking effects. He writes, "We know, now, the supernatural wonders wherewith a priesthood once deluded childlike men into believing that some good god was manifesting himself to them: it was nothing but Mechanism, that ever worked these cheating wonders. Thus to-day again the *super*-natural, just because it is the *un*-natural, can only be brought before a gaping public by the wonders of mechanics; and such a wonder is the secret of the *Berliozian Orchestra*."[58] In Wagner's polemic, the striking effects of Berlioz's orchestrations are struck down like false gods after being unmasked as technical tricks. However, Wagner's critical words also betray the lesson he learned—if you want to outdo Berlioz, you must hide the machinery.

In contrast to the desire to unmask *technê* in *Opera and Drama*, the later Wagner formulated an aesthetic theory wherein he believed himself to have overcome *technê* by simply making it irrelevant. Nowhere is this clearer than in his "Beethoven" essay. As if in response to the anonymous author of the *Musikalische Eilpost*, Wagner's essay begins with a nocturnal scene of distant sounds. On a sleepless evening in Venice, he is suddenly struck by the song of a gondolier. As the sound echoes across the canals, it solicits a response by another distant voice. Like the scene on the Danube, with its distant and resonant Alpine horns, the song of the anonymous gondoliers is autochthonic, "as old as Venice's canals and peoples."[59] The penumbra of darkness, which separates eye from ear, discloses essences obscured by the power of sight. "Whate'er could sun-steeped, color-swarming Venice of the daylight tell me of itself, that that sounding dream of night had not brought infinitely deeper, closer, to my consciousness?" Wagner calls this state—a metaphysics of blindness in which essences are disclosed to the ear alone—"sympathetic hearing." Unlike the author of the *Musikalische Eilpost*, who acknowledges the tenuousness of musical transcendence, Wagner's metaphysics of blindness is surefire. For the sympathetic listener, the distracting visual elements of the concert hall are of no consequence since "our eyesight is paralyzed to such a degree by the effect of the music upon us, that with eyes wide open we no longer intensively see." This experience of musical blindsight is produced anytime the music "really touches us" despite the fact that "the most hideous and distracting things are passing before our eye," such as "the highly trivial aspect of the audience itself, the mechanical movements of the band, [and] the whole peculiar working apparatus of an orchestral production." Wagner argues from the fact that we are ordinarily inattentive to such a spectacle that absorbed listening puts us into "a state essentially akin to that of hypnotic clairvoyance."[60]

The aesthetic of the "Beethoven" essay, which clearly displays Schopenhauer's influence, is at odds with Wagner's earlier views. According to Carl Dahlhaus, "Two decades after *Opera and Drama*, [Wagner] had long since stopped believing that music was, or needed to be a function of drama.... The Feuerbach enthusiast, who accentuated the physical existence of the human being (e.g., the visible action within the drama), had become an adept of Schopenhauer, hearing the 'innermost nature' of the music drama's action in the 'orchestral melody.'"[61] Wagner's surefire claims about the nature of musical transcendence in the "Beethoven" essay, of music's somnambulant and hypnotic subversion of vision by hearing, cannot be easily asserted of a medium like opera, where the physical presence of singing bodies onstage would make blindsight unacceptable.

Wagner confronts the necessary visibility of opera by rigorously controlling what the eye sees through the use of architectural *technê*. The solution is for the mechanism of the orchestra to be literally concealed, not simply neutralized in blindsight, in order to regulate and discipline the attention of the audience in the production of musical transcendence.

> The reader of my previous essays already knows my views about the concealment of the orchestra and...[my condemnation of] the constant visibility of the mechanism for tone-production as an aggressive nuisance....I explained how fine performances of ideal works of music may make this evil imperceptible at last, through our eyesight being neutralized, as it were, by the rapt subversion of the whole sensorium. With a dramatic representation, on the contrary, it is a matter of focusing the eye itself upon a picture and that can only be done by leading it away from the sight of any bodies lying in between such as the technical apparatus for projecting the picture.[62]

Wagner's compromise for the sake of drama evinces the degree to which he struggled with the problem of *technê*. Rather than simply ascribe to the power (or *physis*) of music the capacity to overcome the issue of technical mediation—as he does in the "Beethoven" essay—the architectural space of the performance must be designed to aid in the production of music's hypnotic clairvoyance. The theoretical arrogance of the "Beethoven" essay is challenged by the material facticity of operatic performance. In public and in his writings, Wagner's self-assurance concerning musical transcendence may appear unflappable, but in private, his struggle with transcendence and *technê* is more acute. Cosima Wagner transcribed this statement in September of 1878: "I cannot stand all this costume and grease-paint business! And when I consider how these figures such as Kundry will have to be masqueraded—I immediately think of these repulsive artists' carnivals, and, after having invented the invisible orchestra I would like to create the invisible theater."[63]

The kinship of the acousmatic situation and musical phantasmagoria—as deployed by Wagner, Wackenroder, Borromeo, Schopenhauer, the theosophists, Marsop, or the *Musikalische Eilpost*—relies on the supplementary relationship of *physis* and *technê*. In addition to the unique historical and social circumstances that characterize and specify the deployment of the acousmatic situation by each author and the kind of transcendence effected in each case, a common figure is inscribed. We can identify them as the conditions of acousmatic phantasmagoria:

1. To separate the eye and the ear, one requires *technê*, whether in the form of bodily techniques or architectural constraints.
2. The separation of the eye and ear dissociates the musical effect from its source, cause, or site of production, thus affording an understanding of the musical effect in a phantasmagoric form.
3. Phantasmagoria treats the distance between a sound and its source as a gap or rift between the transcendental and the mundane; when the source is dismissed, transcendence is installed under the name of music's nature or essence—its *physis*.
4. Transcendence is never guaranteed, though its proponents may underestimate its precariousness.

In the excursus that follows this chapter, I trace these conditions in Schaeffer's theory in order to pose the question: "Must *musique concrète* be phantasmagoric?" To demonstrate the pervasiveness of these conditions and the various kinds of tenuous transcendences they support, I conclude with two examples, one pitiful and one frightful.

First, the pitiful: On January 20, 1908, the English composer Joseph Holbrooke premiered his second symphony, *Apollo and the Seaman*, in Queen's Hall, London. The work was no ordinary symphony, but rather, as indicated by the large print above its title, "An Illuminated Symphony."[64] Holbrooke based his work on a poem by Herbert Trench. In the poem, Apollo, disguised as a merchant, converses with a seaman on the sublime topic of man's immortality. Unlike tone poems of the past, Holbrooke had an idea to make the poetry and music work together more precisely and effectively. In the score, he recommends that the hall be darkened and the orchestra "should, as far as possible, be invisible, behind a screen of plants, palms, or foliage—or thin, extremely lofty, decoratively hung festoons and columns of dark, richly-colored veilings designed not to destroy the sound." Most important, the orchestra was to remain behind a large screen upon which the words of the poem and other images were to be projected. Holbrooke specified how many lines of poetry should be projected at a time, the size of the projected script on the screen, and the care that should be given in changing the slides to avoid awkward jerks from stanza to stanza. The aim was clear: "The object to be attained is an effect of dignity, mystery, and solemnity, by a combination of poetry and music simultaneously concentrated upon the same idea."

Despite Holbrooke's fastidiousness, the technical apparatus did not quite provide the desired effect. A review from March 1908 describes the scene: "While the music was playing, the lines of the poem were thrown in eight-inch type on a screen in a darkened hall. Unfortunately the audience was unable to distinguish the lines that had suggested music from those that had been rejected by Mr. Holbrooke."[65] That was not the worst. Before the appearance of Apollo, the reviewer drolly notes that a long organ pedal was held, to arouse the expectations of the audience; then "a gong was struck... and lo, a 16-foot head of Apollo appeared on the screen... the suddenness of Apollo's appearance aroused the laughter of the flippant." Nor did the projectionist skillfully manage the poem. "The operator of the magic lantern sometimes failed to synchronize with the composer's strains...." The effect was so incongruous, so "totally inappropriate," that the result was general confusion, "for the brain was receiving suggestion of one sentiment by sight, and another by hearing."[66]

Now, the frightful: Between November 1939 and March 1944, the *Wiener Symphoniker* performed a series of darkened concerts, or *Dunkelkonzerte*. The musicologist Bryan Gilliam describes how "the darkened Wiener Konzerthaus was transformed into a sacred space," in particular, a space in which to hear the transcendent spiritual meaning of Anton Bruckner's symphonies.[67] At the time, Bruckner had undergone a revival (and revision) under the Nazis, becoming a figure of reverence for the leadership. An "infamous" photograph from Regensburg 1937 captures Hitler, dressed in full military garb, respectfully gazing up at a bust of Bruckner atop a pedestal emblazoned with an iron eagle and swastika. According to Gilliam, in order to "annex" Bruckner, Goebbels and others involved in Nazi cultural propaganda downplayed Bruckner's Catholicism in favor of themes more agreeable to the party. In Goebbels's Regensburg address, he emphasized Bruckner's rustic peasant roots and his victimization at the hands of music critics (like the Jewish critic Eduard Hanslick). In a true act of revisionism, Goebbels described Bruckner's conversion to a new symphonic style after encountering Wagner, claiming that "from that moment onwards the church musician at once retreats almost entirely, and out of him emerges the distinctive symphonist."[68] Bruckner the Aryan, the composer of *Blut und Boden*, is substituted for Bruckner the devout Catholic since, "in order to comprehend him one must look to the roots of his existence, the elemental forces of blood and race that propelled his humanity." The comprehension of Bruckner's music, Goebbels argued, could only occur when situated in the correct (i.e., Nazi) context, for his style is subject to a "complete misunderstanding" when lumped "under the rubric of religious art" or characterized as Masses without texts. After his conversion to Wagner, Bruckner's religious faith had broken free of all Catholic confines and become purely German: "it has its roots in the same heroic feeling for the world from which all truly Great and eternal creations of German art blossom." All the better that Bruckner, like Hitler, was a "son of Austrian soil" but was called to the "intractable intellectual and spiritual common fate that envelops our entire German people."[69]

The Viennese *Dunkelkonzerte* always featured a symphony by Bruckner, along with other works with strongly spiritual themes. Gilliam reproduces a program from November 15, 1940, that opened with Mozart's *Regina coeli*, performed in a "half-darkened hall" (*Im halbberdunkelten Saal*) and, after the intermission, closed with Bruckner's Ninth Symphony in a "completely darkened hall" (*Im gänzlich berdunkelten Saal*). The effect, according to reviewers, was "excellent," the *Dunkelkonzerte* being especially effective for works with a "mysterious romantic character."[70] But it was not quite romanticism the organizers of the concerts were after. Rather, the conservatism of the concert hall reformers from the turn of the century, like Marsop, who sought to aesthetically shape the listening public into a new social body, was outdone by the organizers of the *Dunkelkonzerte* and their "national aestheticism"—to borrow a phrase from Lacoue-Labarthe.[71] A new social body was to be formed through the communing of listeners in the darkened auditorium, hearing in Bruckner's symphonies the spiritual mission to which they were called. The *Dunkelkonzerte* employed all the conditions of acousmatic phantasmagoria in its production of transcendence. The eye and ear were separated by the obscure darkness; the music was separated not only from the source of the orchestra but from the context of Bruckner's religious faith; the sound of the orchestra became a vehicle for sounding out the holy art of the German people, a transcendental sonic message of spiritual destiny and fate. But what about the receivers of this message? Even if, according to the review, the effect was excellent, who knew exactly what message was being received? In Gilliam's words, "Whether or not contemporary German audiences believed the Nazi propaganda, whether or not they sensed their common soil upon hearing a rustic scherzo, communed with God during an adagio, or even perceived Teutonic heroism in a fugal finale is...a large, complex issue yet to be sorted out."[72] Complex indeed, but not without its conditions and history.

INTERLUDE

Must *Musique Concrète* Be Phantasmagoric?

The conditions of musical phantasmagoria are present in Schaeffer's *Traité*. The first condition, the separation of eye and ear through the use of *technê*, is satisfied by Schaeffer's use of modern forms of audio, such as the loudspeaker and the tape recorder, and also in the demand for a certain form of disciplined listening, *écoute réduite*, that trains the listener to audition the sound object while disregarding its technical basis and architectural setting. The second condition, where the separation of eye and ear is recast in phantasmagoric form, is satisfied in Schaeffer's various negative demands: "The sound object is not the instrument that was played.... The sound object is not the magnetic tape.... The sound object is not a state of the mind."[1] Similarly, Schaeffer occludes the means of production when he claims that electroacoustic tools, including the tape recorder, are not actually instruments.[2] The third condition, the transformation of phantasmagoria into musical transcendence—that is, into music's supposed *physis*—is satisfied by Schaeffer's claims about the eidetic nature of the sound object; its indifference to its modes of presentation; the "negligible difference" between the loudspeaker and the Pythagorean veil; the reactivation of an ancient originary horizon in the acousmatic situation; and the "inaugural experiments" of *musique concrète*, the *cloche coupée* and the *sillon fermé*, which disclose a new field of investigation.

But then there is always the stubborn fourth condition. Although Schaeffer and his students describe themselves as the inheritors of a Pythagorean tradition of acousmatic listening, in fact, there is good evidence to argue that they are really inheritors of the tradition of musical phantasmagoria, which was consolidated in the 19th century. Like Wagner (and many others) before him, Schaeffer must expel *technê* in order to reach his musical-aesthetic goals. Yet, if Schaeffer's commitment to the sound object and the phenomenological character of the acousmatic reduction compel phantasmagoria by disallowing a genuine thinking of *technê* and promoting the occultation of production, what is the alternative? Is all *musique concrète* doomed to phantasmagoria?

MUSIQUE CONCRÈTE AND THE PRODUCTION OF PHANTASMAGORIA

To answer that question, we must start in 1958, a significant year in the history of *musique concrète*. Not only was it the year that the *Groupe de recherche musicale* (GRM) was founded, it also brought the acrimonious departure of Pierre Henry and the arrival of Luc Ferrari. With the formation of the GRM, Schaeffer tried to integrate the production of new compositions with collective research projects that would consolidate and systematize the work that had been done over the previous decade. That shift could be seen in the replacement of the term *musique concrète* with competing terms like *musique expérimentale* or even Peignot's suggestion, *musique acousmatique*.

Schaeffer had not worked as a composer while overseas, but 1958 brought commissions for new works that could demonstrate his research on the sound object. Given the opportunity to present at the World's Fair in Brussels, Schaeffer and the young Luc Ferrari worked closely together to prepare works for the French Pavilion. The program first presented on October 5, 1958, featured Schaeffer's *Étude aux allures* and *Étude aux sons animés*, Ferrari's *Étude floue*, *Étude aux accidents*, and *Étude aux sons tendus*, Schaeffer and Ferrari's collaboration *Continuo*, and Xenakis's *Diamorphoses*. (Schaeffer had also intended to present the *Étude aux objets*, but did not complete it in time for the exhibition.) Over in the Philips Pavilion, one could hear Varèse's *Poème électronique* and Xenakis's *Concret PH*; it was a banner year for *musique concrète*.[3]

The new études were dramatically different in character from the *Cinq études de bruit* of a decade earlier. Rather than utilizing trains and pianos, they featured sounds that were coy about exposing their sources, demanding a form of concentrated listening focused on morphological qualities shared among sound objects. Even the titles revealed this change: The iconicity of the *chemin de fer* was now replaced by generalized, non-specific *objets* and *allures*. In Schaeffer's own terms,

> My three new studies [*Étude aux allures*, *Étude aux sons animés*, and *Étude aux objets*] are based on a triple asceticism: an intentionally limited number of sounding bodies [*corps sonores*], manipulations focusing more on "montage" and no longer on deformation, and finally, a compositional bias that consists of submitting itself to the object rather than torturing and modulating it according to preconceived considerations. It is the object, I believe, that has many things to say to us, if we know how to let it speak, and how to assemble it according to its family resemblances and concordance of characters.[4]

Apart from the implicit critique of dodecaphonic music ("torturing and modulating...according to preconceived considerations"), the passage also shows Schaeffer's growing phenomenological commitment, with its keynote of ascetic reductions and a return to "the things themselves." This commitment led toward an aesthetic that was similarly reduced, designed to solicit from the auditor a listening focused on the qualities of the sound objects alone, bracketing all associations and references. Constraining the compositional palette, such pedagogical *études concrètes* followed a dual strategy: On the one hand, the sound objects selected must not be strongly marked by *indexical features* that encourage recognition of the source or cause; on the other hand, the sound objects must share some *immanent feature* in varying degrees, affording comparison and contrast, in order to produce a musical discourse organized around degrees of tension, similarity, difference, and resemblance. For example, in the *Étude aux allures*, Schaeffer employs a limited palette of sound objects, all of which possess "allure," that is, a typo-morphological feature of internal beating, vibrato, or pulsing that is found in sustained sounds.[5] Such études use minimally processed, "appropriately limited sound material to create authentic structures, which will bring out for others the criteria which the composer, in following his own personal schema, endeavors to make audible."[6]

The challenge of treating allure compositionally is that listeners acquire information about the source of a sound from its allure. A listener can often tell if a sound source is natural or mechanical based on the degrees of regularity and irregularity heard in its internal beating or pulsing. Thus, allures have the potential to make the listener overly aware of the sources, instead of being auditioned for their immanent morphological qualities. Or, to put this into the framework of *Traité*, the allure of a sound source offers the possibility of promoting the mode of listening known as *écouter*—an indexical listening for sources. Strategically, Schaeffer tries to diminish this possibility by concatenating (or "montaging") sounds from a variety of sources, but unifying them around the pole of a shared immanent effect. In order to hear Schaeffer's own personal schema in the *Étude aux allures*, the listener must take a step back from the sources used and employ the strategy of reduced listening—*entendre*.

To demonstrate, I ask the reader to perform an experiment. Listen to the opening half-minute of Schaeffer's *Étude aux allures* while attending to the various sources of the sounds heard. When auditioned via *écouter*, the study seems to be little more than a surreal mélange of sounds: a bell, a whistle, a piano chord with the attack removed, a gong, a buzzer, various sustained percussion sounds, a spring, Turkish finger cymbals, a piano with the damper pedal depressed. By overloading the number of sources, Schaeffer denies the auditor any fixed point of orientation available via *écouter*. As a result, he compositionally forces the auditor to shift attention onto morphological characteristics like allure, grasped under the mode *entendre*, in order to make sense of the work. When the mode of listening shifts attention toward the morphological features present and away from information about sources—that is, as the listener shifts from *écouter* to *entendre*—the work suddenly takes on a "musical" quality. One hears allures presented in various degrees of difference and resemblance. Faster pulses are layered over slower, and a subtle rhythmic counterpoint emerges within layers of sustained tones and chords. Like a set of Husserlian "imaginative variations," Schaeffer wants us to hear allure as an invariant that persists across the concatenated series of sound objects. To put it another way, allure is an effect inherent in a variety of sources, and Schaeffer is doing all he can *compositionally* to make that effect audible.

The études of 1958, by focusing their compositional intent on morphological qualities of sound objects, differ from the strategies employed in other canonical works of *musique concrète*. Generally speaking, *concrète* compositions (or parts thereof) are often organized around one of three features inherent in any given sound: the source, the cause, and the effect. For instance, the famous *Étude aux chemins de fer* is organized around a single type of sound source—trains—that produces multiple qualitatively distinct auditory effects. The unity is not morphological, but relies on the listener's recognition of a source. The source is exposed through the presentation of its various aspects: wheels thumping rhythmically on the rails, steam blasts from exhaust valves, and distinctive train whistles that functioned (at least in 1948) as auditory indicators for the French listeners to Schaeffer's broadcast. Although the *Étude aux chemins de fer* deploys this material in a way that makes audible various rhythmic motives inherent in the recording of these sources, the work cannot simply be reduced to a rhythmic study. If one were simply interested in rhythm, a composer of *musique concrète* would be under no demand that all the sounds come from a single source. Although the physical causes of these sounds are all distinct (friction, steam, rattling), the auditory effects, although wildly dissimilar, are related by their secure grouping around a numerically identical source (or type of source). The *Étude aux chemins de fer* is indeed just what it says—a train study.

Another strategy, deployed by Bernard Parmegiani in his *Capture éphémère* (1967), is to focus neither on some numerically distinct sound source nor on a common morphological effect of a set of sound objects, but rather on the manner in which the sound is produced—a sound's "causality," as opposed to its source or effect. In other words, a composer could construct a work of *musique concrète* that presented a series of sounds that were all caused in the same way—a concatenation of *rubbed* sounds, *plucked* sounds, or *struck* sounds, regardless of their actual source. In the opening minute of *Capture éphémère*, Parmegiani presents a variety of unrecognizable sources, which overlap temporally but are unified by their perceived mode of causal production. The passage affords the listener a sequence of sounds that invokes the sensation of objects having been shot or fired, like a missile or arrow, across the virtual musical space, with corresponding spatialization in the stereo field and Doppler effects to give a sense of velocity.

These alternative strategies were not employed in Schaeffer's works for the French Pavilion. Nor were they present in Ferrari's contributions, other than in the *Étude aux accidents*, with its use of selectively edited recordings of a prepared piano. The collaboration piece between Ferrari and Schaeffer, *Continuo*, presents a continuous three-minute block of sound, where a repetitive rumbling and resonating background is counterpointed by a harmonically rich sound whose source is quite unidentifiable (a reversed piano with the attack removed?). This unusual sound, due to its spectral richness, is somewhere between a note and a chord, with a high degree of internal beating and grain—perhaps reminiscent of a motorcycle engine. Over the span of the piece, the sound starts and stops, ascends registers, until it begins to gliss dramatically while being panned across the stereo field. Ferrari's *Étude floue* is similar in vein, beginning with sustaining, highly reverberated pitches—also glissing, warbling, and oscillating in amplitude. Again, the sound sources are impossible to gauge, the indistinctness of the sounds being registered by the title *floue*, with its connotations of indistinct blurs and blobs.

The similarities in Schaeffer and Ferrari's aesthetics in 1958 are not difficult to explain. After the departure of Pierre Henry, Ferrari served as Schaeffer's assistant on the works intended for the French Pavilion.[7] But the confluence of their projects was not to remain intact for long. Ferrari soon began to depart from Schaeffer's research-driven aesthetic, describing himself as a "deviant" within the world of *musique concrète*.[8] He became interested in "anecdotal" sounds, a use of recorded sound that revealed its origins, and even the banality of its origins.[9] Anecdotal sounds broke the taboo on the recognition of the sound source, instituted by Schaeffer's demand for *écoute réduite*. The work *Hétérozygote* (1963–1964) employed a dual strategy: While the piece returns again and again to a texture not unlike that of the *Étude floue* or the *Étude aux allures*, with their sustained and indistinct registral bands of tones, such passages are interrupted by sounds with clear, real-world sources—slide whistles, violins, horns, splashes of water, and most important, speaking voices. According to Ferrari, *Hétérozygote* "used sounds that weren't concrete, that didn't belong in the musical world, but in the world of noises."[10] However, the use of anecdotal sounds was not intended as a provocation or as an explicit refusal of Schaeffer's aesthetic and pedagogical proscriptions, but followed from considerations about the technical tool par excellence of *musique concrète*, the microphone. "I was in the frame of mind that the function of the microphone was to register sounds or to record them on tape. So, whether in the studio, in society, in the street, or in private, it [the microphone] records in the same way. It followed, for me, to introduce anecdotal sounds into musical discourse."[11]

When Schaeffer first heard the work, he gave it a chilly reception, calling it incoherent, formless, and noisy.[12] According to historian Évelyne Gayou, it was Schaeffer's dismissive response that led to Ferrari's eventual departure from the GRM in 1966.[13] But Ferrari's comments about *Hétérozygote* are telling. "I thought he would like it. I didn't do things to please him, but out of a sense of personal necessity. Even so, I was surprised by his reaction, by his violence.... It shocked me deeply coming from him, because I thought it was close to the attitude of *Symphonie pour un homme seul*. I was really taken aback."[14] Schaeffer's reaction reveals the degree to which he had committed himself to the aesthetics of the acousmatic situation and reduced listening, privileging *entendre* over all other modes of listening.

ALMOST NOTHING

While *Hétérozygote* still relied on the juxtaposition of anecdotal sounds with textures that were more akin to Schaeffer's aesthetics of the late '50s and '60s, Ferrari would foreground the "function of the microphone" in his famous piece *Presque rien n°1 ou Le lever du jour au bord de la mer* (1967–1970), perhaps the most extreme "anecdotal" music he ever produced. *Presque rien* presents the sounds of a Dalmatian seaside village with little perceptible editing and no obvious electroacoustic manipulation. Subtitled "dawn at the seaside," the piece on first listening sounds like a documentary or field recording, capturing the sounds of the seaside from early dawn into the busyness of the day. The unsullied presentation of motorboats, lapping water, stridulating insects, footsteps on wooden planks, and singing voices encourages a mode of anecdotal listening. Ferrari's compositional logic is unique in the tradition of *musique concrète*: In contrast to the *Étude aux chemin de fer*, *Presque rien* does not rely on a unified sonic source from which its sounds are derived; indeed, the sources, which are quite diverse, are related to one another as naturally and socially affiliated, temporally contiguous, constitutive parts of a given environment. In contrast to the morphological emphasis of the *Étude aux allures*, the ear finds coherence in *Presque rien*'s environmental situatedness. If one were to approach *Presque rien* under the mode *entendre*, seeking a set of emergent morphological qualities, the result would be a disorganized mélange: No morphological characteristic is shared between voices, motorboats, footsteps, and lapping water.

Eric Drott, in a fine essay on *Presque rien*, notes that the work has consistently generated controversy since it first appeared. "The piece... is generally characterized as a gesture of aesthetic transgression—though there is some disagreement as to what particular principle the work transgresses."[15] On the one hand, the work appears to be a "rupture with the then-dominant aesthetic of French electro-acoustic music," that is, Schaeffer's theory of acousmatic sound and its privileging of *entendre*.[16] "By presenting clearly recognizable sounds, which have undergone little if any overt alteration, the piece marks what Michel Chion and Guy Riebel describe as a 'return of the repressed.' Audible traces of reality, hitherto barred from *musique concrète*, are encountered at every turn in *Presque rien*."[17] On the other hand, *Presque rien* is understood as "a tacit repudiation of the work concept central to Western art," where the dividing line between art and life is effaced through the use of recording. As Drott notes, "the use of magnetic tape to capture a slice of life, and thereby transform it into an object of aesthetic contemplation, places *Presque rien* within a tradition...that stretches from Marcel Duchamp to John Cage and beyond."[18]

In both cases, Drott's characterization of *Presque rien*'s critical reception is right on the nose. In the first camp, the environmental holism of *Presque rien* has encouraged critics and composers to categorize it as a piece of soundscape composition.[19] Soundscape theorist and composer Barry Truax considers *Presque rien* as exemplary of a certain subgenre of soundscape composition, due to the presence of two criteria. First, *Presque rien* emphasizes the use of "found sound" or the "*objet trouvé*."[20] According to Truax, a soundscape composition that emphasizes the use of found sounds is characterized by the fact that "no transformations were used, only editing and sometimes mixing."[21] The "found" character of sound is presented maximally when audible manipulation by the composer is held to a minimum. Second, Truax claims that *Presque rien* is organized around a "fixed spatial perspective emphasizing the flow of time." This is in contrast to other compositional strategies that present, for example, the moving perspective of a journey or a series of discontinuous and variable spatiotemporal perspectives.[22]

One might support Truax's two criteria by appealing to technical features of the production of *Presque rien*. By the mid-'60s, the invention of the Nagra portable tape recorder allowed *concrète* composers to leave the Parisian studio behind and record sounds in the environment. In an interview with Beatrice Robindoré, Ferrari stated, "I was the first composer to use the Nagra portable tape recorder, which had just been invented, and to take it everywhere....I spent a year with a Nagra on my shoulder....I recorded everything I could."[23] Naturally, the use of portable tape recorders facilitates a composer's encounter with potential found sounds. The second trait, the "fixed spatial perspective," is more complicated. The perspectival features of *Presque rien* are facilitated by Ferrari's use of a stereo microphone. Normally, works of *musique concrète*, even multi-channel works, were recorded in mono and subsequently mixed and spatialized—like the sounds that are panned across the stereo field in Schaeffer and Ferrari's *Continuo*. But Ferrari himself notes the importance of the stereo microphone to *Presque rien*, claiming that "with this wonderful discovery of stereo, sounds were present in the depth of space, not just within the left-right axis."[24] The use of the stereo microphone could facilitate a realistic audio reconstruction of environmental space. This helped to encourage *écouter*, the mode of listening that situates its auditory objects within the spatial environment.

But perspective and fixity are not mutually entailed. In fact, the sense of fixity in a recording is not technologically determined because the portable tape recorder alone, even with the supplement of a stereo microphone, does not dictate to the listener a single kind of spatial position. One can use the Nagra to record from a fixed location just as well as while in motion. The unity of place presented in *Presque rien* is a compositional decision. In an interview with Dan Warburton, Ferrari speaks about his strategies for recording *Presque rien*:

> I was in this Dalmatian fishing village, and our bedroom window looked out on a tiny harbor of fishing boats, in an inlet in the hills, almost surrounded by hills—which gave it an extraordinary acoustic. It was very quiet. At night the silence woke me up—that silence we forget when we live in a city. I heard this silence which, little by little, began to be embellished.... It was amazing. I started recording at night, always at the same time when I woke up, about 3 or 4 a.m., and I recorded until about 6 a.m. I had a lot of tapes! And then I hit upon an idea—I recorded those sounds which repeated every day: the first fisherman passing by same time every day with his bicycle, the first hen, the first donkey, and then the lorry which left at 6 a.m. to the port to pick up people arriving on the boat. Events determined by society. And then the composer plays![25]

The fixity of the microphone's location in Ferrari's window provides a unity of place. The lack of audible manipulations gives the sounds a found character. There is no obvious mixing, splicing, or editing—nothing that seems to resemble the careful manipulation of recorded sounds like the *cloche coupée*. The traces of the composer's hand are erased. The use of a stereo microphone fixed in a single spatial position encourages the listener to listen to the piece itself as a kind of listening. Identification is formed between the two sets of ears, the listener's and the composer's. The stillness of the auditor and the stillness of the microphone encourage a blending of horizons. We hear what Ferrari is hearing, and we assume that he is hearing in the mode of *écouter*.

But we don't hear what Ferrari is doing. In fact, the whole question of composition—Schaeffer's problem of bringing the composer's "own personal schema" into audibility—is diminished in *Presque rien*. Perhaps Ferrari is doing very little, almost nothing...*presque rien*? According to Ferrari, the compositional events are determined by *society*, not the hand of the composer. If this is indeed the case, we might think that *Presque rien* moves in the shadow of Cage's 4'33", opening the ear to an expanded field of listening that embraces all sounds, especially the social sounds of the Dalmatian seaside, as worthy of audition. Reduced listening is overturned in the name of a wide, embracing openness.

Such an understanding would fit squarely with Drott's second characterization of *Presque rien*'s reception as an anti-work par excellence, intended to efface the difference between music and noise, art and life. This view can be summarized in a pithy formulation: *Presque rien* as 4'33". Brandon LaBelle, in *Background Noise: Perspectives on Sound Art*, clearly describes *Presque rien* in the way that Drott characterizes it:

> [*Presque rien*] moves outside the confines both of the concert hall and the music studio to confront the random and ambient murmurings of everyday life in such a way as to undermine the Schaefferian sonic investigation, for it positions Ferrari more on the side of a Cagean nonintentionality whereby the composer "becomes a member of the audience," composing as a "contextualized" listener.... Ferrari's "anecdotal" work brings to the surface the split between associative or referential material and an ideal sonorous object by veering toward a concern for the sound

source and its referent as autobiography and individual psychology: the diaristic acoustical mapping of an individual over the course of a single day and how such sonic snapshots may, in turn, reveal conditions of real life.... Ferrari's work "tells stories" by harnessing the "bodily real," the quotidian environment in all its seemingly banal details, thereby invading the cinematic intensities of acousmatic dreaming with the hard edge of actual environments.[26]

As a commentator, LaBelle does not come from the tradition of *musique concrète* or electroacoustic music, but his account dovetails with those of insiders from that world like Truax. Behind the description is a schematic binary that reproduces the Schaefferian opposition of *entendre* and *écouter*, played out on various levels: On the formal register, LaBelle juxtaposes Schaeffer's "given, appropriately limited, sound material" with Ferrari's "random...ambient" indeterminacy; on the intentional register, Schaeffer's communication of the composer's "personal schema" is contrasted with Ferrari's non-intentionality; on the register of compositional material, the "ideal sonorous object" is contrasted with a "concern for the sound source and its referent"; on the register of representation, the "cinematic intensities" of *musique concrète* are juxtaposed against the "hard edge of actual environments."

The more one lingers on aspects of *Presque rien* that are not easily subsumed by this schema, the more the account founders. The claim that the composer becomes a member of the audience or commits to a moment of Cagean non-intentionality encourages the listener to imagine that they are hearing what Ferrari is hearing when listening to *Presque rien*. But, as I stated earlier, such accounts cannot acknowledge what Ferrari is doing—and I would argue that Ferrari is, in fact, doing quite a lot. Although the events may be determined by society, we should not treat lightly the statement that "the composer plays." Things may not be as unintended as LaBelle thinks. In fact, Ferrari's fingerprints are all over the environment—a point to which I return shortly.

Neither of the two positions characterized by Drott can account for what I would argue is the central aspect of *Presque rien*—not only that *it is recorded* (as is all *musique concrète*), but also that it brings this recorded character into audibility. Both positions perpetuate a phantasmagoric lack of consideration for the means of production, the recording device itself. On the one hand, if the work is understood as simply an overturning of Schaeffer's "dominant aesthetic," whereby the privileging of *entendre* over *écouter* is reversed, the recorded character of *Presque rien* is lost. While *écouter* can only expose the source of the recorded sound, *entendre* can only expose its morphology. Schaeffer's phenomenological organization of modes of listening accommodates no mode in which the recording itself can be considered or its recorded character auscultated. The means of production, the recording machine, always remains phantasmagorically outside the scene of listening. To hear the *technê* of *musique concrète*, one requires a mode of listening wholly outside of Schaeffer's schema. On the other hand, interpreting Ferrari's work in terms of "Cagean non-intentionality" also blocks reflection on the basic condition of *musique concrète*, namely, that it is recorded. It is not inessential that 4'33" is *performed* while *Presque rien* is recorded; but when the two pieces are identified, this difference is effaced and a phantasmagoric expulsion of the means of production reigns. *How* does *Presque rien*'s recorded character *necessarily* differentiate it from a work like 4'33"?

ÉCOUTER, ENTENDRE, IMAGE, AND VESTIGE

To develop this mode of listening, one can turn to recent work by Jean-Luc Nancy. Continuing in the wake of Derrida's critique of phenomenology, Nancy has put forth an analysis of *écouter* and *entendre* that is "post-phenomenological" in character, moving far from the ambitus of Schaeffer's thought.[27] Nancy's discussion of *écouter* and *entendre* begins on the very first page of his slim volume *Listening*, in the form of a question: "hasn't philosophy, forcibly and in advance, superimposed or substituted upon listening something that might be more on the order of *understanding*?" [... *la philosophie n'a-t-elle pas d'avance et forcément superposé ou bien substitué à l'écoute quelque chose qui serait plutôt de l'ordre de l'entente?*].[28]

The English translation loses the terseness of the French. One might read the word "understanding" as a translation derived from *comprendre* rather than *l'entente*. The latter comes from the Old French noun meaning "intent," which is itself derived from the verb *entendre*, "to direct one's attention," which echoes the Latin, *intendere*—"to stretch out, to lean toward, to strain." If one were to mistake *comprendre* for *entendre* in this passage, as Charlotte Mandell's translation is prone to encourage, we miss the phenomenological context that is being evoked in Nancy's opening statement. Moreover, we lose the tension between *écouter* and *entendre* that animates Nancy's argument afterward. After all, *tendre* means to stretch or tighten.

This tension promotes a philosophical question, or at least a question about the philosopher: "Isn't the philosopher someone who always hears...but who cannot listen...[who] neutralizes listening within himself, so that he can philosophize?" [*Le philosophe ne serait-il pas celui qui entend toujours...mais qui ne peut écouter...qui neutralise en lui l'écoute, et pour pouvoir philosopher?*].[29] The philosopher finds him- or herself in a situation of a tension, of balance or oscillation, between "a sense (that one listens to) and a truth (that one understands)" [... *entre un sens (qu'on écouter) et une vérité (qu'on entend)*...].[30] As the French makes explicit, the struggle between sense and truth is a struggle between *écouter* and *entendre*. The ear is the common thread upon which the tension travels, an ear that oscillates between sense organ and sense maker. Perhaps the English translation is doomed to sever the thread that ties the listening ear to the hearing ear, *écouter* to *entendre*, by unloosing sensation from understanding and encouraging the reader to falsely cast the difference in terms of faculty psychology—sensibility versus understanding—rather than an oscillation of difference within the same.

On first glance, one might argue that Nancy and Schaeffer resemble each other because of their shared characterization of *listening* as fundamentally non-indexical and non-signifying. For example, Nancy appears to reject indexical listening by claiming that music "makes sound and makes sense no longer as the sounds of some things, but in their own resonance."[31] Musical listening is irreducible to an ecological listening concerned with the size, speed, source, and location of sounds. Nancy also rejects a listening aimed at signification when he writes: "If *listening* is distinguished from *hearing*... that necessarily signifies that listening is listening to something other than sense in its signifying sense" [*Si l'écoute se distingue de l'entendre... cela signifie forcément que l'écoute est à l'écoute d'autre chose que du sens en son sens signifiant*].[32] Despite these apparent similarities, one must be attentive to the great divergence in Schaeffer's and Nancy's approaches in order to differentiate their shared desire to find a listening that is reducible to neither the index nor the sign. The language used—in particular, the selection of verbs employed—marks their divergence. Nancy selects *écouter* as the axis for his interrogation of listening because of his sensitivity to the etymology and implications of the verb *entendre*. Listening, as *entendre* or as intention, provides one with the structure necessary for a Husserlian epistemology: a subject, possessing the capacity for attention, who wills the directing; and an intentional object toward which this attention is directed and from which it attains its meaning. *Ego, cogito,* and *cogitatum.*

Nancy is critical of this epistemology. In his essay "The Forgetting of Philosophy," he argues that signification always involves two registers, the sensible and the ideal, creating a relationship of perfect conjunction:

> Signification... is the presentation of meaning. Signification consists in the establishment or assignment of the presence of a factual (or sensible) reality in the ideal (or intelligible) mode (which is what one calls "meaning"); or else, and reciprocally, it consists in the assignment of the presence of an intelligible determination in the sensible mode (a particular reality and/or the materiality of the sign itself). From Plato to Saussure, signification is, properly speaking, the conjunction of a sensible and an intelligible, conjoined in such a way that each presents the other.[33]

It doesn't matter if we start with the sensible and establish an intelligible meaning (as in Kant's description of the processing of the manifold by the categories and forms of intuition), or if we begin with the intelligible and trace its manifestation in the sensible realm (as in Hegel's self-exteriorization of the Absolute Spirit). In either case, "*Signification is...the very model of a structure or system that is closed upon itself....*Before the terrifying or maddening abyss that is opened between the possibility that thought is empty and the correlative possibility that reality is chaos... signification is the assurance that closes the gaping void by rendering its two sides homogeneous."[34] Nancy reads the creation of signification, the adequacy—or *adéquation*—of the sensible and the intelligible as a will-to-truth, a decision made in recoil from the difference that threatens such willful homogenization. The agent of this will-to-truth is none other than the subject, "capable of presenting the concept and the intuition together, that is, the one through the other."[35]

Although Schaeffer seems skeptical toward signification generally—for instance, he directs his investigation away from the mode *comprendre* toward *écouter* and *entendre*—in the end, the Schaefferian "sound object" meets Nancy's criteria for signification. It is the presentation of an ideal, intentional object within a sensible mode: audition, whether real or imagined. It conjoins the sensible and the intelligible such that each presents the other. It matters little if listening to some piece of *musique concrète* is conceptualized as the sensible manifestation *of* the sound object, or as a perceptual act of grasping the sound object within audition—the economy has been closed and a perfect adequacy is delivered.

The English translation of *Listening* misses an opportunity to bring out the important ways in which Nancy's modes of listening differ from Schaeffer's phenomenological characterization. However, the difference is clearly present in Nancy's opening question, if one is attentive to Nancy's French:

> Isn't the philosopher someone who always hears [*entend*] (and who hears [*entend*] everything), but who cannot listen [*écouter*], or who, more precisely, neutralizes listening [*l'écouter*] within himself, so that he can philosophize?

Not, however, without finding himself immediately given over to the slight, keen indecision that grates, rings out, or shouts between "listening" [*écoute*] and "understanding" [*entente*]: between two kinds of hearing [*entre deux auditions*], between two paces [*allures*] of the *same* (the same *sense* [*sens*], but what sense [*sens*] precisely? that's another question), between a tension [*tension*] and a balance [*adéquation*], or else, if you prefer, between a sense [*sens*] (that one listens to [*qu'on écoute*]) and a truth (that one understands [*qu'on entend*])....[36]

There is a tremendous amount of work being done here. Nancy aligns *entendre* with *adéquation* and thus invokes his critique of signification. Recall, this critique is aimed at the whole history of Western metaphysics, as he says, from "Plato to Saussure" with phenomenology as the culminating move in the sequence. Similarly, *écouter* aligns with *tension*, but not *intention*, rather a tension that is to be distinguished from phenomenological intentionality—a putting *in-tension*, by a stretching or directedness that is not directed by the phenomenological subject, or projected across the "terrifying or maddening" abyss that haunts signification. At the same time, the passage makes clear that Nancy is not really choosing sides; the philosopher is not simply being critiqued, for who is Nancy if not a philosopher? Rather the "keen indecision" that affects the philosopher marks the spot of an aporia, one that will be important for developing a non-phantasmagoric mode of listening. To hear the recorded character of the recording in *Presque rien*—to hear its *technê*—one must negotiate the aporetic relation between *écouter* and *entendre*, the oscillation between tension and *adéquation*, the spacing or "*différance*" between sense and sensation.

Nancy situates his critique of signification in terms of "presentation" (*Darstellung*), a term that can be traced back to its origin in the aesthetic theory of the German enlightenment.[37] Hegel's definition of art as "the sensible presentation of the Idea" has precisely the structure of signification elaborated above: "the assignment of the presence of an intelligible determination in the sensible mode" or "the conjunction of a sensible and an intelligible, conjoined in such a way that each presents the other."[38] Nancy takes Hegel's definition of art, "the sensible presentation of the Idea," as the basic premise of Western aesthetic theory. "No other definition escapes from this one sufficiently to oppose it in any fundamental way. It encloses, up until today, the being or essence of art."[39] Art, in its various formulations, centers itself on a certain form of presentation, where the invisibility of the idea is made sensuously visible or the inaudibility of the idea is made sensuously audible. It is "ideality made present."[40] The Hegelian definition of art would hold for Schaeffer's project: The sound object, which functions as an eidetic object, is made sensuously available in the *concrète* work.

For Hegel, the necessity to sensuously present the idea has come to an end; philosophy supplants art. With philosophy, spirit can come to know itself in a medium like language, one that has sloughed off the brute materiality of paint, stone, and inarticulate sound. The inadequacy that leads to the "end of art" is on the part of sensible presentation, not the idea. Yet, for Nancy, the question of presentation remains front and center, and he pursues this through two lines of thought. First, why is it that language is better able to present the idea? Or, to put it otherwise, why is it that philosophy, if it has overcome the brute materiality of things, still needs language in order to manifest itself? Why is it that "the Idea cannot be what it is...except through, in and as this sensible order that is at the same time its outside?" Why is it that "the Idea must go outside itself in order to be itself?"[41] Second, what remains of sensuous presentation after the idea has departed? Why does sensuous presentation maintain itself in this new milieu, and why does it still capture our attention? Was there something to sensuous presentation that was always in excess of the idea, something that could only come to the fore once the idea had departed? If the Hegelian aesthetic disallowed the possibility of thinking presentation as anything other than an ornament to the idea, how can we now understand the question of presentation today?

After the idea has departed, it would be very easy to think that the purpose of sensuous presentation is to represent precisely this disappearance. Art could take up the task of revealing to its audience the great lack that is now at the center of things—of trying to present the unpresentability of the departure of the idea in the form of a sublime incomprehensibility. But Nancy is eager to resist this move, for it re-inscribes sensuous presentation back into its previous role of presenting something other than itself, an idea, albeit it a negative idea. Even the attempt to present the nothing is still a type of presentation. The meaningfulness of the presentation remains somewhere other than right at the surface.

What remains after the idea has departed is "the vestige," and to understand it we must contrast it with "the image." Nancy writes,

> It would be necessary to distinguish, in art, between image and vestige...to distinguish that which operates or demands an *identification* of the mode or the cause, even if it is a negative one, from that which proposes—or exposes—merely the thing, *some thing*, and thus, in a sense, *anything whatsoever*, but not in any way whatsoever, not as the image of the Nothing...(Nancy 1996: 96).

Whereas the image is always defined within the economy of the sensible presentation of the idea, and thus always functions as a site for the homogeneous relay of a sensible particular and the idea, the vestige comes into focus when one considers the sensible remainder that persists after the idea has departed. The word *vestigium* literally means the sole of the foot, and like a footprint, it describes the mark left behind after some event has occurred. The vestige, according to Aquinas, "represent[s] only the causality of the cause, but not its form.... [It] shows that someone has passed by but not who it is."[42] To put it a bit more simply, perhaps we should think of the *vestigium* as it is often translated—as a trace. The challenge is to think about the trace not as something that leads us back to the source or idea that produced it or would subsume it. Rather, it is to try and think of the trace *as a trace*, as surface, as being right there at the surface and opaquely present in all of its sensibility. This cluster of Nancian terms—vestige, exposure, surface, and trace—spurs us to think of the artwork differently: not as the artwork intended to represent nothing, but as the artwork that has nothing to represent. Strategically, Nancy introduces the vestige and the image to force a distinction that, after the "end of art," curbs the possibility of thinking of the artwork within the closed economy of signification. The vestige is not a sensuous presentation of the departure of the idea, not a sublime picturing or representation of some vast nothing beyond comprehension. The vestige is simply the sensible remainder that cannot be ignored when the closed economy of signification has been evacuated. The possibilities are clearly drawn: "...either the 'nothing,' in an obstinate and I dare say obsessional manner, is still understood as negative of the

Idea [i.e., as an image]... or else it can be understood otherwise. This is what I would like to propose under the name of the *almost nothing* that is the vestige."[43]

I would like to play on this fortuitous conjunction of "almost nothings" or *presque riens*—on the one hand, what Nancy proposes "under the name of the *presque rien*," and on the other, what Ferrari composes under the name of *Presque rien*. To hear Ferrari's piece as a vestige opens the possibility of hearing it non-phantasmagorically: to hear the work as indifferent to the idea, as neither a concatenation of sound objects nor its opposite, as a series of indexical traces that reference some particular moment at the Dalmatian seaside; instead, to simply hear what remains on the surface after the idea has departed—the recorded character of the recording.

HEARING ALMOST NOTHING

It is tempting to regard *Presque rien* as a soundscape recording, indeed, almost irresistible. However, close attention to Ferrari's compositional decisions significantly challenges this temptation.[44] One way to hear Ferrari's intervention in *Presque rien* is to focus on his idiosyncratic mixing of various sounds and strata. In a soundscape recording, the listener relies on aural cues for the reconstruction of spatial relations, evaluating distances according to their volume, reverberation, and spectral attenuation. A well-mixed soundscape can give us the illusion of depth, and we will hear *through the recording* the intended spatiality: the distant water lapping, the closeness of footsteps on the floorboards, the passing of a motor, a singing voice reverberating off the hard surfaces of the street. In other words, we receive *an image*.[45]

Presque rien does not present this kind of soundscape. If one listens closely to the mix, the listener may notice that everything is pressed up to the surface and presented with nearly equal audibility and clarity. The cicadas are loud, just as loud as the sounds of lapping water, a sputtering engine, hammering, footsteps on wooden planks, or a speaking voice. There is no differentiation and thus there are massive incongruences: When have you ever experienced an auditory environment in which motors, insects, and lapping waves are all *equally audible*? Ferrari's mixing resists a realistic reconstruction of the environment, effacing the difference between foreground and background. Everything is selected, and hence nothing is selected. Flatness is foregrounded. Selection, differentiation, depth, hearing-in, spatiality, causality, signification—all must move beyond the auditory surface in order to generate adequacy between surface and projection. Perhaps attending to this flatness can help us hear *Presque rien* as a vestigial art that is right there at the surface.

The flatness of the mixing suggests other kinds of flatnesses tied to audition—the flatness of the magnetic tape on which *Presque rien* is recorded, as well as the flatness of the eardrum stretched across the auditory canal. As a means of reproducing sounds, the magnetic tape inherits a line of technological developments that can be traced to the gramophone, the telephone, Bell's phonautograph, and even Helmholtz's work on the decomposition and analysis of sound.[46] This line relies on a change in the way that sound is conceptualized. Jonathan Sterne has argued that modern forms of sound reproduction articulate "a shift from models... based on imitations of the mouth to models based on imitations of the ear."[47] In this new regime, sounds become conceptualized as effects that can be reproduced independently of their sources. The eardrum is central in this conceptualization: If one can transmit the proper vibrations to the eardrum, then one can (in essence) conjure the source from its effects alone. It should be mentioned that early attempts at sound-reproducing machines, like Bell's phonautograph, literalized this goal of reverse engineering the eardrum by actually using the severed ear from a cadaver as a primitive recording diaphragm.[48] Detaching sounds from their causes, tympanic technologies operated by "contemplating and constructing sound as a kind of *effect*."[49]

Yet, for all the reverse engineering, tympanic technologies only modeled the physiology of hearing, not the psychology of listening. As Friedrich Kittler wrote, "The phonograph does not hear as do ears that have been trained immediately to filter voices, words, and sounds out of noise; it registers acoustic events as such."[50] One *requires* a psychology of listening to filter and process the inscription of the acoustic event. Schaeffer's theories of *musique concrète* recoil from the acoustic inscription afforded by tympanic forms of sound reproduction. The acoustic inscription must be reintegrated into an ideal context where it can be conceptualized and auditioned as either the sign of the sonic source or cause, or as the adumbration of this or that sound object. The inscription is phantasmagorically recuperated under *écouter* or *entendre*. Rather than compose a series of adumbrations that "image" a set of sound objects, as Schaeffer did in his études of 1958, *Presque rien* presents a series of acoustic inscriptions or "traces" that resist this recuperation. Paradoxically, one could think of these auditory vestiges as adumbrations without objects. *Presque rien* is *musique concrète* that articulates the technical limits of *musique concrète*. It points back at itself, articulating its own technical condition by bringing the recorded character of the recording into audibility.[51] By pressing everything up to the surface, by turning away from the sonorous object back toward the facticity of its own recording, it rearticulates the condition of possibility of *musique concrète*, or what (following Kant) one might call its transcendental condition—namely, that *musique concrète* is recorded.

Of course, *musique concrète* is recorded. However, it is not an exaggeration to say that this condition was evaded in Schaeffer's phantasmagoric phenomenological project. If the law of *musique concrète* was to make recordings, this law was always augmented by an aesthetic demand that the content of the recording be something worth hearing, that it possessed the potential for the auditory inscription to be transmuted into a sound object that would not be confounded with the sound's cause or source. The law *thou shalt make recordings* was constrained by a corollary: *thou shalt make recordings...such that your recording meets this or that aesthetic criterion*. In Schaeffer's case, the new law might read: *thou shalt make recordings such that they explore the morphology of this or that sound object*. In other words, the *transcendental condition* of recording had to be effaced for the sake of the sound object's *transcendence*. But in *Presque rien*, Ferrari treats recording no longer as simply the means of production for exploring an aesthetic criterion, but brings the recorded character of the recording to the fore. Ferrari excises the aesthetic augmentation of the law and exposes the transcendental condition of *musique concrète*. *Thou shalt make recordings* is revealed as simply meaning *record whatever*.[52]

To *record whatever* implies an absolute aesthetic indifference toward the auditory events recorded—precisely the relation that the recording machine has to the environment. It is not about taste, or delectation, or the sensitivity of the ear to hear remarkable sounds. The microphone does not care, it just *records whatever*. This is precisely what Ferrari says he was doing during the period of *Presque rien*, though his commentators seem to have been deaf to it. Recall that Ferrari states quite clearly: "I was in the frame of mind that the function of the microphone was to register sounds or to record them on tape. So, whether in the studio, in society, in the street, or in private, it [the microphone] records in the same way."[53] The logic of the microphone is primary in *Presque rien*, not the logic of concatenating this or that feature of a sound object. The microphone records sounds, whatever they may be. To hear that aspect of the piece requires one to swerve away from both *entendre* and *écouter*—to care neither about the specific things recorded nor the morphology of the sound objects invoked, both of which are of absolute indifference to the microphone.

But as an aesthetic position, the law *record whatever* leads to a paradox because, to borrow the words of art historian Thierry de Duve, it "does not prescribe anything determined."[54] Since a recording is always a recording of some particular thing, "...[it] is forever fatally and excessively overdetermined. It is impossible to [record] anything whatever while avoiding that it be *this* thing by the same token."[55] Perhaps one could state the paradox like this: While trying to meet the transcendental condition of *recording whatever*, the recording is also stuck in the immanent condition of always being a recording of some particular thing.

This is the paradox of *Presque rien*. It is *almost nothing* because it is *almost anything whatsoever*. And to be *almost anything whatsoever* means that while the recording records this particular morning at this particular seaside in this particular Croatian seaside town, it is also indifferent to this fact. It could be replaced by something else, by some other recording, but this replacement would still encounter the same paradox. As a trace or vestige, *Presque rien* is held in tension by the law *record whatever*, suspended between the infinite ideal of *the whatever in general* and the overdetermination of being this particular thing.

This tension, this paradox of the law *record whatever*, deposits us back within the framework of Nancy's opening question. Perhaps Nancy is correct to chide the philosopher—and who is Schaeffer in the *Traité* if not a philosopher?—for being a poor listener, while placing his finger on a moment of keen indecision that troubles the superimposition of *entendre* over *écouter*. But the same cannot be said for the composer or listener who grips this keen indecision like a microphone with which to auscultate the world. On the other end of this device, one discovers a listening sensitively balanced between immanence and transcendence, or (to let Nancy's words resound in a new context), "between two kinds of hearing, between two paces [*allures*] of the *same*...between a tension and a balance...between a sense (that one listens to [*qu'on écoute*]) and a truth (that one understands [*qu'on entend*])...."[56]

Between *entendre* and *écouter* a theory of acousmatic listening must go.

5

Kafka and the Ontology of Acousmatic Sound

In the physical world, sounds are produced when one object activates another. A bow rubs against a string, air is forced through a vocal tract and shaped by a mouth, a raindrop collides with a windowpane. Objects emit sounds. Whether emitting the sounds of instrumental music, household noises, or speaking voices, objects have an inherent potentiality to produce sounds when struck, rubbed, percussed, or bowed. Sounds are emitted at the intersection of an action and a body. Thus, one might posit a simple law: Every sonic effect is the result of the interaction of a source and a cause. Without this interaction, there is no emission of sound.

In the phenomenological world, things are more complicated. Phenomenology begins as a description and analysis of structures of consciousness. It seeks to establish a presuppositionless philosophy grounded on the indubitable evidence of first-person experience alone. In a phenomenology of sound, the knowledge and assumptions we import from the natural sciences like acoustics are suspended, so that the immanent structure of sound *as experienced* can be described. Thus, for the phenomenologist, the acoustical relation of sonic source, cause, and effect cannot simply be presupposed. Real sounds offer evidence as indubitably as imagined sounds, although the thoroughgoing phenomenologist will note their different modes of presentation, degrees of clarity, and distinctness. Moreover, the case of imagined sounds presents a challenge to the presumed co-presence of source, cause, and effect. A listener might hear imaginary sounds that do not possess a source or cause—sounds that seem to be simply sui generis, autonomous, or without location in the physical world. Imaginary sounds dissolve the unity of sonic source, cause, and effect as parts of a single physical event or process, and draw an ontological line between the effect and its source or cause. The effect can be taken as an object in its own right, a "sound object" in Schaeffer's parlance, severed from its originating physical body or causal event. The nature of this severance is *ontological* because the sound object and the physical source or cause of the sound are not only understood as different in kind, but as different kinds of being.

After establishing this ontological distinction, phenomenological accounts of listening all follow a certain trajectory. Because they are made on the basis of first-person experience, such accounts invariably begin with the perceived sonic effect and work their way back to its source and cause—rather than assuming the sonic effect as a result of the interaction of source and cause. This is the basis for Schaeffer's statement that "One forgets that *it is the sound object, given in perception, which designates the signal to be studied*, and that, therefore, it should never be a question of reconstructing it on the basis of the signal."[1]

Schaeffer is not alone. Other phenomenologists concerned with analyses of sound, listening, and the senses follow the same trajectory. For instance, Hans Jonas wrote a brief comparative essay on the phenomenology of the senses in 1954, entitled "The Nobility of Sight."[2] Jonas begins with a contrast between seeing and hearing in order to defend the superiority of sight in terms of its direct access to external entities. While sight instantaneously presents a world of coexistent objects, detached from yet disposed before the beholder, hearing is doomed to access this world only through the medium of sounds. He writes,

> What the sound immediately discloses is not an object but a dynamical event at the locus of the object, and thereby mediately the state the object is in at the moment of that occurrence.... The immediate object of hearing is the sounds themselves, and then these indicate something else, viz. the actions producing those sounds; and only in the third place does the experience of hearing reveal the agent as an entity whose existence is independent of the noise it makes.... The object-reference of sounds is not provided by the sounds as such, and it transcends the performance of mere hearing. All indications of existents, of enduring things beyond the sound-events themselves, are extraneous to their own nature.[3]

Jonas chides hearing for being twice removed from the sound's objective source: first, by its mediate attention to the state of the external object—the cause of the sound—at the moment of hearing; and second, by the ear's immediate apprehension of sonic effects, the "sounds themselves," which are distinct from all worldly ties. The sound itself is insufficient for establishing reference back to a source; the referentiality of sounds depends on an act of the listener, who supplies a knowledge of the workings of the physical world in order to reason about potential physical sources and causes. The listener must transcend the sound itself to move back one step to the sonic cause or "the actions producing those sounds," and finally to the sonic source. Thus sounds, for Jonas, reveal a chain of mediations.[4]

However, Jonas finds a silver lining. Because of the mediate relationship of sound to the physical-causal objects from which they are emitted, "sound is eminently suited to constitute its own, immanent 'objectivity' of acoustic values as such—and thus, free from other-representative duty, to represent just itself."[5] Jonas distinguishes two forms of objectivity: the immanent objectivity of the sonic effect, which could be characterized as a "sound object," and the external objectivity of a sound considered as the emission or production of some physical source or cause. As an example, Jonas considers the sound of a barking dog: the immanent objectivity of the bark qua sound can be distinguished from the bark as signaling the presence of some particular dog. By describing the sound itself as immanently objective, Jonas repeats the gesture of drawing an ontological line, separating the sound's index from the sound itself.

Jonas's decision to describe the sound itself as an "immanent objectivity of *acoustic* values" is perhaps an unfortunate formulation. The scientific and physical connotations of the word "acoustic" abrogate the difference between the causal context and the sound itself, which Jonas had just established. Perhaps Jonas would have used the word "acousmatic" instead, had it been available. Acousmatics (*l'acousmatique*), which in Schaeffer's work designates an experience of sound that has undergone the test of the *epoché*, functions as the counterpart to sounds grasped under the natural standpoint, which would encompass the science of acoustics (*l'acoustique*). Regardless of the terminology, both Schaeffer and Jonas stake their theories of listening on the ontological separation of sounds from sources, and both rely on a shared set of phenomenological procedures aimed at disclosing the immanent objectivity of sound as an intentional object. Both are led to posit a regional ontology of sounds themselves.

One could turn to Erwin Straus's work for more evidence of the phenomenological commitment to the ontological separation of the source from the sound itself. Straus, a pioneer in the phenomenology of the senses, first established his mature views on sound in an essay entitled "The Forms of Spatiality."[6] He begins with the premise that sounds must be essentially distinguished from their sources because "it is of the essence of sound to separate itself from the sound source."[7] Aside from the methodological procedures that encourage ontological separation, Straus is motivated to draw a distinction between sound and source to account for the experience of music as an autonomous art. He argues that *music exploits the autonomy of sound itself and turns it to advantage*. Straus writes, "The sound that detaches itself from the sound source can take on a pure and autonomous existence; but this possibility is fulfilled solely in the tones of music, while noise retains the character of indicating and pointing to."[8] Music, unlike the rest of the arts, has a specific and unique claim to autonomy, because it exploits the essential separability of the sound itself from its source. "There is no visual art that is analogous to music, and there can be none because color does not separate itself from the object as tone does. In music alone tone reaches a purely autonomous existence. Music is the complete realization of the essential possibilities of the acoustical."[9] Jonas, perhaps unsurprisingly, also argues for music's autonomy along the same lines as Straus. "In hearing music," Jonas writes, "our synthesis of a manifold to a unity of perception refers not to an object other than the sensory contents but to their own order and interconnection."[10] Music's autonomy, argues the phenomenologist, is grounded in sound's essential separability from its conditions of production and external sources. The only thing that matters is the sounds themselves and their organization in time, pitch, duration, and timbre. Music, as an art, relies on sound's immanent properties alone.

Of course, this view about music's autonomy is not a product of the phenomenological tradition alone, but is closely tied to the history and rise of the autonomous musical work in the 19th century—a tradition that phenomenology tacitly accepts while mistaking it as simply given in the phenomena. The separability of the sound itself from its source also grounds two positions that are closely tied to the rise of the autonomous musical work: phantasmagoria and formalism. In musical phantasmagoria, as described in chapter 4, the separation of the sound itself from its source severs music from its conditions of production, making the latter dispensable or inessential and reifying the former into an ontology of the *tone*. In musical formalism, the ontological separation of the sound itself from its source encourages attention onto the formal configuration of tones alone. While musical formalism has never been univocal in its aims, the separation of tones (and their formal configuration) from the fully aspectual totality of sound (which would include its source, cause, and signification) is a methodological given in musical thinkers from Eduard Hanslick and Heinrich Schenker to Milton Babbitt and Allen Forte.

Nor has commitment to the ontological separation of sound from source perished since deconstruction and critical theory hastened the demise of phenomenology. Perhaps there is no clearer instance of the commitment than in Roger Scruton's *Aesthetics of Music*. Claiming that musical sounds are intentional and not material objects, Scruton argues,

> The person who listens to sounds, and hears them as music, is not seeking in them for information about their cause, or for clues as to what is happening. On the contrary, he is hearing the sounds *apart* from the material world. They are detached in his perception, and understood in terms of their experienced order.... What we understand, in understanding music, is not the material world, but the intentional object: the organization that can be heard in the experience.[11]

What is surprising about this statement is not its venerable commitment to the phantasmagoric separation of musical sound from its conditions of production. Rather, given Scruton's ultraconservative diagnosis of modern music, it is surprising that he affirmatively cites Pierre Schaeffer as the thinker of this position. Surprising indeed, but not misplaced, for Scruton is correct to situate himself in a tradition of phenomenological thinking about music even if he arrives at a very different set of aesthetic valuations about particular works. Scruton's conservative defense of tonality and the great mainstream of musical works may contrast with Schaeffer's aesthetics of *musique concrète*, but both operate with a similar ontology.

Beyond their agreement concerning ontology, important differences should be noted. Most significantly, Scruton conflates the acousmatic reduction and reduced listening. He writes, "in listening, Schaeffer argues, we spontaneously detach the sound from the circumstances of its production and attend to it as it is in itself: this, the 'acousmatic' experience of sound, is fortified by recording and broadcasting, which complete the severance of sound from its cause that has already begun in the concert hall."[12] When Scruton uses the word "acousmatic," he really means both the acousmatic reduction and reduced listening. He jumps immediately to the latter and situates it within a horizon of listening practices originating in the concert hall. One no longer needs the use of screens, scrims, covered orchestral pits, or loudspeakers in Scruton's world, for any time one hears music, one is already listening within the acousmonium. Reduced listening becomes simply coextensive with musical listening, and the whole complex is called acousmatic. Nowhere is this more boldly stated than when Scruton writes, "The acousmatic experience of sound is precisely what is exploited by the art of music."[13]

The conflation of the acousmatic reduction and reduced listening masks an important distinction between the two—a distinction that has not been appropriately appreciated in writing on acousmatic sound. As described in chapter 1, these two distinct yet interlocked reductions perform different operations that should be distinguished in any robust account of acousmatic sound. The first reduction, the acousmatic reduction, is Schaeffer's equivalent of the phenomenological *epoché*. By separating seeing from hearing and barring visual access to sonic sources and causes, the acousmatic reduction *does not advocate any particular mode of listening*. All modes of listening are available, depending on the attention of the listener. Indeed, Schaeffer's small typology of modes of listening (*écouter, entendre, ouïr, comprendre*) emerges upon the ground of the acousmatic reduction. On the ontological level, there is no claim regarding the separation and difference between the sonic source, cause, and effect. In fact, the purpose of the *epoché* was as a methodical corrective, to help the philosopher avoid all presuppositions based on thetic positings, which assume the world around us to be something factually given. The point of the *epoché* is to "parenthesize everything which that positing encompasses with respect to being."[14] In other words, the *epoché* transforms the philosophical subject into a kind of ontological agnostic, one who does not presuppose or force a pre-given ontology onto their experience of the world.

The second reduction, the eidetic reduction, does just the opposite. It discloses the sound object, the noematic correlate of reduced listening. The eidetic reduction, which discovers the essential or invariant morphological features of a sound object, can only occur in the mode *entendre*. Thus, the acousmatic reduction and eidetic reduction have distinct ontological consequences. While the acousmatic reduction is agnostic, reduced listening is committed, ontologically distinguishing the sound object from its source or cause. Undoubtedly, Schaeffer saw the acousmatic reduction as a preparatory step in establishing the sound object as the foundation of musical research, and reduced listening as the proper mode for auditioning *musique concrète*. But that does not mean that one cannot revisit the acousmatic reduction by reasserting its agnosticism and challenging the subsequent ontological separation of source, cause, and effect authorized by reduced listening.

In what follows, I want to focus on the relationship of sonic source, cause, and effect articulated in the acousmatic reduction before any ontological separation has been asserted. My intention is to expose an unexpected aporia that inhabits the relationship of source, cause, and effect in the acousmatic reduction. If this aporia has been neglected, perhaps it is due to the quick set of moves that aligns the acousmatic reduction with reduced listening and the sound object, and discourages consideration of the acousmatic reduction apart from Schaeffer's modes of listening and ontology of the sound object. My "counter-theorist" to Schaeffer (and the phenomenological tradition generally) will also be unexpected, namely, Franz Kafka. In particular, I focus on a late, unfinished tale titled "The Burrow," which presents the reader with a series of patient analyses of an acousmatically auditioned sound. Kafka's reflections will help to expose the inner logic of the acousmatic reduction in a discourse that is far removed from the phenomenological tradition.[15] Kafka, unlike Schaeffer, Jonas, Straus, or even Scruton, has no desire to ontologically separate sonic effects from their sources. Kafka's rich literary imagination allows the reader to inhabit imaginary worlds that disclose the precise logic of acousmatic sound apart from its actual sonorousness.

INTO THE BURROW

"The Burrow" is a story about acousmatic listening.[16] Written in the winter of 1923–1924, months before Kafka's death, it is a tale of an unidentified animal inhabiting an impenetrable burrow—some kind of mole or badger. The narrator, using the first person throughout, describes the lavish care spent on the construction and defense of his burrow. Kafka's mole relies on the sense of vision the least. The burrow, comprising various chambers and a central, ration-stocked castle keep, is heard, tasted, smelt, and felt—but not seen. Bathed in darkness, the narrator's ear dominates the other senses, listening for intruders or identifying tiny insects—"small fry"—that wriggle their way through the soil. After a brief sojourn aboveground, the mole returns to its burrow only to discover an unidentifiable high-pitched sound, "an almost inaudible whistling noise."[17] The sound continues "always on the same thin note, with regular pauses, now a sort of whistling [*Zischen*] but again like a kind of piping [*Pfeifen*]."[18] The continuity of the piping is a strong counterpoint to the narrator's rapid succession of changing attitudes. As the mole investigates, positing unverifiable hypotheses, it becomes impossible to determine if the sound comes from one or many places; who or what could be causing it; if it comes from near or far; or if it is not simply imagined. Walter Benjamin elegantly described the mood of the narrator: "as [the burrower] flits from one worry to the next, it nibbles at every anxiety with the fickleness of despair."[19]

Given the current interest in the sonic aspects of Kafka's work, it is surprising that "The Burrow" has not received the same treatment as the other late stories, like "Josephine the Mouse Singer" and "The Investigations of a Dog."[20] Perhaps this is because the other late stories explicitly concern music rather than sound, the former featuring a singing mouse, the latter dancing dogs that produce a strange music from their coordinated motions. In addition, the narrator watches and auditions the musical spectacle with rapt attention.

But "The Burrow" makes no mention of music, only sound—a sound anxiously auditioned by a worried creature. Many exegeses of "The Burrow" tend to ignore the story's sonorousness, focusing primarily on the mole's elaborate descriptions of the burrow's construction. The original German title "*Der Bau*" encourages this reading, with its emphasis on the construction of the burrow and its potential associations—most importantly, an association between the burrow and writing.

In the 1960s, Heinz Politzer identified "The Burrow" specifically as a story about Kafka's own literary production. "In an almost allegorical way," Politzer writes, "'The Burrow' is identical with Kafka's own work.... While the narrator...describes the hole it has dug in the soil, Kafka explains in a multitude of hardly veiled hints that he is about to discuss the very nature of his own writing."[21] For Politzer, the nature of that writing is a confrontation of author and other that can only be articulated in the form of parables and paradoxes. By the 1990s, after Derridean theories of writing had impacted literary studies, the identification of the burrow with writing remained, albeit in a new form, in which writing migrates from the biographical to the impersonal. For example, Rosemary Arrojo draws a connection between the labyrinthine burrow and a theory of textuality, reading "The Burrow" as "a poignant illustration of Nietzsche's notions of the text and the world as labyrinth."[22] Playing on a double entendre, the *passages* of the burrow must be "constantly reviewed because of the 'manifold possibilities' of their uncontrollable 'ramifications.'..."[23] The proliferating meanings of Kafka's passages disseminate wildly, operating at a register uncontainable by the presence of the author/builder's intentionality or by any transcendental signified.[24] Similarly, Stanley Corngold makes this casual observation: "That a story of Kafka's called 'Der Bau'—which means, literally, 'The Building' or 'The Construction'—alludes to Kafka's literary enterprise will come as no surprise."[25] The nonchalance of Corngold's assertion evinces its ubiquity.[26]

This text-centered reading of "The Burrow" depends on a correspondence between the production of the burrow and Kafka's literary production. Yet, by focusing on the construction of the burrow rather than the sounds heard inside, it elicits a nagging question: What is the meaning of the *sound* in the second half of the tale? Kafka scholars have proffered diverse hypotheses: It is the sound of the existential self, whose threatening judgment is ignored by the narrator; or the sound of something entirely alien and ominous, whose terrible force comes to destroy the narrator; the infernal, obsessive compulsion that drove Kafka as a writer; a representation of the narrator's fear and anxiety in the face of solitude; hallucinations and phantasms of paranoia and mental illness; an abstraction from the terrifying, blind acoustic experience of soldiers involved in trench warfare during WWI; or simply the sound of Kafka's tubercular cough, a symptom of the disease that ultimately took his life.[27] Taken individually, the readings are each defensible, but taken together, they present an astonishing multiplicity of irreconcilable interpretations.

Most critics treat the sound in "The Burrow" symbolically, rather than sonorously. Sound acts as a metaphor for some other form of experience—moral, political, philosophical, or psychological. Deleuze and Guattari offer an alternative to these symbolic readings in *Kafka: Toward a Minor Literature*, a book not only ostensibly concerned with the sonorousness of sound in Kafka's work, but also explicitly invested in "The Burrow" as a central text. In fact, no text may be quite as relevant as "The Burrow," for in its passages, Deleuze and Guattari find the literary equivalent of their master concept, the rhizome. Just read their opening sentence: "How can one enter into Kafka's work? This work is a rhizome, a burrow."[28]

Many of the central concepts from Deleuze and Guattari's other works—becoming-animal, territorialization, lines of flight—appear in the course of their reading of Kafka. Sound is described as "pure and intense,"[29] the kind of force requisite to open up a deterritorializing line of flight. But the same does not hold for music, which is explicitly differentiated from sound by Deleuze and Guattari. Music is unable to reach sound's pure intensity because of its relation to conventional systems of signification: "It isn't a composed and semiotically shaped music that interests Kafka, but rather a pure sonorous material."[30] By escaping from music's power of signification, sound, being the stuff from which music is made, also contains the potential to evade music, to slither from music's grasp and, in so doing, undermine music by abolishing its signifying order.[31] Deleuze and Guattari write, "What interests Kafka is a pure and intense sonorous material that is always connected to *its own abolition*—a deterritorialized musical sound, a cry that escapes signification, composition, song, words—a sonority that ruptures in order to break away from a chain that is still all too signifying. In sound, intensity alone matters, and such sound is generally monotone and always non-signifying."[32] Perhaps the high-pitched whistling or piping heard in the burrow would be just such an example of pure and intense, non-signifying, sonorous material—just like the raspy speech of Gregor Samsa, the voice on the telephone in *The Castle*, or even the silent singing of Kafka's sirens.[33]

Sound's deterritorializing function shares many similarities with another one of Deleuze and Guattari's master concepts, one uniquely suited to Kafka's stories: the concept of becoming-animal. Both are ruptures, insubordinate to any other territorialization or plane of consistency they may encounter. The language used to describe sound and becoming-animal is strikingly similar, even to the cursory reader. Both are pure intensities, non-signifying bits of unformed matter or material, involved in breaking away or discovering paths of escape. Deleuze and Guattari write, "To become animal is...to stake out the path of escape...to find a world of pure intensities where all forms come undone, as do all the significations, signifiers, and signifieds, to the benefit of an unformed matter of deterritorialized flux, of nonsignifying signs."[34] Both sound and becoming-animal encourage a line of flight that evades signifying orders. Often they are found together, in that the process of becoming-animal changes the production of sounds from signifying to non-signifying. In language that recalls the sonic—language of vibrations, intensities, thresholds, and movements—one can follow the crossing of sonic production and becoming-animal in this passage:

> Kafka's animals never refer to a mythology or to archetypes but correspond solely to...zones of liberated intensities where contents free themselves...from the signifier that formalized them. There is no longer anything but movements, vibrations, thresholds in a deserted matter: animals, mice, dogs, apes, cockroaches are distinguished only by this or that threshold, this or that vibration, by the particular underground tunnel in the rhizome or the burrow. Because these tunnels are underground intensities. In the becoming-mouse, it is a whistling that pulls the music and the meaning from the words. In the becoming-ape, it is a coughing that "sound[s] dangerous but mean[s] nothing"...In the becoming-insect, it is a mournful whining that carries along the voice and blurs the resonance of words.[35]

The animals' sounds escape the signifying chains of territorialized language and liberate themselves from music and meaning.[36] Similarly, one would assume that the same disruptive conjunction of sound and becoming-animal would hold for "The Burrow" too.

Examining the account in more detail, it is surprising to see that Deleuze and Guattari *do not* explicitly identify the "monotone" and "non-signifying" sound heard in the burrow with any kind of pure intensity or line of flight.[37] The sound heard in the burrow is never explicitly addressed, nor is any conjunction posited between becoming-mole and the resonating high-pitched sound. Unlike Kafka's other creatures, whose sonorous productions conjoin with their animality, the mole is a listener rather than a performer. Josephine's whistling, Gregor Samsa's whine, and the ape's coughing are all actively produced—auditioned by others but made by those who have become-animal. Sound may deterritorialize, but where is listening? For Deleuze and Guattari, it is only in the burrow's tunnels (or passages) that "underground intensities" are to be found.[38] Again, it is the production of the burrow, not the sounds that inhabit it, that ultimately fascinates Deleuze and Guattari. Their reading of the burrow does not radically differ from text-centered accounts, which treat the burrow as the figure of endlessly creative, textual production.

FRANTIC HYPOTHESES

Rather than posit another hermeneutic (or anti-hermeneutic) interpretation of the sound in the burrow, or read the text to support some pet theory concerning sound in general, I will simply reassert that "The Burrow" is a text about acousmatic sound and acousmatic listening. Even if Kafka did not know of the term "acousmatic," the sound in the burrow clearly meets the definition of acousmatic sound that Schaeffer cites from Larousse: "a sound that one hears without seeing the causes behind it." Moreover, the specifically acousmatic focus of "The Burrow" differentiates it from Kafka's other late tales, "Josephine the Mouse Singer" and "The Investigations of a Dog," which thematically concern listening and sound.

In "Josephine," the eye and the ear are tightly bound in the reception of musical performance. The narrator declares that there is nothing special about Josephine's singing. On its own, her voice is indistinguishable from a mere "piping" (a *Pfeifen*—the same word used in "The Burrow" to describe the high-pitched monotone), but when reconnected to the body from which it is emitted, the sound is transformed: "to comprehend her art it is necessary not only to hear but to see her...when you sit before her, you know: this piping of hers is no piping."[39] Kafka posits a necessary connection between the eye and the ear in shaping auditory experience. Heard acousmatically, the power of Josephine's singing would be simply annihilated. One could not, for instance, understand Josephine's art by listening to a recording, for the inseparability of voice, gesture, and spectacle would be broken.

In contrast, "The Investigations of a Dog" challenges the close intertwining of audition and vision by slightly displacing the two domains. Here, the canine protagonist is stunned by the music he encounters coming from a pack of seven dogs. "I could not recognize how they produced it.... They did not speak, they did not sing, they remained generally silent, almost determinedly silent; but from the empty air they conjured music."[40] Although the mechanics of causal production are uncertain, the visual dimension is by no means reduced. This presents the unusual situation where the auditory effect and the source are both known, both equally manifest, while the cause remains mysterious. The music *seems* to be emitted from the very movements of the pack: "Everything was music, the lifting and setting down of their feet, certain turns of the head, their running and their standing still, the positions they took up in relation to one another, the symmetrical patterns which they produced."[41] The visual gestures, while distinct, correspond to a music that comes from nowhere in the visual scene but rather, like an *acousmate*, "from the air." The two modalities, the eye and the ear, have become detached in Kafka's descriptions.

This kind of sonic experience encroaches on the domain of acousmatic sound. Larousse's oft-cited definition *nearly* fits the situation. The narrator sees the source, the pack, but cannot identify the music's cause. The visual presence of the source and the palpability of the auditory effect operate in tandem, but across a gulf not bridged by any mechanical cause. Although the visual source is not obscured behind some figural Pythagorean veil, a strange puzzle remains: the simultaneous co-presence of spectacle and sound, both in absolute correspondence, but seemingly without worldly connection.

Scruton's claim that, in the concert hall, "we spontaneously detach the sound from the circumstances of its production and attend to it as it is in itself" could be applied to Kafka's canine observer. For the dog's experience of music prolongs a traditional view about music, originating in 19th-century musical aesthetics and perpetuated in Scruton's conservative work. One finds a similar correspondence-yet-detachment between the worlds of vision and sound in Wagner's writings: "besides the world that presents itself to sight, in waking as in dreams, we are conscious of the existence of a second world, perceptible only through the ear, manifesting itself through sound; literally a *sound world* beside the *light world*, a world of which we may say that it bears the same relation to the visible world as dreaming to waking."[42] There can be no causal relation between these two worlds. Wagner insists, "The dream organ cannot be roused into action by *outer* impressions."[43] The sound world and light world remain distinct, situated upon opposite shores of a "mystic gulf" as fact is divided from essence.

If the "Investigations" presented a *weak* acousmatic thesis, where the eye and the ear are co-present yet displaced, the "Burrow" meets the most stringent requirements. In its dark interior, vision does no good—the cause and source are lost to subterranean obscurity. Kafka's choice of a mole for the protagonist was not simply fortuitous; it was likely intended to recall various folk tales and received wisdom concerning moles and their notable sensory powers. Pliny the Elder, in *Natural History*, recounts moles' acute sense of hearing and even their ability to comprehend speech: "…moles hear more distinctly than other [animals], although buried in the earth, so dense and sluggish an element as it is; and what is even more, although every sound has a tendency upwards, they can hear the words that are spoken; and, it is said, they can even understand it if you talk about them, and will take to flight immediately."[44] Alexander Pope, in the *Essay on Man*, locates the mole at the absolute bottom of the scale of visual acuity: "What modes of sight betwixt each wide extreme, The mole's dim curtain and the lynx's beam."[45] Although the association of blindness with a dim curtain is not likely a direct allusion to the mythic Pythagorean veil, the couplet depends upon an association of curtains, screens, and veils with blindness.

One might be tempted to correlate this total separation of eye from ear with a separation of the auditory effect from its source and cause. Orthodox Schaefferian theory would make precisely such a move, arguing that sounds, when acousmatically reduced, take on an aspect of intensified profundity; and they do so in proportion to their ontological severance from worldly sources and causes. But Kafka's mole finds solace neither in a phenomenological shift away from the natural attitude toward sound taken as a pure "sound object," nor in any kind of worldly bracketing or ontological separation. The mole's attention is constantly preoccupied with the mysterious source of the sound and its possible meanings, moving through a crescendo of frantic hypotheses—seven, to be exact.

1. When the "almost inaudible whistling noise" is first heard, it is "immediately recognized" as having been caused by the burrowing of some "small fry," which, in wriggling though the burrow, exposed a "current of air."[46] The sound is rationalized away as a non-intentional trace, a leftover residue that signifies nothing at all.
2. Attempting to confirm that the sound is due to a current of air, the mole notes that the uniformity of the sound continues at every location in the burrow. If

the sound were to come from a single location, proximity to the source should correspond to perceived volume. Confounded by the uniformity of the sound's volume, the narrator posits that there must be two sources quite widely spaced; as one diminishes in its perceived volume, the other increases, giving the effect of uniformity.[47] Two sources have now replaced one.

3. As soon as the two-source hypothesis is entertained, the mole replaces it with another: "it is a noise produced by the burrowing of some species of small fry."[48] The source is now attributed to the digging of the creatures themselves, and no longer to their resultant air channels. The source has moved from the non-intentional traces of action to causal ascriptions of action itself. There is a degree of intentionality in the sound, in the sense that the digging is purposive; but the sounds are hardly full of significance in the way that, for example, a speaking voice would be. It is the sound of action, not communication. But this hypothesis is also quickly negated, for the appearance of small fry is nothing new to the burrow, so why would they have suddenly become audible? "One could assume, for instance, that the noise I hear is simply that of the small fry themselves at their work. But all my experience contradicts this; I cannot suddenly begin to hear now a thing that I have never heard before though it was always there."[49]

4. The rejection of the third hypothesis leads to a new position: that the sound comes from "some animal unknown" to the narrator, a "whole huge swarm that has suddenly fallen upon my domain, a huge swarm of little creatures."[50] But if this were the case, why has the narrator never encountered them? The only possibility is that this swarm is composed of creatures "far tinier than any I am acquainted with, and that it is only the noise they make that is greater."[51]

5. After finding a moment of respite from the sound by huddling at the moss-covered entrance to the burrow, the narrator returns only to move rapidly through three more hypotheses. First, the original view is reinstated, that the sound in the burrow is the sound of air caused by channels dug by the small fry. This return leads the mole to reflect on the fruitlessness of endless hypothesizing: "One could play with hypotheses… one is not at liberty to make *a priori* assumptions, but must wait until one finds the cause, or it reveals itself."[52]

6. Next, immediately overturning the appeal to reason, the narrator wonders if the cause might not be a "water burst" that "seems a piping or whistling," but "is in reality a gurgling." This hypothesis, like the first, is also a non-intentional trace—a sound, like those described by Deleuze and Guattari, that seems ominous but signifies nothing.

7. Recalling the work done to drain the sandy soil in which the burrow was built, the narrator jumps from non-intentional trace to the view that the sound must be coming from a single source, a beast, "dangerous beyond all one's powers of conception."[53] The source grows more and more distinct in the mole's imagination, to the point where the mole can even hear a *Doppelgänger* in the sound. Recalling a memory of a similar sound from the early days of constructing the burrow, the mole concludes that the noise comes "from some kind of burrowing similar to my own; it was somewhat fainter, of course, but how much of that might be put down to the distance one could not tell."[54]

The difficulties of locating the source are increased because the sound seems to come from no particular location or, equally, from all places at once. The mole never seems to "be getting any nearer to the place where the noise is, it goes on always on the same thin note, with regular pauses, now a sort of whistling but again like a kind of piping." "The noise can be heard everywhere and always at the same strength, and moreover uniformly, both by day and night."[55] The topography of the burrow complicates and obscures judgments about the distance and location of sounds. Because of its labyrinthine construction, and the resonances generated by such involutions, the burrow blends sounds that originate from the outside with those that originate inside the passages. The German title of the tale, "*Der Bau*," accentuates this fact. Because of its strange denotation, the word is nearly impossible to translate into English. As Mladen Dolar suggests, "It can mean the process of building, construction; the result of building, the edifice; the structure, the making (of a plant, of a novel...); a jail, a burrow, a hole in the ground, a mine. The oscillation is not only between the process and the result...but also between erecting an edifice and digging a hole."[56] Oscillating between above and below, the status of *der Bau* similarly oscillates between inside and outside. Rather than read the passages of the burrow as metaphors for Kafka's textual production, it might be more fruitful to recognize something *organic* about the burrow's topology. Like an ear, the burrow leads from a single soft and protected entrance into a series of tunnels and passageways of differing (but specialized) size and function. And, as with the ear, sound does not simply travel through the burrow, but penetrates it from various points. Just as vibrations travel through the bones of the skull to be received inside the ear, the burrow is similarly permeable, combining signals from both inside and out into a single resonance.

UNDERDETERMINATION

In his book *Individuals*, philosopher P. F. Strawson imagines what it would be like to inhabit a "purely auditory world," a world that is known through no other sensory modality than the ear.[57] Strawson notes, surprisingly, that this purely auditory world would be a world without space. He argues,

> Where sense experience is not only auditory in character, but also at least tactual and kinaesthetic as well...we can then sometimes assign spatial predicates on the strength of hearing alone. But from this fact it does not follow that where experience is supposed to be exclusively auditory in character, there would any place for spatial concepts....Sounds of course have temporal relations to each other, and may vary in character in certain ways: in loudness, pitch and timbre. But they have no intrinsic spatial characteristics.[58]

Spatial predicates rely on visual, tactual, or kinesthetic contributions that supplement audition, on a multimodal perception of the world. Strawson's space-less auditory world is indeed uni-modal: hearing without seeing (or touching or moving). While space-less, this auditory world is hardly without content; it forms an immanent sphere of sounds related to each other (and these relations would be absolutely intrinsic) but not related to other entities or predicates imported from non-auditory sensory modalities.

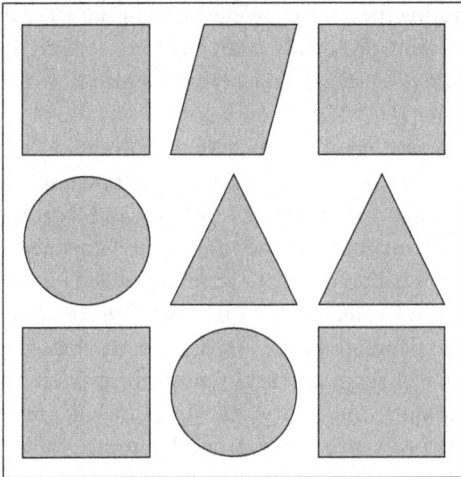

Figure 5.1 After P. F. Strawson, *Individuals*, 22.

This purely auditory world has some unexpected features; in particular, it raises a strange issue concerning identity. Strawson argues that, in a purely auditory world, one can only establish *qualitative* identity but not *numerical* identity. That is, without reference to a spatial framework, one can only establish that two sounds sound alike (i.e., that they are qualitatively identical), but we cannot definitively establish that they are the same numerically distinct individual. For example, we may hear two pipings in succession that, as far as we can discriminate, are qualitatively identical, but we could never be sure if they are two numerically distinct sounds or one numerically identical particular sounding twice.

This may seem an odd problem to fret about, but it is consequential for understanding the logic at work in acousmatic listening. If we contrast the purely auditory world with the spatial world, we can begin to see just how odd this problem really is. Strawson sets up the question of numerical and qualitative identity by giving a diagram similar to figure 5.1.[59]

The diagram helps to disambiguate two senses of the concept "the same." According to Strawson, when we say "The figure in the top-left-hand corner of this diagram is the same as the figure that has a parallelogram to its right and a circle beneath it," we are speaking about the same *numerically identical* object. Whereas, if we say "The figure in the top-left-hand corner of the diagram is the same as the figure in the bottom-right-hand corner," we are speaking of two objects that are the same only in terms of their *qualitative identity*.[60] Qualitative identity means that we are talking about the same *type* or *universal* (i.e., each token or particular possesses the quality); numerical identity means that we are talking about one and the same re-identifiable token or particular. In the spatial world, we can easily distinguish between these two kinds of identity because of the role that location plays in re-identifying numerically distinct particulars.

But in a purely auditory world, there are serious worries about the status of numerical identity. How can we tell the difference between two qualitatively identical particulars and one numerically identical particular played twice? Can any particular in a purely auditory world be re-identifiable as numerically identical? Strawson writes, "A note could be re-identified, or a sequence of notes or a sonata. But what sense could be given to the idea of identifying a particular sound as the same again after an interval during which it is not heard?"[61] Re-identification could only happen at the level of qualitative identity—at the level of the *type*, like a pitch class, a pattern, a musical work, or a sound object—but not at the level of individual tokens or material things. A purely auditory world, surprisingly enough, turns out to be a world where types or universals, rather than particulars, are primary.

Strawson's distinction between numerical and qualitative identity is useful for analyzing an aspect of Kafka's tale in a theoretically rigorous manner. As the mole flits from fickle worry to fickle worry, running through its seven hypotheses, the mysterious sound is ascribed to different *numerically distinct* sources and causes: a whistling crack, the gurgling of water, some small fry, a swarm of small fry, or a large creature much like the mole. At the same time, the sound remains *qualitatively identical* throughout. It is the same sound again and again or, at the very least, the mole recognizes the sonic effect as qualitatively the same: "always on the same thin note, with regular pauses, now a sort of whistling but again like a kind of piping." Only the qualitative identity of the sound can be guaranteed; its numerical identity is always insecure. Kafka brings this condition to the fore by having his narrator hypothesize so many different, yet logically possible, attributions. The anxious uncertainty that underscores the narrator's fickle hypotheses registers an insight into the logic of acousmatic sound: namely, that the auditory effect, when unaccompanied by contributions from other senses, underdetermines ascriptions of source and cause. To put it another way, given a certain qualitatively identical sound object, its numerical identity cannot be secured without the contributions of the other senses to report on the state of its cause and source.

The frantic hypotheses of the mole rely on acousmatic underdetermination to permit their constantly changing attributions. The auditory effect cannot be simply reunited with its source and cause. Nor can it be absolutely separated. By keeping alive the search for the source, Kafka's narrator does not posit an ontological separation of the sound object from its source. Rather than develop the difference between the auditory effect and its source in ontological terms, Kafka deploys acousmatic sound in order to emphasize the *spacing* of source, cause, and effect, without simply permitting their separation.[62] This spacing of source, cause, and effect provokes a feeling of anxiety. The constant reappearance of the mysterious sound, with its underdetermination of source and cause, is the motor that drives the imagination of the narrator through a series of anxious hypotheses.[63]

CERTAINTY AND UNCERTAINTY

This analysis of acousmatic sound, in which an acousmatically auditioned sonic effect necessarily underdetermines ascriptions of source and cause, contrasts with Schaeffer's own analysis. In the *Traité*, the acousmatic reduction functions as a relay on the path to reduced listening. The latter treats the auditory effect as a sound object detachable from causal and worldly affiliations. When Schaeffer speaks as the theorist of the sound object, he speaks from a position of Husserlian detachment and eidetic perfection. The anxious narrator of Kafka's tale would appear to occupy a position concerning acousmatic sound that is wholly unrelated to Schaeffer's style of thinking. How could the anxious listening of Kafka's mole be situated vis-à-vis Schaeffer's theoretically detached reduced listener? In a few brief instances in the *Traité*, Schaeffer addresses the anxiety that disturbs acousmatic sound, although he is quick to dismiss it. At one point, he writes:

> For the traditional musician and for the acoustician, an important aspect of the recognition of sounds is the identification of the sonorous sources. When the latter are effectuated without the support of vision, *musical conditioning is unsettled*. Often surprised, often uncertain, we discover that much of what we thought we were hearing was in reality only seen, and explained, by the context.[64]

Schaeffer acknowledges that there is something unsettled, or unsettling, about acousmatic sound. However, the feeling is attributed to the challenge of overcoming "musical conditioning," of overturning longstanding habits that supplement pure audition with visual and tactual information. Perhaps Schaeffer misrecognizes what precisely is unsettling about acousmatic sound. The uneasiness is not the discomfort of acquiring a new habit, of teaching the ear—that old dog!—a new trick.

It is not the overcoming of habit that is unsettling, but rather a structural feature of acousmatic sound that is disturbing, namely, that the sound object is never quite autonomous; that this nearly-*but-not-quite*-autonomous auditory effect *necessarily* underdetermines attributions of source and cause; that the autonomous effect, when heard acousmatically, is pursued by the shadow of its source and cause, a shadow that it cannot escape because without it, the acousmaticity of a sound simply dissipates. The tension inherent in acousmatic sound depends on the possibility that it may become *dis-acousmatized* when identified with a body, or pigeonholed away as a sound object.

By emphasizing underdetermination and uncertainty in acousmatic listening, my theory of acousmatic sound differs quite dramatically from Schaefferian theory. In Schaeffer's work, the acousmatic reduction separates the eye from the ear; this anticipates the separation of the sound from its source. Schaeffer's division of the sensorium must be contrasted with the *epistemological* approach that Kafka authorizes. In the latter, one can talk about a phenomenon like *acousmaticity*, the degree to which a sound's source or cause can be ascertained. Were acousmatic sounds truly autonomous, as they are in Schaeffer's theory of the sound object, they would possess none of their gripping tension and mystery. For example, in Kafka's text, the narrator might have discovered the origin of the sound, which would have reunited source, cause, and effect while dissipating the sound's acousmaticity. However, the same dissipation would have occurred if the mole was simply resigned to treating the sound in the burrow as a sound object, secure in its ontological severance from its source and cause. How can the Schaefferian theory explain the mole's profound anxiety as anything other than pathological—a fixation on *écouter*? Without an account of acousmaticity, a theory of acousmatic listening is severely limited in its explanatory force.

Shifts in the degree of a sound's acousmaticity might be crucial for a listener. For instance, territorial listening depends upon the interception of a sonic effect that precedes the proximity of the source. Roland Barthes, in his essay "Listening," writes,

"It is doubtless by this notion of territory... that we can best grasp the *function* of listening, insofar as territory can be essentially defined as the space of security... listening is that preliminary attention which permits intercepting whatever might disturb the territorial system."[65] The thing intercepted is a signal, and yet not the source or cause. The security at work in territorial listening depends on the rapid reduction of a sonic effect to its potentially predatory source, but acousmatic underdetermination forecloses the easy attainment of such security. There are always degrees of acousmaticity. In successful territorial listening, the acousmatic source is identified ahead of its arrival. The first appearance of the sonic effect affords the territorial listener an advantage, in that the signal permeates the territory in advance of the predator. Yet, that advantage does not guarantee success; by the time the predator is seen, it is already too late; the territory has been breached. Listening, lacking a perspective but possessing a perimeter, is the first line of defense for the anxious animal.

However, the spacing of source, cause, and effect also encourages a different kind of anxious listening, which is perhaps best displayed in the cinematic figure of the *acousmêtre*, as theorized by Michel Chion.[66] Like Schaeffer's acousmatic reduction, which exploits the split between vision and audition, film is similarly divided between the projected image and the soundtrack. The *acousmêtre*—the shadowy figure whose voice can be heard, but whose body cannot be located—exploits the acousmatic underdetermination of the source by the sonic effect. It makes a sound that comes from no particular location. The acousmatic voice floats or drapes itself around the onscreen characters. According to Chion, the *acousmêtre* has four main powers: "the ability to be everywhere, to see all, to know all, and to have complete power. In other words: ubiquity, panopticism, omniscience, and omnipotence."[67]

It is precisely the spacing of the auditory effect from its source or cause that grants the *acousmêtre* and acousmatic sound their strange power. Thus, one might arrive at the laconic formulation that acousmatic sound is constituted by a structural gap. To be more precise: When source, cause, and effect are simultaneously present, acousmatic sound *is not*. Or, similarly, when the effect becomes an "essence," detached from the cause and effect, acousmatic sound *is not*. Thus, the very acousmaticity of sound—its quality of being acousmatic—depends on the spacing of source, cause, and effect. Acousmatic sound exists structurally between these two possibilities. This neither-heteronomous-nor-autonomous sound can neither be reduced to its source nor reified as an object in its own right. It only *is* when source, cause, and effect are spaced. But even to use the word *is* is itself an infelicity, for the *being* of acousmatic sound *is* to be a gap. Acousmatic sound is neither entity nor sound object nor effect nor source nor cause. It flickers into being only with spacing, with the simultaneous difference and relation of auditory effect, cause, and source. With tongue planted firmly in cheek, one could refer to acousmatic sound's ontology as a non-ontology, a *nontology*.

The simultaneously heteronomous and autonomous auditory effect raises a series of problems for the conceptualization of acousmatic sound, for this strange auditory "effect" is neither directly related to its source or cause, nor is it an object in its own right. According to Mladen Dolar, the acousmatic voice "always displays something of an effect emancipated from its cause. There is a gap between the source and its auditory result, which can never be quite bridged."[68] In fact, Dolar goes so far as to argue that the acousmatic voice can never truly be effectively latched back onto a body, that "*there is no such thing as disacousmatization.*"[69] No matter how drastic one's curiosity or how fervently one tears down the veil to see the speaker behind the curtain, there is no de-veiling of the voice. For Dolar, the source of the voice *always remains veiled by the body of the speaker*, that "there is always something totally incongruous" between a person's "aspect" and "his or her voice." According to this view, the spacing of cause and effect can never be closed, for "Every emission of the voice is by its very essence *ventriloquism*."[70] The voice is neither body nor language, but a phantom effect in excess of both fields. Like music itself, a site of interference between the acoustic and the eidetic, the voice is an illusory *akousma*, reducible to neither the materiality of the body nor the ideality of language. Situated between acoustic and eidetic registers, for Dolar, the acousmaticity of the voice is a terminal condition.

There is much more to say about Dolar's claim in chapter 7. For the time being, one must take his claim with certain qualifications. Dolar exaggerates by making this split between source, cause, and effect permanent; I, on the other hand, do not think that acousmaticity is a permanent condition, but rather a special situation. When a sound is heard in its full acousmaticity, it brings into audibility the incongruous spacing of source, cause, and effect, and (because they are implicated in the determination of source, cause, and effect) the eye and ear. This spacing can be overcome, but at that moment, the acousmaticity of the sound is gone. Some may find that state of dis-acousmatization a relief, given the anxiety, uncertainty, and underdetermination of acousmatic listening.

One can understand the impulse for reduction, whether to the eidetic intuition of the sound object or to the materiality of the source, as recoiling from an acousmatic sound's constitutive unsettledness. Schaeffer follows the former route by positioning the experience of acousmatic sound as a prologue to reduced listening. *Musique concrète* may capitalize on the spacing of source and cause afforded by radio and sound reproduction, but it diminishes the unsettling aspects of such spacing by demanding that the listener hear the sonic effects as self-generated, as autonomous sound objects bracketed from worldly connection. The privilege that Schaeffer gave to reduced listening in the theory and practice of *musique concrète* has set the terms of a great debate within sample-based electronic music ever since: to refer or not to refer? Theorists and historians of electronic music have traced the aesthetic battles over the issue of reference and identification of source and cause, often positing a roster of composers who break with Schaeffer's "puritan position" on reduced listening.[71] The roster of composers who reassert the significance of sonic sources and causes usually begins with Luc Ferrari and includes figures such as Trevor Wishart, Michel Chion, R. Murray Schafer, Hildegard Westerkamp, and Denis Smalley.[72] On the opposite side, Francisco Lopez is often singled out as a lone defender of reduced listening. Such affiliations are undoubtedly correct, insofar as acousmatic sound is construed as a compositional or aesthetic problem concerning a sample's reference or significance alone. But the problem of acousmatic sound has a larger scope that raises ontological questions about the relationship of sonic sources, causes, and effects. Beyond the compositional dilemma of acousmatic sound (to refer or not to refer?), one can detect the presence of a *decision*, motivated neither by love nor distaste for reference nor by dispassionate philosophical reasoning, but by a recoil from the unsettled (and unsettling) relationship of sonic sources, causes, and effects inherent in acousmatic sound. To decide in favor of reduced listening is one way of negotiating sonic incongruousness—by demanding that it simply go away. To decide in favor of the source by ignoring the structural gap of acousmatic sound and reducing the effect to the source and cause is another. Neither addresses the central problem.

Since the former route has been addressed in chapter 1 and in my review of the phenomenological tradition that Schaeffer shares with figures like Jonas and Straus, I will give some extended consideration to the latter route. The desire to lift the "acousmatic veil" and dis-acousmatize the sonic effect by reattaching it to its source and cause has motivated a variety of related theoretical positions in sound studies, philosophy, and film.

1. R. Murray Schafer's writings and soundscape recordings (in association with his World Soundscape Project) are designed to encourage listeners to reconnect sounds with the environment—an environment pathologized by increasing noise levels, the preponderance of recorded and broadcast sounds, and lack of aesthetic care for sound design.[73] On the ontological register, Schafer tries to overcome Pierre Schaeffer's sound object by positing a new entity as part of the soundscape, known as the "sound event." Unlike the sound object, which holds itself deliberately to "physical and psychophysical terms" and avoids "considering [sounds'] semantic or referential aspects," the sound event places the sonic effect back into a situated spatial and cultural context.[74] "When we focus on individual sounds in order to consider their associative meanings as signals, symbols, keynotes or soundmarks, I propose to call them *sound events*, to avoid confusion with *sound objects*, which are laboratory specimens."[75]

The *decisive* aspect of Schafer's thinking is legible in his critique of *schizophonia*. For Schafer, technologies of sound reproduction have created a new sonic order where all sounds can be separated from their origins in order to be electronically transmitted, reproduced, or broadcast. The portability of sound allows for any sound to be heard anywhere. Schafer bemoans the interchangeability of all sonic environments afforded by schizophonia, and thus the dedifferentiation and lack of attention paid to the specificity of any single sonic environment. This modern state of affairs contrasts with a myth of sonic origins that grounds Schafer's thinking. "Originally all sounds were originals. They occurred at one time in one place only. Sounds were then indissolubly tied to the mechanisms that produced them. The human voice traveled only as far as one could shout. Every sound was uncounterfeitable, unique."[76] In comparison to this Edenic vision of sonic effects and sources indissolubly identified, modern schizophonia appears as a hell on earth—a condition where all sounds are doomed to circulate endlessly without attachment to their sources.

One way to challenge Schafer's account is to compare it with the theory of acousmatic sound developed thus far. First, it should be noted that acousmatic sound cannot be simply identified with mechanically reproducible or electronically transmitted sound. As I have already argued, the utilization of acousmatic situations in the form of bodily techniques and architectural practices was already well established in the 19th century. Those practices may well be schizophonic—in the sense that they encourage a separation of source, cause, and effect—but then schizophonia can no longer be identified with sound reproduction and broadcast media. Second, environmentally situated listeners always have the potential to encounter acousmatic sounds. If we take Kafka's "The Burrow" as an example, the narrator is both environmentally situated and an auditor of acousmatic sound. There is no reason to hold with Schafer that "every sound was uncounterfeitable, unique." Even in Schafer's Edenic sonic myth, acousmatic experiences would be available. From a biological perspective, territorial listening would be the norm, not an experience of sonic plenitude. Thus there is no reason to accept that the natural or mythical condition of sounds affirms the unity and uniqueness of source, cause, and effect.

2. Schafer's work on the soundscape encourages a counter-reduction where the sonic effect is reduced back to its source or cause, as opposed to the endless tokens of types that follow from the eidetically reduced ontology of sound objects. This form of counter-reduction is often marked by its affirmative claim for sonic uniqueness. This is as apparent in the soundscape recordings made by Schafer and his World Soundscape Project, like those of the harbor in Vancouver, as it is in the work of philosopher Adrianna Cavarero. In *For More than One Voice*, Cavarero offers an ambitious rereading of the history of Western metaphysics by addressing the theme of the voice.[77] In her view, the philosophical tradition, from Plato to Derrida, has systematically suppressed the role of the voice in order to support a videocentric metaphysics, which privileges the access to truth by means of vision, *eidos, theoria*, and such. In philosophy's videocentrism, the voice has been made a refugee; thus the resonating *phoné* has been subsumed under the guise of silent logos, creating a situation where the Said, the propositional content of one's utterance, is privileged over the act of Saying. In tracing the historical disenfranchisement of *phoné*, Cavarero argues for a "vocal ontology of uniqueness," a claim that every voice indexes a unique individual. Uniqueness grounds an ethical ideal whereby individuals will no longer have their voices stolen away into the impersonality of logos, but will engage in a reciprocal exchange of sonorous vocalizations.

Cavarero surveys numerous moments from the history of Western literature and philosophy in her critique. But for the purposes of my argument, I focus on only two: her chapter on Italo Calvino's tale "A King Listens" and the appendix on Shakespeare's *Romeo and Juliet*. Both are significant in that they invoke acousmatic situations. Cavarero's affirmation of a "vocal ontology of uniqueness," like Schafer's emphasis on the uniqueness of every sound event, can be understood in relation to the unsettled character of acousmatic sound, and as motivating the insistent reduction of the sonic effect to the materiality of its source or cause.

Cavarero's appendix offers both a reading of *Romeo and Juliet* and a response to Derrida's brief essay on the play, "Aphorism Countertime."[78] (As one might imagine, any philosopher who is writing affirmatively about the voice must deal with Derrida's work, which is predicated on a foundational critique of Western metaphysics as logocentric and phonocentric.) For Cavarero, Derrida's critique of speech in the name of writing miscasts the history of metaphysics and thus misses an opportunity to separate the role of *phoné* from logos. The history of metaphysics should be understood as a history of "the devocalization of logos, instead of as a triumph of phonocentrism."[79] Derrida is incorrect to argue that philosophy has perpetuated a metaphysics of presence that privileges acts of speech over writing and, in contrast, Cavarero asserts vision's more historically significant role: "[Derrida's] thesis on metaphysical phonocentrism supplants the far more plausible, and philologically documentable, centrality of videocentrism."[80]

To counter Derrida's proscription of the *phoné*, Cavarero analyzes the famous balcony scene where the lovers converse under the veil of night. In that scene, Juliet poses the question "What's in a name?" in order to sensitively consider the relationship between the signifier "Romeo" and the individual designated by this name. The chain of proper names—Romeo and Montague, Juliet and Capulet—leads to a harrowing order of social, economic, and familial bonds that keep the lovers apart. Yet behind the names are the unique individuals who reciprocate each other's love. Derrida argues that although the proper name and the bearer are non-coincident, the two cannot be definitively separated; since the name of the bearer is given at birth and will persist after the bearer's demise, it cannot be simply renounced or changed at will; it is not coincident with the presence that bears it, yet that presence cannot be individuated, designated as the same, without some proper name; the name and its bearer are related through *espacement*.

Cavarero argues otherwise. The difference between the name and the unique individual who bears the name is articulated through an exchange of sonorous voices; thus the acousmatic situation of the balcony scene affords the lovers the opportunity to recognize each other's vocalic uniqueness, an exchange of pure *phoné* apart from any chain of signification. According to Cavarero,

> The crucial point is that this [unique] ontological status or, better, the singularity of the human being loved by Juliet, is manifested as voice in the balcony scene. Recognizing Romeo's voice, the young girl recognizes the uniqueness of the loved one, separable from the proper name, which is communicated to her vocally. Thus the essential bond between voice and uniqueness—theatrically underscored by a nocturnal darkness that empowers the exclusive role of the acoustic sphere—comes to the fore.... Shakespeare could have set the scene in the light of day. By day, the sense of the dialogue, the request to separate Romeo from his name, would not have changed. But he set the scene at night—not, or not only, because the penumbra foreshadows their death, but above all because the voice of Romeo, unseen and therefore unidentifiable through the gaze, is the immediate, sonorous revelation that is proper to that embodied uniqueness that Juliet wants to separate from the name.[81]

In addition to affording the recognition of vocalic uniqueness, the darkened setting of the balcony scene (where the lovers are "unseen and unidentifiable through the gaze") is significant. By diminishing the role of vision in the balcony scene, by short-circuiting the gaze or videocentric metaphysics, Romeo and Juliet are able to discover the "exclusive role of the acoustic sphere," which is "empowered" by "nocturnal darkness." The prohibition of visual access between Romeo and Juliet resonates with Cavarero's critique of metaphysics as fundamentally videocentric.

But what establishes "the essential bond between voice and uniqueness"? Why should the uniqueness of the beloved be any more or less unique when accessed via the ear, rather than via vision? Why should *any* sense modality be privileged in this way? (To argue that the voice is privileged because it is unseen is circular.) Since the acousmatic situation underdetermines ascriptions of source and cause—or, one might say, it never guarantees numerical identity—there is good reason to be skeptical of Cavarero's leap toward uniqueness. What guarantees that Juliet is indeed speaking to Romeo? What if someone were imitating his voice? Derrida, without providing an explicit analysis of acousmatic sound, notes precisely this aspect of the situation: "She [Juliet] is speaking, here, in the night, and there is nothing to assure her that she is addressing Romeo himself, present in person."[82] Given the ontological agnosticism of the acousmatic situation, Cavarero's vocal ontology of uniqueness, equal and opposite to Schaeffer's ontology of the sound object, is ultimately a philosophical *decision*.

Cavarero introduces her thesis concerning the "vocal ontology of uniqueness" through a reading of Italo Calvino's short story "A King Listens."[83] Part of an unfinished collection of tales about the five senses, the story describes a king who sits immobile on his throne, listening to the sounds of his kingdom from beyond the castle walls. Unable to leave the throne, in fear that another may accede to it, the king's only access to his realm is auditory. Calvino's king is an allegorical figure for the act of listening. This is noted in the topography of the king's castle. Like the underground domicile of Kafka's "The Burrow," the king's palace is constructed like "a great ear, whose anatomy and architecture trade names and functions: pavilions, ducts, shells, labyrinths."[84] That is not the only similarity shared by "A King Listens" and "The Burrow." In both tales, the listening is characterized as a fickle act; both king and mole listen with uncertainty to the sounds that penetrate deep inside their enclosures. Paranoia runs deep, and both remain fixed in acousmatic situations that foil attempts at attaining certainty. In one instance, the king considers the sound of the changing sentries outside the palace. Wondering when he last heard the change of guards, he asks, "How many hours has it been since you heard the changing of the sentries? And what if the squad of guards faithful to you has been captured by the conspirators?"[85] If lack of sound produces a worry about security, the presence of sound cannot assuage the anxiety by bringing certainty. The king muses,

> Perhaps danger lurks in regularity itself. The trumpeter sounds the usual blast at the exact hour, as on every other day; but do you not sense that he is doing this with too much precision?...Perhaps the troops of the guard are no longer those who were faithful to you....Or perhaps, without their being replaced, they have gone over to the side of the conspirators....Perhaps everything continues as before, but the palace is already in the hands of usurpers....[86]

The example could be recast in Strawson's terms. The sound of the guards is qualitatively identical to the sound expected by the king, but qualitative identity cannot guarantee the numerical identity of its source and cause. Mimicry is possible.

In a related example, the king listens intently to a beating tom-tom, unsure if the sound is real or imaginary. Does it come from the underground dungeons, where someone is tapping out a message, or is it a paranoid figment of the king's imagination? Calvino's king wonders, "You want absolute proof that what you hear comes from within you, not from outside? Absolute proof you will never have."[87]

Given the similarities between Kafka's worried mole and Calvino's paranoid king, "A King Listens" seems an odd place for Cavarero to mine a thesis concerning the voice and its ontological uniqueness. Cavarero's reading centers on a single episode in the tale, when the king overhears a singing voice transported to the palace on a breeze. Upon hearing the unknown woman's song, the king's anxieties are momentarily assuaged. The woman's voice effects a moment of transformation; it is inassimilable to the worried posture of the king's paranoid listening. Upon hearing the voice, the king notes that "it is no longer fear that makes you prick up your ears."[88] The king, attracted by the sound of the voice, articulates a sentiment that is central for Cavarero's account. Calvino writes,

> A voice means this: there is a living person, throat, chest, feelings, who sends into the air this voice, different from all other voices. A voice involves the throat, saliva, infancy, the patina of experienced life, the mind's intentions, the pleasure of giving a personal form to sound waves. What attracts you is the pleasure this voice puts into existing: into existing as a voice; but this pleasure leads you to imagine how this person might be different from every other person, as the voice is different.[89]

The king experiences a moment of transcendence as he is ravished by the singing voice. Cavarero reads this episode as attesting to the "simple vocal self-revelation" of a unique individual, a uniqueness that is never available to the semantic register, which uses a shared and impersonal lexicon to make its statements, but only available to the fleshy singularity of vocal emission.[90] "When the human voice vibrates," Cavarero asserts, "there is someone in flesh and bone who emits it."[91] Although Cavarero admits that the "corporal root of uniqueness is also perceptible by sight," the invisibility of the voice is privileged because it is hidden and interior. Vocal emission corresponds with "the fleshy cavity that alludes to the deep body, the most bodily part of the body. The impalpability of sonorous vibrations, which is as colorless as the air, comes out of a wet mouth and arises from the red of the flesh. This is also why, as Calvino suggests, the voice is the equivalent of what the unique person has that is most hidden and most genuine."[92]

By placing emphasis on the scene of overheard singing, Cavarero uses Calvino's text to develop her central thesis concerning the vocal ontology of uniqueness. In a thetic sentence, Cavarero asserts, "Every voice 'certainly comes from a person, unique, unrepeatable, like every person,' Calvino assures us. He calls our attention to what we might call a vocal phenomenology of uniqueness. This is an ontology that concerns the incarnate singularity of every existence insofar as she or he manifests her- or himself vocally."[93] In positing this thesis, Cavarero neglects to cite the rest of Calvino's sentence, an omission that changes the sentiment substantially. Here is the full passage:

> That voice comes certainly from a person, unique, inimitable like every person; a voice, however, is not a person, it is something suspended in the air, detached from the solidity of things. The voice, too, is unique and inimitable, but perhaps in a different way from a person: they might not resemble each other, voice and person. Or else, they could resemble each other in a secret way, not perceptible at first: the voice could be the equivalent of the hidden and most genuine part of the

person. Is it a bodiless you that listens to a bodiless voice? In that case, whether you actually hear it or merely remember it or imagine it makes no difference.[94]

Calvino's sentiment is far less thetic and self-assured than in Cavarero's reading, for the statement that the "voice comes certainly from a person, unique, inimitable like every person" appears in the midst of a rapidly shifting series of hypotheses, queries, and worries. Immediately after asserting the uniqueness of a voice, Calvino seems to deny its immediate association with the fleshy materiality of the person from whom it is emitted. The voice is detached or separated from its source. The king entertains a thought much closer to Dolar's position that there is always something incongruous between a person's voice and aspect than Cavarero's "simple truth of the vocal," where the voice immediately communicates "the elementary givens of existence."[95] Perhaps voice and body are related, sharing a deep, secret resemblance, but nothing guarantees such resemblance. Recoiling from such uncertainty, the king moves in a direction where he posits the voice as disembodied, fully detachable and distinct from the source of its emission. The king's final thought should recall Schaeffer's eidetic reduction. The voice becomes a sound object, severed from source and cause. Moreover, as in Schaeffer's eidetic reduction, where imaginative variation is used to reveal the essence of a sound object, the mode of presentation is indifferent—it makes no difference whether the voice is real, remembered, or imagined.

Another scene from Calvino's tale further challenges Cavarero's claim concerning the uniqueness of the voice. In an attempt to find the singing woman, the king entertains the possibility of staging a singing contest in which all female subjects would be ordered to sing for the king. Upon hearing her voice, the king would be able to declare, "She is the one!" and discover the unique source of the voice. Just as before, the king begins to doubt his plan: "But are you sure that, for the steps of the throne, it would be the same voice? That it would not try to imitate the intonation of the court singers? That it would not be confused with the many voices you have become accustomed to hearing...."[96] It is not clear how Cavarero's vocal ontology of uniqueness responds to the possibility of vocal dissimulation. Thus, in the midst of such a perpetually shifting context, it seems unwise to select any particular assertion of Calvino or the king as definitive. Rather, the *indecision* itself is central; the acousmatic situation, whether heard deep in the burrow or in the throne room, underdetermines ascriptions of sonic effects to sources and causes. There are many ways to recoil from that situation. To select, as Cavarero does, one assertion from out of the flow of Calvino's text as supposed proof for a vocal phenomenology betrays a lack of close attention to what Calvino has to say about listening. It is deaf to the nuances of Calvino's text and a far inferior form of listening to that of the king.

3. The counter-reduction of Schafer and Cavarero, motivated by acousmatic uncertainty, is really a drive to dis-acousmatize sound. But while music and philosophy have developed two characteristic strategies for dis-acousmatizing sound, by recasting the auditory effect into a sound object ontologically distinct from its source or cause or by reducing the auditory effect onto the materiality of its source and cause, film seems to have only the latter option. The role of dis-acousmatization in film has been thoroughly explored by Michel Chion, whose pioneering work on film sound is often centered on the issues of acousmatic sound. In *The Voice in Cinema*,

Chion offers numerous instances where an *acousmêtre* is introduced early in the film, in order to motivate its eventual dis-acousmatization.[97] Chion notes,

> An entire image, an entire story, an entire film can thus hang on the epiphany of the acousmêtre. Everything can boil down to a quest to bring the acousmêtre into the light. In this description we can recognize *Mabuse* and *Psycho*, but also the numerous mystery, gangster, and fantasy films that are all about 'defusing' the acousmêtre, who is the hidden monster, or the Big Boss, or an evil genius, or on rare occasions a wise man.[98]

A comic use of dis-acousmatization appears in the final scene of *Singin' in the Rain*. Throughout the film, Kathy Selden (played by Debbie Reynolds) has been acting as the voice double for the selfish and talentless Lina Lamont (played by Jean Hagen). Lamont, who has been suppressing Selden's voice by using it as her own, finally gets her comeuppance when she is exposed as a fraud on the opening night of her new musical picture. Just before Lamont is exposed, there is a remarkable scene where the camera reveals both women standing before microphones but separated by a curtain. Jean Hagen stands in front miming the song, while Debbie Reynolds stands behind it singing (see figure 5.2). The moment of dis-acousmatization occurs when the trio of Gene Kelly, Donald O'Connor, and Millard Mitchell lifts the curtain that acts as a mythical Pythagorean veil. In Chion's insightful description, "An astonishing shot reveals the two women, one behind the other, with the two microphones lined up, both singing with this single voice that wanders between them looking for its source. The audience understands and attributes the voice to its true body"[99] (see figure 5.3).

After the curtain is lifted and Debbie Reynolds is revealed as the true source of the voice, both the onscreen and actual audience are treated to a duet version of "You Are My Lucky Star." The symbolic nature of the performance is obvious enough: Kelly and Reynolds sing a duet together that establishes the reciprocity of their love. Furthermore, Reynolds now gets to come out from behind the curtain and receive the first public acknowledgment from her beloved of their veiled relationship, as well as become the rightful recipient of the audience's adoration. Dis-acousmatization has restored everything to its rightful place.

However, the security of *Singin' in the Rain*'s final dis-acousmatization is easily shaken. Reynolds's voice finds its true body only in the symbolic register. In reality, something much more disturbing occurs. In their study of *Singin' in the Rain*, Hess and Dabholkar noted that "Debbie Reynolds had a natural, virtually untrained singing voice that worked perfectly for simple, bouncy songs, but not so well for others."[100] "You Are My Lucky Star" proved to be too much for the young starlet to handle and, in the final duet with Gene Kelly, her voice was dubbed after many takes. Reynolds's actual voice, which was used while Jean Hagen lip-synched to "Singin' in the Rain," is not the same voice issuing from her lips in the final duet. That voice belonged to another little-known voice double named Betty Noyes.[101]

By counterpoising the reflections on the phenomenology of sound that opened this chapter against the three examples taken from sound studies, philosophy, and film, we attain an overview of the two common strategies employed in dodging the constitutive underdetermination of acousmatic sound. Acousmatic sound is unsettling because it depends on a structural spacing of sonic source, cause, and effect

Figure 5.2 Final scene of *Singin' in the Rain*. Courtesy of Photofest NYC.

Figure 5.3 Moment of dis-acousmatization from *Singin' in the Rain*. Courtesy of Photofest NYC.

that is fundamentally insecure. On the one hand, there is the drive to secure certainty by discovering the material source of acousmatic sound, by lifting the mythical Pythagorean veil and seeing the source in all of its nakedness. On the other hand, there is the drive to secure certainty by bracketing everything that is inessential to encounter the sound object in all of its absolute and essential detachment. In contrast to these two reductions, the utility of Kafka's "The Burrow" is clear. Kafka chooses neither of these routes, maintaining the anxiety inherent in acousmatic sound against any reduction. In his text, one touches the root of the acousmatic situation, free of any drive to theorize away the problems of acousmatic sound.

CHIASMUS

Although the protagonist of "The Burrow" is certainly uncertain, the reader of this tale is not free from epistemic worry either. Among the various fickle hypotheses concerning the source of the sound heard in the burrow, attentive readers will not fail to notice one glaring omission. Is it not possible that the sound heard in the burrow is simply produced by the mole? Couldn't the sound's persistence and omnipresence be explained by some kind of physiological or psychic tinnitus, a ringing in the head of the narrator falsely ascribed to the passageways of the burrow? There is nothing in Kafka's text to help rule out this possibility, although the mole never explicitly considers it. Upon reflection, it appears that Kafka may have given the reader a tantalizing clue that this is indeed the case. After attributing various causes to the sound, the mole finally convinces himself that "the whistling is made by some beast, and moreover not by a great many small ones, but by a single big one."[102] In developing this deduction, the mole claims, "I could clearly recognize that the noise came from some kind of burrowing similar to my own...."[103] A double is posited, perhaps revealing a tenuous identification between the narrator and a rival.

The possibility that the burrower misrecognizes his own sonic production for that of another forms a corollary to my earlier claim. If the sonic effect underdetermines attributions of its source or cause, then the location of that source as definitively located inside or outside the listener's own body also becomes uncertain. Kafka, whose attention to the strange paradoxicality of sound is exemplary, is not alone in exploring this phenomenon. In this respect, "The Burrow" could be placed into a wider literary context where such reversals of inner and outer often occur as a function of sonic underdetermination.

Take Poe's famous story "The Tell-Tale Heart." Everyone knows the tale of the hyperaesthesic murderer who kills his elderly victim, chops up the body, and buries it under the floorboards, only to be driven to madness and confess the crime because of the incessant post-mortem beating of the victim's heart. Yet the plot alone does not do justice to the auditory undecidability of Poe's tale. The reader is left with grave doubts about its veracity, told by an unreliable narrator as a way of ostensibly conveying the sanity and reason behind his murderous act.

In the final scene, the murderer sits with the police in the very room where the body is buried to demonstrate his ease and composure. Soon the sound begins and the murderer's bearing becomes unsettled. The beating of the heart may belong to the victim, as the narrator implies, yet it may just as well belong to the murderer himself. The source is underdetermined, a chiastic swap of inner and outer. At first it seems that the police have not heard the sound, yet this turns quickly into paranoia:

> I gasped for breath—and yet the officers heard it not. I talked more quickly—more vehemently; but the noise steadily increased.... It grew louder—louder—louder! And still the men chatted pleasantly, and smiled. Was it possible they heard not? Almighty God!—no, no! They heard!—they suspected!—they knew!—they were making a mockery of my horror!... Any thing was more tolerable than this derision! I could bear those hypocritical smiles no more![104]

The ambiguity is preserved in the final line of the story: "'Villains!' I shrieked, 'dissemble no more! I admit the deed!—tear up the planks! here, here!—It is the beating of his hideous heart!'"[105] But where exactly is the location of this "here"? Does it refer, ostensively, to the location under the planks or to the breast of the murderer? Without a visual gesture to accompany the phrase, the reader will never know to where the murderer's finger points. If it were the murderer's body producing the sound, the dissociation of the subject from his own body would be a sign of the madness that has already begun. This possibility is supported by the subtle fact that the last line of the text is in quotation marks, the only place in the entire narrative where the murderer quotes himself. The reification of the narrator's voice within the discourse depends on an unraveling of the auto-affective circuit of the subject from its body. The quotation, like the sound of the beating heart, cannot be definitively situated or possessed.

Poe's murderer, unlike Kafka's mole, considers the possibility of the sound's subjective source before rejecting it. In the final scene with the policemen, the sound begins as something other than a beating, occupying, like Kafka's *Pfeifen*, a much higher register. Poe writes, "My head ached, and I fancied a ringing in my ears.... The ringing became more distinct.... I talked more freely to get rid of the feeling; but it continued and gained definitiveness—until, at length, I found that the noise was not within my ears.... Yet the sound increased.... It was a low, dull, quick sound—much such a sound as a watch makes when enveloped in cotton."[106]

The change in register is fascinating; as the sound grows in distinctness and definitiveness, it also shifts in pitch from high to low, from inner to outer. The repeated presence of these two registers—one occupied by a high-pitched ringing, whistling, or *Pfeifen*, the other occupied by a low rumbling, beating, or *Zischen*—leads me toward a tenuous speculation. Michel Chion, in *Film: A Sound Art*, introduces the idea of a "fundamental noise" in cinema, usually a continuous or periodic complex sound mass (like the sound of the ocean, gyrating fans, rain, flowing air, breathing, repetitive clicking, or hissing), a sound that other sounds try to cover up, but that gets uncovered again at the end of the film.[107] It acts as a sonic floor upon which the action is situated that, when uncovered, exposes a cosmos indifferent to human agents in its impersonal, mechanical repetition. Chion considers the reason for the presence of fundamental noises in film: that they first emerged as attempts to cover the mechanical sound of the projector; but his discussion also reverses the story, suggesting that we hear in the projector's noise another instance of the incessant, inhuman, cosmological machinery that fundamental noises represent.

Borrowing Chion's terminology, could it be that our two sounds—the high-pitched *Pfeifen* and the low-pitched *Zischen*—are fundamental acousmatic sounds? Of course, these sounds would not have emerged as ways of covering (or prolonging) the sound of the film projector; they would have emerged from a different, less mechanical origin. It is suggestive that these two sounds are also present in John Cage's foundational account of his visit to an anechoic chamber. Cage, upon entering the chamber at Harvard University in order to finally hear silence, expresses surprise at what he discovers: "[I] heard two sounds, one high and one low. When I described them to the engineer in charge, he informed me that the high one was my nervous system in operation, the low one my blood in circulation. Until I die there will be sounds. And they will continue following my death. One need not fear about the future of music."[108] I want to distinguish the punch line of Cage's anecdote—"one need not fear about the future of music"—from an earlier punch line given to the engineer. Imagine the uncanny shock at realizing that two sounds, which were assumed at the outset to be attributed to exterior sources, were actually subjectively produced. Cage's stunning confusion of inner and outer recapitulates the uncanny sonic underdetermination of Kafka and Poe's tales. Inner to outer, outer to inner—once the chiasmus is underway, it matters little in which direction we traverse.[109]

By reading—or should I say listening to—"The Burrow," an aporia at the heart of acousmatic sound is disclosed. The aporia depends on the fact that the auditory effect, when unaccompanied by contributions from other senses, underdetermines ascriptions of source and cause; and, as a corollary, that the chiasmus of sonic ascriptions to inner and outer sounds occurs as a function of this underdetermination. We may be surprised to find this articulated in Kafka's tale, for what can Kafka tell us about listening? But then again, why should we give more credence to Hans Jonas or Erwin Stein or Pierre Schaeffer than Kafka? Why should the philosopher be a more insightful, more systematic researcher than the novelist? If listening to "The Burrow" can shed light on the problems of acousmatic sound—so be it. Who ever said Kafka was only for reading?

PART FOUR

Cases

6

Acousmatic Fabrications
Les Paul and the "Les Paulverizer"

In part III, I articulated a set of conditions that underlies the experience of acousmatic sound. I developed these conditions by pursuing the three objections to Schaeffer's theory of acousmatic sound—concerning phantasmagoria, myth, and the ontology of the sound object. The conditions that I articulated draw upon two central premises: first, a tripartite ontology of sound (where sounds are broken into three indissoluble moments—source, cause, and effect), and second, the supplementary relationship of *physis* and *technê*. The former allows us to describe the strange ontology of acousmatic sound, which *comes into being* (or *is*) only when the source, cause, and effect of a sound are spaced. Spacing manifests itself when the sonic effect is underdetermined by the source or cause. The latter brings into a play a large number of cultural techniques (bodily techniques, architectural devices, technological inventions) involved in the production of acousmatic spacing.

But a theory is only as good as the cases it illuminates, and I have two in mind: one musical, one philosophical. The first concerns the guitarist Les Paul; it focuses on his use of technology to produce "live" acousmatic music and the reasons that led him to do so. The second concerns the acousmatic voice and the philosophical tradition that has heard it resounding in the furthest recesses of the subject. Starting with Edison's talking machine, it ends with a take on the talking cure, in order to sound out the acousmaticity of the question "Who speaks?"

STOLEN IDENTITY

The career of guitarist Les Paul presents an unusual case for exploring the conditions of acousmatic sound and their practical consequences.[1] Paul and his wife, singer Mary Ford, were among the most commercially successful pop musicians of the early 1950s, with numerous chart-topping singles, radio programs, and even a daily five-minute television program. Paul built his reputation during the 1940s as a jazz guitarist, with a style influenced by Django Reinhardt. Backing singers like Bing Crosby and the Andrews Sisters, Paul also released recordings with his own trio, performing frequently on local and national radio. In addition to his musical skills,

Paul was an inventor of electronic audio equipment. He is often credited with playing a seminal role in the development of the solid-body electric guitar, magnetic pick-ups, multitrack recording technology, and various techniques for recording sound in the studio. The marriage of Paul's musical and technical skills led him to experiment with "sound-on-sound" or "multiple" recordings, which are more commonly referred to as overdubbed or multitrack recordings.[2] These early recordings, instrumental records of jazz standards such as "Lover" and "Brazil," were instant hits for the fledgling Capitol Records.

However, it was not simply a love of technology that motivated Paul to develop his "new sound," but something else entirely. In the 1940s, Paul's unique style of virtuosic runs and tender descants was widely recognizable and an object of admiration and even imitation by other guitarists.

> My mother came backstage and she told me that she had heard me on the radio. "Lester," she said, "you were fine." Only I knew it wasn't me; I hadn't been on the radio. Then I found out that it had been [Chicago-based guitarist] George Barnes. No reflection on George, of course, but I figured that my own mother had a right to know when her son was playing.... I decided that I wasn't gonna record or go on the radio or anything until I could work out something so much me, that my mother would know that was her Lester playing.[3]

Here we have an unexpected consequence of acousmatic underdetermination, a case of stolen identity: Am I hearing Les Paul or George Barnes, and how am I to tell the difference? Acousmatic sound provokes the questions: "Who speaks? Whose sound is this?" Barnes, as a mimic, speaks in Les Paul's voice, momentarily appropriating his identity, and expropriating Paul of what he assumed was uniquely his possession—his sound. Radio is a medium where sound is often heard acousmatically. Because radio broadcasts sound without a visual referent, it affords the possibility of acousmatic underdetermination. Although the lack of visual referent can allow the listener a space for imaginative supplementation—a space that, in the past, was occupied by serials, theater, and other specific radio programs—in the case of Les Paul, it allowed for a strange kind of sonic identity theft. Paul became the unwitting victim of an acousmatic robbery.

According to popular music scholar Steve Waksman, "Paul's wry comments about his mother's misidentification uncover a private, familial motivation behind the innovations that brought him mass acceptance."[4] Rather than read Paul's reaction in familial terms, I would argue that it is a case of mimetic rivalry, where one's very identity is challenged by a rival or double that triggers a recoil from the threat of identification.[5] By working in secret in the studio, Paul would try to discover a proprietary sound, one that would allow him to create something "so much me" that his identity could never be stolen again.[6] But, in a strange irony, Paul's strategy would exploit the underdetermination of acousmatic sound in a related medium, sound recording, and the technique of overdubbing it affords.

Making overdubbed recordings was a tedious process. The early disc-cutting (and magnetic tape) technology that Paul used for most of the 1940s until the mid-1950s did not allow for the possibility of correcting mistakes. Once something was recorded, it could not be altered, but only added to. An article from the *Saturday Evening Post* describes the process:

> To make a record, Les has to hold in his musical memory all the parts he wishes to record. Then he records the least important, or background part, first, knowing that some of this will be lost in the final version. He plays back the tape and accompanies it with a second part, which is recorded on a second tape. If the second round is satisfactory, he then transfers Part No. 2 on top of Part No. 1 on the first tape, and so on, recording the melody and very delicate sounds, such as the tinkle of a bell, at the very last.[7]

In addition to layering part upon part, Paul also leveraged some of the affordances of recording to produce special effects—delay, echo, and slapback. But one of Paul's most ingenious tricks was to alter the recording speed when layering parts. By slowing down a recording to half of its original speed, the entire recording is not only slower in tempo, but drops in pitch by one octave. Say, for example, you have a recording of a rhythm guitar playing through the chord changes to some jazz standard. Call this guitar A. If the recording of guitar A is slowed to half of its original speed, its tempo also slows in half and the pitch drops an octave. If a solo guitar part (guitar B) is overdubbed onto the half-speed recording, an unusual effect occurs. When the recording is brought back up to its normal speed, guitar A sounds like it did originally, but guitar B is now one octave higher than before and twice as fast. But not only is the solo part faster and higher, the whole frequency spectrum of the guitar has been shifted, making the guitar sound brighter and clearer.[8]

By altering the recording speed, Paul was able to create guitar lines that were faster and higher than humanly possible—but that still sounded, potentially, within the realm of attainability. The technique is used prominently on Paul's first single for Capitol, which featured covers of "Lover" and "Brazil." Half-speed techniques exploited the affordances of recorded sound in a way that simply layering sound-on-sound in real time did not. Potentially, a talented arranger could create transcriptions of sound-on-sound recordings, giving each of the parts to a different guitarist, and perform them live, so long as no half-speed techniques were used. But this is no longer the case once the tape speed is manipulated. Simply put, there is no way to perform sounds with the heightened speed, transposition, and spectral shift of half-speed recordings live onstage—one can only play them back from the recording.

More significantly, these superhuman runs were unable to be duplicated by other guitarists. They helped to relieve the threat of mimetic doubling and expropriation that so troubled Paul. More than relying on new technological advances, Les Paul's "new sound" was also *his* sound—one that was not vulnerable to being mimed or doubled by George Barnes or anyone else because it was "so much" him. It required an apparatus of homemade machines and secret techniques in order to function—a technological analogue to those inner, hidden secrets that most deeply constitute the self. *Technê* was instrumental in producing the stunning proprietary sound of these recordings, but only if it could be kept secret. The sound was new, strange, and unsettling, yet thrilling. Part of its appeal was that it provoked the question "How does he do it?" This gave Paul great leeway over how to dissimulate the answer.

MISDIRECTIONS

Like the new sound itself, the answer was multiple. Paul, who had a weekly 15-minute radio show in the 1950s, used it to offer various cheeky solutions to the puzzle. Little mention was made of the use of overdubbing. Paul preferred to misdirect the audience by attributing the new sound to unusual, little guitars or to his ability to play multiple instruments at the same time. These two strategies differ in that the former is a false attribution of the source, while the latter is a false attribution of the cause.

Imaginary Sources: "That Little Guitar with the Weird Sound"

In the first episode of the Les Paul Show (which aired on NBC radio on May 5, 1950), the question "How does he do it?" was lightly handled. There was little storyline, simply a few introductions that set up the Les Paul trio as they played through a variety of genres: novelty numbers, ballads, hillbilly tunes, and jazz. The first song they "performed" was "Nola," but in actuality, Paul inserted the recording that Capitol Records had released in 1949. In this arrangement of "Nola," originally a novelty piano solo written by Felix Arndt in 1915, a lead guitar part produced by the use of a half-speed recording is prominently featured, along with other signature effects, like echo, deadened-string pizzicato, and mandolin-style tremolandi. The strangeness of the effect (which gives the impression of a futuristic, mechanical carousel or souped-up theater organ) is left almost untouched, as Paul immediately jumped in to introduce the next number.

> LES: Here's Mary Ford ready to sing...
> MARY (interrupting): Les, I think you ought to tell them about that little guitar you just played with the weird sound.
> LES: Oh. Well, that guitar is about eighteen inches long and it's tuned about an octave higher than a big, standard guitar. And that's how we get that different sound.

Of course, nothing could have been farther from the truth. The actual recording techniques were simply disguised and left unconsidered.

Imaginary Causes: "Mary Sings Three Parts at Once"

After the initial episode, few attempts were made to explain the new sound through deferrals to an imaginary source. A more common strategy was to claim that Les was playing multiple guitars at once, and that Mary was singing multiple parts simultaneously—a false attribution of causality that would amount to something like a musical magic trick.

Figure 6.1 A doctored photograph of Les Paul. Courtesy of Photofest NYC.

In an episode airing seven weeks later, on June 23, 1950, Les and Mary introduced themselves in this way:

LES: That's Mary Ford.... Mary sings three parts at once and does some very fantastic things with vocals on this program.
MARY: And Les plays seven guitars and all the rhythm instruments on this program....

It was followed by a rendition of "The World Is Waiting for the Sunrise." Playing the opening bars rubato, Mary sings in one part, then two, then three—adding a voice at each pause in the melody, with Les counting each time a voice was added. When all three parts were harmonized, Les yelled, "Oh that's great! Now just stay right like that and let me get over here by the guitars and we'll knock ourselves out. Wait a minute! Seven guitars and three voices... now let me get to the guitars. Here we go."

The false attribution that Les and Mary were simultaneously playing multiple instruments and singing multiple parts was easily afforded by radio—where the invisibility of the performing body encourages the imagination to concoct all sorts of correspondingly impossible physical situations. If seeing is believing, then hearing is imagining. However, the idea that Les and Mary possessed special skills to play and sing multiple parts wasn't simply propagated on radio; it was also propagated in images. These images came in two varieties. Through the use of doctored photographs or illustrations, multiple Les Pauls form a string band (figure 6.1) or a

Figure 6.2 Cover of Les Paul's *The New Sound* (Capitol Records, 1950). Courtesy of Universal Music Enterprises, a Division of UMG Recordings, Inc.

single Les Paul plays multiple instruments, as on the cover of his Columbia album *The New Sound*, where he appears as a cross between Ganesh and Giacomo Balla's dog (figure 6.2).[9] In both images, it should be noted, Paul plays exactly seven guitars. While the cover promotes a comic image of Paul playing multiple instruments, the liner notes on the backside of the LP are ambiguous about the attribution, opting for feigned ignorance and advertising lingo:

> Les Paul now brings us a captivating demonstration of his theory that what is good on one guitar is eight times as good on eight guitars—and to prove it, he plays them all himself! How this can be done is Les' secret, and he steadfastly refuses to divulge it... but we do know that the results are bright, gay and intriguing—and filled with good humor.

Similar to the radio pilot episode, the curiosity or unsettledness provoked by the music is as quickly acknowledged as it is dismissed.

PYTHAGORAS MEETS THE HITMAKER

The creation of the new sound also created another problem: It was impossible to produce it live onstage. It could only be *reproduced* by playing along with prerecorded tracks, but it could not be *recreated*. The new sound may have given Paul some proprietary control over his persona, but because of its asynchronous mode of production, it could only fully exist on phonograph discs or reels of magnetic tape. As long as the *technê* that produced the new sound remained hidden away, everything was in Paul's control. However, an essential problem remained: How did one preserve the acousmaticity of the new sound in the visual space of live performance, where the *technê* would be on display for all to see?

Due to the phenomenal success of the duo, there was great demand and financial incentives for Paul and Ford to perform live. On the road, the duo was often augmented with a few additional musicians: Ford's singing sister Carol and her bass-playing husband Wally Kamin. Ford, who was an excellent guitarist, would play some of the rhythm parts. But augmenting the size of the group was not an adequate solution. In an interview with John Sievert, from 1977, Paul described the situation:

> You walk out there with just one voice and one guitar, and you've got a problem. If they yell out, "How High the Moon," you've got to give them something as close as possible [to the recording]. So I came up with the bright idea of taking Mary's sister and hiding her offstage in a john or up in an attic—wherever—with a long microphone. Whatever Mary did onstage, she did offstage. If Mary sniffled, she sniffled. It just stopped everyone dead.[10]

Here, Pythagoras, with his legendary technical trick of veiling the body, joins forces with the hitmakers of the 1950s. One might imagine the whole scenario inspired by the final scene of *Singin' in the Rain*—if only Paul's performances had not predated the film. Yet, unlike the victorious *dis-acousmatization* of Lina Lamont's voice to the benefit of Kathy Selden, Paul did his best to keep Carol a secret.

Of course, there were moments when the secret was almost discovered. According to Paul's biographer Mary Alice Shaughnessy, one night,

> [t]he stage manager playfully kissed Carol's neck, doing his best to make Mary's pretty backup singer giggle halfway through a song. Naturally the audience wondered where the disembodied voice was coming from. But Les went out of his way to conceal Carol's supporting role in the show. The way he figured it, the more mystery surrounding Les Paul and Mary Ford the better. Poor Carol never did get a chance to step out from behind the curtains to soak up some of the applause.[11]

Paul, in order to maintain an air of mystery—a demand formed in reaction to the threat of mimetic rivalry—had no compunction about putting others in situations where *their* identities would be mimetically doubled. When Mary came down with a case of nerves before their October 1951 performances at New York City's Paramount Theater, Paul put Carol, "who closely resembled her sister in visage and voice," into Mary's strapless gown and placed her onstage.[12]

Paul, a great spinner of yarns, tells a tale about the degree to which the audience would go, in an attempt to answer a question marked by acousmatic unsettledness, "How did he do it?"

> People couldn't believe it or figure it out.... One night I hear the mayor of Buffalo sitting in the front row tell his wife, "Oh, it's simple. It's radar." So a couple years

after playing with the extra voice and an orchestra and everything, they began to think that they heard all kinds of things. They put things in there that weren't there.... You know who figured out the trick with Mary's sister? Nobody could figure it out. Life Magazine couldn't. We wouldn't tell anybody; it was a secret for years. Then one night, a man came backstage with his little girl and says, "If I tell you how you're getting that sound, will you give me a yes or no?" I said, "Sure" and the little girl says, "Where's the other lady?" It took a little kid who didn't have a complicated mind. Everybody saw machines, turntables, radar—everything but the simplest thing.[13]

Despite the questionable veracity of this tale, it illuminates something central about how acousmatic sound underdetermines attributions of source and cause. By having Mary onstage, the audience is misdirected to believe that they have certainty about the source of the sound, but they are left to wonder about its technical cause. There must be some kind of mechanical supplement that makes one voice sound like two. But the trick doesn't involve any form of unusual causality; rather it relies on a deception at the source—in this case, two sources, not one. As Paul emphasizes, the effectiveness of the trick depends on its simplicity. With ravenous desire to discover the cause, everyone overlooks the source. The story reaches its fable-like conclusion when the innocent child, the only one simple enough to figure out what is really going on, instinctively and naïvely intuits the most obvious—and least desirous—of solutions.

Finally, in 1953, a long profile on the duo in the *Saturday Evening Post* unveiled the trick for all to see:

On the stage, Les and Mary, with a guitar apiece, are backed up by one visible supporter. Wally Kamin, with a bass fiddle. In the wings offstage—and this has been kept a secret till now—stands Carol, Mary's sister, with a mike. Carol's voice is so similar to Mary's that when she chimes in, a split second behind Mary, with a harmonic contribution, the double sound does seem to be issuing from Mary's gifted throat.[14]

Technically, the trick exploits the way that audio technologies, like the microphone, mixer, and loudspeaker, allow a sound to be displaced from its source and reproduced elsewhere. By close-miking the voice, amplifying it, and diffusing it through a loudspeaker, it could be mixed with other voices and emitted from a single location. Truly, the "double sound" does indeed issue from one place, the cone of the loudspeaker—not Mary's gifted throat.

THE "LES PAULVERIZER"

By July of 1950, an alternative strategy for explaining the new sound had emerged on Paul's radio program. Paul began to characterize himself as someone who liked to "tinker" with electronics, and often described their house as full of electronic gear and gadgets.[15] One piece of gadgetry was supposedly capable of multiplying sounds—plug in "one guitar and make it sound like six," or sing into it and sound like a whole choir. Eventually, the device was baptized the "Les Paulverizer."[16]

Acousmatic Fabrications: Les Paul and the "Les Paulverizer"

When we did the radio show for NBC I had a problem. I was doing everything; I produced, I directed, I wrote the script, I acted and I played. In the script I tried to explain how I'd take Mary's voice and multiply it, but it was all so technical. Then I came up with the idea of my magic box, the Les Paulverizer, which did everything for me, and this worked and it became very popular among listeners. It became part of the show, with me saying things like, "Mary, you sing this song and the Les Paulverizer will multiply you into twelve."[17]

Paul wove the machine into the plot in a variety of ways, usually for the sake of a gag. In an episode entitled "The case of the missing Les Paulverizer," the machine has suddenly vanished for the first few minutes—until Les discovers that Mary has accidently broken it.

> LES: Mary, why did you ever go down in the basement with that gadget in the first place?
> MARY: Well, I thought if the thing could make one guitar sound like six, I could plug in my new Hoover vacuum cleaner and clean the house six times as fast.[18]

In another episode, Les, desperate to find a job, scours through the newspaper only to find an advertisement from a booking agent looking to hire a string orchestra and a glee club. Deciding that this is the perfect gig, Les tries to convince Mary that they need to audition—but only over the phone.

> MARY: Oh no, Les! You're not a string orchestra and I'm not a glee club.
> LES: Yeah, but I can make you sound like one with my Les Paulverizer.
> MARY: What will the fella say who's going to hire us? It's just you and me. Kind of a small organization, isn't it?
> LES: But Mary, he won't see us. We're going to audition on the telephone. All you have to do now is stand over there now by the Les Paulverizer and mumble into the microphone and your one voice will sound like a whole room full of voices.
> MARY: OK, I sure hope it works.[19]

After a variety of tactics to dissuade the booker from coming over to see the act, the duo finally auditions over the phone and gets the gig. But the dishonest trick receives its comeuppance in a joke at the end of the program. The booker, Mr. Fairchild, is so delighted with the sound of the string orchestra and glee club that he decides to send a bus over to pick up the whole troupe and put them on his television show. Les hangs up.

The same year that the *Saturday Evening Post* revealed the tricks Paul and Ford employed for their live performances, the imaginary Les Paulverizer underwent a similar unveiling. On an episode of *Omnibus*, which aired on October 23, 1953, Alistair Cooke had Paul and Ford as guests on the program.[20] Rather than simply interview the couple and have them perform a few numbers, the program was focused on Paul and Ford's technique for making sound-on-sound recordings. In so doing, the segment began with a humorous exposé of the Les Paulverizer. Standing on stage in front of a wall-sized machine with switches, dials, audio jacks, and an oscilloscope, Cooke introduces what he calls a "popular conception" about how Paul and Ford make their records:

I've been told... that they just play a guitar into this vast machine and then they set dials and it comes out with 25 guitars. There is a widespread belief that they use this electronic machine that is about as simple as a...uh...cyclotron, and we have a model here just to show you what this popular conception is. Now, Les, show us how it works.

After strapping on a guitar and plugging it into the machine, Les demonstrates how he can turn one guitar track into a string band. After playing a brief phrase from his composition "The Kangaroo," Paul flips switches, adjusts dials, and waits for a second, and then—with tongue in cheek—hits the machine with his fist to make the sound appear.[21] Next, Mary sings a bit from the chorus of "Don'cha Hear Them Bells," and after similar adjustments and another punch (which causes the sound of thunder), her voice is tripled. Finally, Cooke himself has a turn on the Paulverizer. Before speaking into the microphone, Paul adjusts the machine, cheekily saying he's got to "put a little English on that." After taking a moment, Cooke, with his distinctive British accent, speaks clearly and distinctly into the microphone: "This is Alistair Cooke, who comes up every Sunday on *Omnibus*." Of course, the joke is on him, since what comes out of the Paulverizer is an unrefined Cockney voice: "This is Alastair Cooke, the lad wot comes up every Sunday on *Omnibus*."

After having a laugh about the Les Paulverizer, Cooke turns to the camera and says:

> You see ladies and gentlemen, that this is the final demolition of this popular and ignorant rumor, that the basis of Les Paul and Mary Ford's music is electronics. They make music the way people have made music since the world began. First of all, they are musicians; they have an accurate ear for harmony; they work very hard; they have a lot of patience and they take advantage of the trick which, granted, electronics makes possible, that you can record one part of a song and then you can play it back to yourself and then you can accompany that part and keep on recording.

Remarkably, the rest of the segment is given over to an actual demonstration of how Paul and Ford make their records. On the stage set are two Ampex tape machines, and Paul takes the audience through the process of recording one track and then adding another on top of that, and so forth. Both Paul and Ford take turns building up the opening phrase of "How High the Moon."

Given the popularity of Paul and Ford in 1953, who (according to Cooke) had sold some 15 million records, people were curious to know "how they did it." The necessity to offer such false demonstrations simply in order to demolish the popular conception reveals just how much the acousmaticity of Paul's new sound—that is, the way in which their music underdetermined its sources and, especially, its technological causes—mattered to their audience. The irony is that Paul himself started the popular misconception by his use of the imaginary Les Paulverizer on his radio program. The need to keep his new sound proprietary, to keep the little technical trick hidden away from the public, gave way to its eventual revelation when his gifts as a guitarist and musician were being questioned. Cooke's "demolition" reasserts and reassures the audience that Paul and Ford's music is made the old-fashioned way, the way music has always been made, through the skill and talent of the musicians; the technological advantage of multitracked recording is acknowledged, but it is put back into a subordinate role to their musical skills. In line with the long history of the expulsion of *technê*, the *physis* of the musician's natural gifts is reasserted.[22]

ACOUSMATIC FABRICATIONS

After its disclosure on *Omnibus*, one might have expected the demise of the fictional Les Paulverizer. But that's not exactly what happened. In the span of a few years, the Les Paulverizer was transformed from an imaginary gag into an actual device. Sometime around 1956, Paul constructed a special black box (also called the "Les Paulverizer") that sat on the end of his guitar—an acousmatic fabrication, if ever there was one.[23] He was motivated by the continual problem of live performance:

> Wherever we performed, people kept asking the same thing [i.e., why doesn't the duo sound like they did on the records?], so what I did was sit down and build a box that I called the Les Paulverizer. This sat on my guitar and it started and stopped the tape machines, rewound them, recorded, added the echo and did everything right there on the stage.[24]

It is uncertain to what degree the box was actually capable of doing all of these tasks. Elsewhere, Paul describes it as "a remote control box for a tape recorder, and it's mounted right in the guitar."[25] Indisputably, the box allowed Paul to control a tape recorder so that he could start and stop prerecorded segments from his guitar without the audience suspecting any hidden machinery. It seemed as if the little black box was making the guitar sound like a dozen instruments, rather than the somewhat disappointing realization that he was simply playing along with prerecorded tracks. Or, in Paul's words:

> When I told Mary we were going to use the Les Paulverizer, she said, "I'm not going on the stage with this thing! It's never been tried." I said, "It'll work...." The tape machine would be hidden behind a curtain, so everyone would still think the Paulverizer was this magic box.[26]

According to Paul, the newly fabricated Les Paulverizer was first used in a command performance for President Eisenhower at the White House. The degree of deception involved in this performance was much higher than in the fictional phone call to the booker, Mr. Fairchild, and this time the joke was on Eisenhower:

> Well, we went down to Washington, and there we were, performing for Eisenhower, Nixon and all the bigwigs, and through the first five songs everything went great. Then Nixon leapt up, put his arms around me and said, "Maybe the President has a favorite song. Why don't you ask him?" I said, "That's a great idea." I could have killed Nixon.

I said, "Mr. President, Vice President Nixon came up with an idea here—I'd like to ask if you have a favorite song that Mary and I can play for you." I was thinking, "Oh my God, what are we going to do?" because we really had to play the next number on the tape. Anyway, Eisenhower couldn't think of a favorite song, so he asked Mamie and she said, "Well, when we were leaving Denver and you got pains in your chest, we pulled over to the side of the road and I turned the radio on and we heard 'Vaya Con Dios.'..." So help me God, that was the next number on the tape!

I still have the letter that Richard Nixon sent me, describing how Eisenhower had stopped him down in the tunnels beneath the White House and said, "You know, that Les Paul is bothering me. I still can't figure out the Les Paulverizer." I've also got a letter that Eisenhower wrote to Pat Nixon, saying, "I'll never figure out how that guy could do what he did. It was the most amazing thing I've ever seen." And of course it was amazing. If it hadn't been for Mamie picking that song, we'd have been dead. It's just lucky that Nixon didn't suggest they ask for a second song.[27]

Like his fellow politician, the mayor of Buffalo, Eisenhower is the unwitting dupe of Paul's misdirection. But the nature of the misdirection is different. Unlike the case of Mary's sister hidden behind the curtain, where the audience overlooks the doubled source of the voice by trying to discern the technical cause behind it ("It's radar!"), here, one piece of technology veiled another. Paul pointed to the Les Paulverizer as if it were the technical gadget that created the effect, while keeping its real function at bay. Like a magician, Paul maintained an aura of mystery by holding the actual workings of the box in reserve. The audience sees Les and Mary playing and singing, but they hear a string orchestra and glee club. Isn't that the epitome of a *black box*—an input, an output, and a mechanism where one is never sure what happens inside?

Fundamentally, the Les Paulverizer presents the listener with a paradox: On the one hand, it acts as a proxy for the real cause of the sound (i.e., overdubbed recording); on the other hand, the gadget remains mysterious enough to leave the details about the sound's production unanswered. If the acousmaticity of sound is ultimately created from the altered relationship between seeing and hearing, from the underdetermination of the sound's source and cause by its effect, then one could articulate the paradox as follows: The device simultaneously *dis-acousmatizes* and *re-acousmatizes* sound. It manifests a sonic cause while keeping the real causality invisibly veiled.

Beyond this paradox, there is a bitter irony. If the motivation for developing the new sound stemmed from the threat of a mimetic rival, Paul's solution for appropriating and securing his identity—his sound—confined this security to the space of recording. In so doing, it cut out Paul's role as a live performer. People wanted to hear what was on the recordings, and that demand could be met in no other way than to simply play them more recordings. That was the basic fact that could never be acknowledged. The fabrication of the Les Paulverizer gave Paul a role to play in live performances—he could press buttons and claim responsibility for the invention of the amazing Paulverizer, which did all sorts of astonishing electronic transformations, while in reality, it simply played back what was already there. But it did something else as well: If the original threat motivating the new sound came from George

Barnes's ability to mimic Les Paul, in the end, the Les Paulverizer made Les Paul into something of a mimic. In order to secure his identity, the fabrication forced Paul to ventriloquize himself. Miming along with a recording, the duo pretended as if the sound was being created spontaneously, but in actuality, Les and Mary lip-synched or added an additional part to a prerecorded track. Acting as both ventriloquist and dummy, the duo gestured along to a voice thrown onto recordings and back onto their bodies.

This is a disappointing conclusion. Yet if we draw our attention away from the problem of live performance and back to the confines of the recordings, perhaps a different conclusion emerges. For within the preserve of acousmatic boundaries, Paul's attempt to appropriate a sound that was "so much me" was far more successful. Although other singers in the 1950s made overdubbed recordings, the most similar being Patti Page, there remains something gripping—uncanny and unsettling—in Les Paul and Mary Ford's music.[28]

Despite all the overdubbing in Page's recordings—like "Confess," "Tennessee Waltz," and "Old Cape Cod"—there is little uncanniness. The recordings mimic natural acoustic space through their mixing and panning. The lush orchestral arrangements supporting Page's harmonies are balanced so that her voice stands out front and center, as in her recording of "Tennessee Waltz." Often, Page sings backup harmonies that contrast with a solo line, allowing the listener to imaginatively differentiate between distinct sounding bodies. The contrast of the solo voice against the chorus helps to diminish the potentially uncanny effect of overdubbing.

Paul's arrangements were radically different. Surrounding Ford's voice with an assortment of effects—half-speed recordings, echo, delay, and slapback—Paul placed the voice into a setting that did not reproduce the orthogonal axes of physical space. The voice was closely miked, with Ford singing quietly and directly into it, producing an aural image that lacked the virtual distance of Page's recordings. It is pulled from the body and onto the tape, panned and reverberated anywhere in the recording's space.

In Paul and Ford's version of the "Tennessee Waltz," two Mary Fords croon softly into the listener's ears—so close, in fact, that they seem to inhabit no acoustic space at all. Rather, they seem to sound from within the head of the listener. One loses all grasp of the direction from which the voice is emitted. Like an *acousmêtre*, Ford's voice comes simultaneously from everywhere and nowhere. Her nearly affectless presentation of the melody is similarly inhuman; it lacks all traces of what Roland Barthes identifies as "the grain of the voice," the audible quality of the "materiality of the body emerging from the throat."[29] In this recording, the palpability of Ford's body is all but absent, yet the voice remains, lingering like a spectral vestige. Underneath the voice, a metallic guitar plays tremolando, miming an ominous shudder. In contrast to Ford's voice, Paul's tinny background strumming is artificially reverberated and echoed, giving it a distant and lonesome quality. The entire setting is constructed to present something other than habitable, *heimlich* physical space. Rather, Paul and Ford's recording of the "Tennessee Waltz" presents a series of acoustical anomalies, designed to convey the loneliness and heartbreak of the lyrics through a sonic analogue of broken and evacuated interiority.

Some of the strange, multiplicative power of the Les Paulverizer is conveyed by an unpublished photograph from that era (see figure 6.3). Through the trick of multiple

Figure 6.3 A multiple exposure photograph of Les Paul and Mary Ford. Photograph by Arthur Rothstein, 1956. Courtesy of the Arthur Rothstein Archive.

exposures on the same photographic film, three Mary Fords are pictured singing in harmony, while one Les Paul strums along. (It should be noted that multiple exposure photography presents a good analogy to the layering of sound-on-sound that Paul and Ford pioneered.) The figures emerge from a black background that frustrates all sense of spatial orientation. In the center of the image, Mary lays her left hand gently on her husband's shoulder. Each Mary is differentiated by facial expression, head position, and posture. Yet her multiple bodies are somewhat less than corporeal. In the lower left of the photograph, a collision of arms links a pair of clasped hands; on the left side, a phosphorescent glow marks the intersection of one Mary's shoulder with another Mary's arm. Even the hand gently lying on Les's shoulder is ghostly, silhouetted against the absorptive black ground. The formal arrangement supports this sense of indistinctness. By overlaying multiple images in an ever-expanding pattern, Mary appears to be produced by Les, strummed into being, emanating from the sound holes of his guitar, and radiating outward like a sound wave toward the edge of the frame. Just as the ontological status of sound is difficult to determine—is it an event or an object?—Mary's image seems to inhabit a netherworld, the *intermundia*, insecurely fixed between the material and immaterial. These radiating figures, one intersecting the next, construct a singular topography of unnatural spatiality.

Compare this image with the cover illustration of *The New Sound*. If the earlier image conveys a feat of physical mastery, the latter captures the technologically tinged effect of the Les Paulverizer's phantom multiplications. Whereas the first image, a childish cartoon, keeps the question of source, cause, and effect in an abstract realm, "full of good humor," the latter image exploits photography's indexicality to produce an effect that is more troubling in its technological mediation of reality. Like those remarkable multiple recordings, in which voices and guitars emit sounds from no particular location and space is disoriented into a simultaneous near and far, the photographic space of the image is similarly dislocated—the unreal inversion of the lived space of the real world, while still claiming a hold upon that world. This silly little image is perhaps the best visualization we can have of acousmatic sound, one that truly preserves its acousmaticity. It manages to capture the ontological uncertainty that marks the effect of acousmatic sound—quasi-real, nearly-but-not-quite autonomous, and uncanny.

7

The Acousmatic Voice

> [The phonograph] can't record an eloquent silence, or the sound of rumors. In fact, as far as voices go, it is helpless to represent the voice of conscience.[1]
> —Auguste Villiers de L'Isle-Adam

§1. THE PHONOGRAPHIC VOICE

On any average day in 1906, you could walk into the doors of an Edison phonograph dealer and hear the commodity speak. When the stylus was dropped and its voice emerged from the horn, this is what it said:

> I am the Edison phonograph, created by the great wizard of the New World to delight those who would have melody or be amused. I can sing you tender songs of love. I can give you merry tales and joyous laughter. I can transport you to the realms of music. I can cause you to join in the rhythmic dance. I can lull the babe to sweet repose, or waken in the aged heart soft memories of youthful days.[2]

As the advertising cylinder spins on, the tone wavers. Its list of indispensible skills modulates into a plea for companionship: "I never get tired and you will never tire of me.... The more you become acquainted with me, the better you will like me. Ask the dealer." The power of the advertisement relies on our willingness to hear the voice emitted from the phonograph as if it were the phonograph's voice, imploring us to purchase it, take it home, and let it be our companion. At the same time, we are quite certain that the voice is not its own, that it emerges only with a cranked spring and properly placed stylus. We know very well...but nevertheless....

Philosophical Jokes and Linguistic Tricks

Before the talking machine, there was the writing machine. In the late 18th century, the horologist Pierre Jacquet-Droz made his name by designing ingenious watches and clocks that integrated mechanical singing birds, animated natural scenes, and musical elements. The feats of mechanical engineering that went into these inventions were impressive indeed, capturing the attention of European royalty. Yet they cannot compare with the three astonishing automatons he and his son built between 1769 and 1774. The androids, which are still on display today in Jacquet-Droz's home city of Neuchâtel, captivated the public's attention with their uncanny ability to imitate complex human actions. In the first, the Musician, a girl of about 10 years of age, sits on a stool and plays a harmonium; the music sounds as her fingers actually depress the keys. In the second, the Draughtsman, a young boy, sits and draws with a pencil on a sheet of paper, sketching rococo images of animals, portraits of famous persons like Louis XV and George III, and even a picture of Cupid riding a chariot pulled by a butterfly. It employs variations in pressure to create shading effects, drawing some lines stronger and some lighter, occasionally removing its hand to get a better view of its work.

The Scribe is perhaps the most fascinating automaton of the group. Seated at a small writing table, it looks like a young boy, only a few years old, with a mop of curly hair. Barefoot, dressed in breeches and a red coat with a cravat and ruffles on the sleeves, the child slowly and deliberately writes on a sheet of paper, pulling his pad along as he goes, occasionally refreshing his quill in a nearby inkpot. In the 1780s, an exhibition of the automata was held at Covent Garden, and an advertisement for the event contains a wonderful description of the Scribe that enumerates its various gestures:

This figure dips its pen in the ink, shakes out what is superfluous, and writes distinctly and correctly.... It places the initial letters with propriety, and leaves a suitable space between the words it writes. When it has finished a line it passes on to the next, always observing the proper distance between the lines: while it writes, its eyes are fixed on its work, but as soon as it has finished a letter or a word, it casts a look at the copy, seeming to imitate it.[3]

On the back of the Scribe, covered by its garments, a panel provides access to its central mechanism. Inside sits a removable disc, fitted with a special set of adjustable wedges that determine the characters to be written. This allows the android to be essentially programmed to output a variety of statements, as long as they are fewer than 40 characters in length—all that the disc can hold.[4] Nowadays, the Scribe typically writes short self-promotional snippets: "*Les automates Jacquet-Droz à neuchatel*," or "*Soyez les bienvenus a neuchatel*." Yet, of all the sentences the Scribe has written, one stands out. "I think," writes the automaton, "therefore I am." We might note that the success of this "eerie philosophical joke," to borrow a phrase from Gaby Wood, depends on two noteworthy features of the word "I."[5] First, the word is always an indicator of self-reference, whereby a speaker (or writer) designates him- or herself as the subject of the statement. Second, "I" is a signifier and, like all signs, it has the possibility of functioning in the absence of its referent. The written word "I" can be understood even when the subject who utters it is not present. Derrida refers to this feature of the written sign as its "iterability," noting that "the ideal iterability that forms the structure of all marks is that which undoubtedly allows them to be released from any context, to be freed from all determined bonds to its origin, its meaning, or its referent...."[6] Such iterability is made especially clear in the case of written language, where texts can circulate in the absence of their author's presence.

The Scribe exploits both of these features simultaneously. The child's inscription, "I think, therefore I am," designates the writer as the self-referential subject of the statement. At the same time, the iterability of the word "I" offers a sign of self-reference that functions regardless of the presence of its referent. The author that animates this act of writing seems to be absent, for we know that the automaton—this elaborate horological construction of cams, gears, springs, and flywheels—cannot actually be referring to itself. But presence is a tricky thing, for not only are *we* present at the act of writing so is the machinery that writes. We see the ink flow from the pen, yet we can only make sense of the self-reference of the "I" as simulated. The two features of the word "I," its self-referentiality and its iterability, reach a limit with the Scribe, for they cannot both be asserted at once. We either dismiss the sentence altogether, demoting the word "I" to the meaninglessness of an iterable signifier—one long vertical stroke or series of marks, depending on the language (*I, je, ego, ich*)—or we accept the unacceptable meaning of the sentence and take the "I" as indicating the self-reference of the automaton.

In linguistics, the word "I" belongs to a class known as "shifters." Otto Jespersen, in his 1921 volume entitled *Language: Its Nature, Development and Origin*, coined the term to designate a category of words "whose meaning differs according to the situation."[7] Jespersen included examples such as "father," "mother," and "enemy"—words that can pick out different referents depending on the context and the speaker—but noted that the most important class of shifters is found in pronouns (such as "I," "you," "we") and spatial and temporal adverbs (such as "now," "here," "there," etc.). Shifters can present challenges to children in the process of acquiring language since they may encounter the word "I" coming repeatedly from the mouths of various sources, such as their mother, father, uncles, aunts, grandparents, or siblings. Some people obviate the child's potential confusion by referring to themselves in the third person: "Mommy," "Daddy," or "Granny." Thus, there is a view that the acquisition of the use of "I" is a real achievement on the child's part. The Scribe might exemplify just such a case, a child who has not only learned to refer to himself but to write down philosophical propositions about himself in a neat, well-spaced, and legible script. Another example comes from the philosopher Fichte who, according to Jespersen, celebrated not his son's birthday, but rather the day upon which he first referred to himself as "I."[8] Jespersen does not share Fichte's deep philosophical investment in the constitutive role played by the "I" and "Non-I" in the formation of subject and object, slyly noting that "Germans would not be Germans, and philosophers would not be philosophers, if they did not make the most of the child's use of 'I,' in which they see the first sign of self-consciousness." Despite the various philosophical systems built on the achievement of such self-reference, Jespersen soberly assures us that "a boy who speaks of himself as 'Jack' can have just as full and strong a perception of himself...as one who has learnt the little linguistic trick of saying 'I.'"[9]

The Phonographic Voice

A little more than a century after the births of Jacquet-Droz's mechanical children, another very famous child learned to effectively harness the power of this "little linguistic trick." This child was the "latest born of Edison," the phonograph, which by 1888 had reached the ripe age of 11.[10] Not only had the phonograph matured since its invention, but Edison had begun producing machines that employed wax cylinders (or "phonograms"), which provided a more durable recording surface with better sound quality than the original tin sheets used in the phonographs of 1877.

In an article from 1888 titled "The Perfected Phonograph," Edison acknowledged that the phonograph "may still be in its childhood; but it is destined to a vigorous maturity."[11] To help ensure that goal, Colonel George E. Gouraud, Edison's principal overseas representative, masterminded a publicity campaign to reintroduce the new and improved phonograph to England.[12] Gouraud, a colorful character with a streak for theatrical salesmanship, demonstrated the "perfected phonograph" to the English public in the 1888 Handel Festival at the Crystal Palace and at more intimate gatherings at his house atop London's Beulah Hill, known as "Little Menlo."

Gouraud came prepared with a variety of spectacular gimmicks, including a recording of Wordsworth's "To a Cuckoo" (with the telling line "Shall I call thee a bird, or but a wandering voice?"), as well as cards that invited the public "To meet Prof. Edison / *Non presentem, sed alloquentem*! [Not in present, but in voice]"[13] One of the most fascinating recordings played for the English public was composed and recorded by Gouraud's brother-in-law, the poet and minister Horatio Nelson Powers, entitled "The Phonograph's Salutation."[14] Powers's poem, recited in a fluid elocutionary style, manages to succinctly capture many of the most salient and culturally powerful tropes of phonography: the power of the machine to reproduce all modes of human speech, to capture fugitive sound in a permanent inscription, to overcome the mediation of the sign by harnessing the living power of the voice, and to embalm and preserve the dead. Gathering them all together, "The Phonograph's Salutation" uses the "little linguistic trick" of the shifter to great effect:

> I seize the palpitating air. I hoard
> Music and speech. All lips that breathe are mine,
> I speak, and the inviolable word
> Authenticates its origin and sign.
>
> I am a tomb, a paradise, a throne;
> An angel, prophet, slave, immortal friend;
> My living records, in their native tone,
> Convict the knave, and disputations end.
>
> In me are souls embalmed. I am an ear,
> Flawless as truth, and truth's own tongue am I.
> I am a resurrection; men may hear
> The quick and dead converse, as I reply.
>
> Hail, English shores, and homes, and marts of peace!
> New trophies, Gouraud, yet are to be won.
> May "sweetness," "light," and brotherhood increase!
> I am the latest born of Edison.

Powers's poem is often invoked to underscore the uncanny power of the phonograph for the generation that witnessed its birth. Frances Dyson reads the poem as a demonstration of the "'haunted' nature of audio media" for its early audiences, which saw no contradiction in the phonograph's powers to both inscribe the soul and reanimate the dead.[15] John M. Picker describes the poem as an attestation to the "macabre power" of the machine, noting how the phonograph turns out to be a "self-contained contradiction"—at once master and slave, tomb and resurrection, organ of hearing and organ of speech.[16] Undoubtedly, the poem presents such uncanny contradictions. But the nature of that contradiction can be productively analyzed by pursuing the question posed by use of the shifter: Who is speaking? On first blush, the poem emphasizes the phonograph's technical powers to reproduce the voices of others: "All lips that breathe are mine." However, the conceit of Powers's verse is to present the personified phonograph speaking for itself, in its own *phonographic voice*.[17] Numerous times does the phonograph speak of itself through the use of shifters like "I," "me," and "mine." Such emphasis on the self-reflexive powers of the phonograph harmonized with Gouraud's theatrical sales pitch, which presented the newly perfected phonograph as something of a debutant, offering a "coming-out" or social introduction of the adolescent to the English public. In Gouraud's presentations, "The Phonograph's Salutation" followed a recording in which the phonograph spoke directly to the press, as a proxy for the absent Edison:

> Gentlemen, in the name of Edison, to whose rare genius, incomparable patience, and indefatigable industry I owe my being, I greet you. I thank you for the honor you do me by your presence here to-day. My only regret is that my great master is not here to meet you in the flesh, as he is in the voice. But in his absence I should be failing in my duty, as well as in my pleasure, did I not take this, my first opportunity, to thank you and all the press of the great city of London, both present and absent, for the generous and flattering reception with which my coming to the mother country has been heralded by you to the world.[18]

The use of the first person makes it seem as if the phonograph is addressing the public directly, showing off its skills in the arts of eloquence, flattery, and poetic recitation. The "little linguistic trick" of the shifter is employed to ascribe all kinds of subjective states to the speaking machine. It refers to its honor, pleasure, and duty; it acknowledges and respects its maker, Edison, to whom it owes its being. Consequently, it is not unlike Jacquet-Droz's writing automaton. In both cases, to understand the meaning of the machine's statements (whether spoken or written), we must take the "I" to be a sign of the machine's impossible self-reference. Yet the phonographic voice articulated in "The Phonograph's Salutation" and its introductory text exceeds the powers of the Scribe. It seizes the once fugitive aspect of the sonic signifier ("the palpitating air") and makes it iterable. The vocal sign could no longer be exempt from the condition of writing, for it became reproducible even in the absence of its speaker. This is not lost on Edison, nor on Gouraud, whose discourse and presentation to the English press constantly emphasized the play of presence and absence at the heart of phonography: from Gouraud's card ("Meet Prof. Edison, not *present* but in the voice") to Edison's writings ("reproduction...without the *presence* or consent of the original source") to the introduction to Powers's poem ("I thank you for the honor you do me by your *presence* here to-day...in [Edison's] *absence*...[I] thank you and all the press of the great city of London, both *present* and *absent*..."). Such play is exemplified in the phonographic voice, the recording that speaks of itself and, in doing so, demands the impossible. To understand its discourse, one must fulfill the absent self-referential subject of the "I" with a mechanical presence that confounds the very notion of the subject.

Perhaps the best way to explain the paradoxical phonographic voice is to situate it within the history and horizon of acousmatic sound, by showing its intelligibility in terms of the acousmatic spacing of source, cause, and effect. The effect is presented at playback, in the sound of the voice that says "I." Instead of hearing this effect as autonomous (like a sound object), the potential autonomy of the effect is challenged by the meaning of the word "I," which points equivocally back toward its source. No matter how emphatically the phonographic voice says "I," the source is underdetermined since "I" could refer to *either* the soul that animates the voice of Powers's recitation *or* the soul that animates the voice of the machine itself. As listeners to "The Phonograph's Salutation," we are quite willing to play the game of pretending that the source of the voice comes from the phonograph itself, that it speaks to us about itself from the depths of its own machinery. To understand the conceit of this poem, we are compelled to play along. The reproducing and recording machine, while remaining faithful to the utterance ("seizing the air" as if it could capture the voice without mediation), encourages the fantasy that it speaks spontaneously, *as if* it were the machine itself that produces language. Of course, this subjunctive aspect is all important, because the phonograph, at the moment it speaks, also displays its machinery to the viewer. The very ambiguity of the shifter, its indication of a source that it simultaneously underdetermines, affords such paradoxical fantasies.

The "eerie philosophical joke" of Jacquet-Droz's Scribe returns in the "I" of the phonographic voice. Yet the joke cannot simply be laughed off, for the possibility of a machine that speaks of itself poses a challenge to the philosophical identification of the voice with an animating soul or subject. Before the age of phonography, one might have been able to convincingly argue that the voice possessed a privilege over the written sign, that due to its animation, the voice could never truly resound separate from the presence of the speaker, from its source. Even in cases of the acousmatic voice, such as Borromeo's singing nuns, the underdetermination of the source of the voice opened up the imaginative possibility of substituting a choir of nuns for a choir of angels. Yet the introduction of the phonographic voice encourages two irreconcilable assertions. First, by making the vocal sign iterable and reproducible in the absence of the source, it reveals the vocal sign as a form of writing and disallows the voice to guarantee presence. Second, when the phonographic voice says "I," it emphasizes the *spacing*, and not the severance, of the voice from its source. Hence, the paradox: Exposing the iterability of the sign, its separability from specific contexts, and thus the vocal sign's autonomy, the phonographic voice simultaneously forces attention back toward the source of the voice by underdetermining it yet demanding that it be (impossibly) present.

What is the philosophical response to the phonographic voice? If the sound of the voice is no longer adequate to establish it as the marker of the soul's presence, to what can the philosopher appeal in order to secure the voice? This chapter attempts to sketch a response to this problem by outlining the philosophy of the voice in the wake of Edison. My account focuses on two aspects. First, I argue that the philosophy of the voice after Edison unfolds as a concatenated sequence of responses, wherein each philosopher introduces a new kind of voice in relation to the last: The *phonographic voice* will be contested by the *phenomenological voice* of Husserl; the *phenomenological voice* will itself be contested by Heidegger's discovery of an *ontological voice* in *Being and Time*; and, finally, the *ontological voice* will be reworked into a Lacanian *psychoanalytic voice*, as described in the work of Slavoj Žižek and Mladen Dolar. Second, I argue that each of these voices (*phonographic, phenomenological, ontological, psychoanalytic*) is deeply affected by the paradox of the acousmatic voice: the voice that speaks from an underdetermined source.

Before turning to those texts, I should note that only Žižek and Dolar explicitly speak of the "*voix acousmatique*," having discovered the term in the film theory of Michel Chion. However, the paucity of the word "acousmatic" in the philosophy of the voice should not be taken as evidence for its irrelevance. Given the rarity of the word before Schaeffer, one would not expect writers of the first half of the 20th century to be speaking explicitly of acousmatic sounds or voices. Yet the experience of the acousmatic voice, a speaking voice whose source remains underdetermined, remains central to Husserl and Heidegger's analyses. It is *that* experience, and the attempt to come to grips with it, that is my concern here. As I argued earlier, even under a different set of conditions, one in which the word acousmatic was never coined, we could still investigate a history of voices without sources. At the same time, the explicit appearance of the *voix acousmatique* in Žižek and Dolar's writings is not to be taken lightly. Once the term became widely available (via Schaeffer's *Traité* and Chion's film theory), it is not coincidental that Žižek and Dolar addressed it. Their use of the acousmatic voice is a response to the philosophy of the voice they inherited. Via Žižek and Dolar, an attentive reader can hear the echo of the acousmatic voice in the texts that preceded theirs.

§2. HUSSERL AND THE PHENOMENOLOGICAL VOICE

In 1901, thirteen years after Gouraud played "The Phonograph's Salutation" for the British public, Husserl published the *Logical Investigations*. Consistent with much of Husserl's later writings, the *Logical Investigations* was focused on providing a philosophical critique of naturalistic and psychologistic theories of logic. The work is a set of "investigations" that include various topics, such as semantics, the theory of signs, mereology, grammar, judgment, and consciousness. To this list, we could add "the voice," discussed in Husserl's First Investigation on "Expression and Meaning." Jacques Derrida, in *Speech and Phenomena*, offers a close reading of this investigation. Derrida argues that Husserl's theory of the voice operates within a long tradition of the metaphysics of presence, that is, the philosophical tradition of granting a privileged status to presence as a source of truth, goodness, and value. Husserl conceives of the voice as a medium for presence that, in contrast to writing, guarantees the meaningfulness of spoken language. But, as Derrida argues, Husserl's theory of the voice fails to live up to this metaphysics of presence, for the condition of the possibility of all signs, spoken and written, is iterability; thus speech cannot ultimately be differentiated from writing in terms of presence.

Although my reading of Husserl is influenced by Derrida's account, I approach the problem of the Husserlian voice from a slightly different angle. My aim is to demonstrate how Husserl's account of the voice—which I refer to as the *phenomenological voice*—can be understood as a philosophical response to the phonographic voice. In particular, Husserl grounds his theory of the voice in the spontaneity and intentionality of the living, speaking subject and in the immediacy of the auto-affective circuit generated between the speaking tongue and the listening ear. These features are unavailable to the phonographic voice, which can only simulate the spontaneity, intentionality, and auto-affectivity of live speech. By making these features into voice-defining traits, Husserl ultimately transforms the voice into something non-sonorous, unable to be phonographically recorded, but not exactly inaudible. To show why this is the case, we must delve into the details of Husserl's account.

Expression and indication

Husserl begins the First Investigation by noting that the term "sign" is ambiguous. It covers two distinct concepts: *indication* and *expression*. Indications are signs that point beyond themselves toward something else. For example, a brand on livestock indicates which ranch the animals belongs to; a flag on a ship indicates a country of origin; a canal on Mars indicates the presence of water, or of intelligent life; a knot in a handkerchief acts as a reminder to do some task. In each case, the sign operates as a site of transfer. The person who encounters the indicative sign is led from the sign itself, present to the observer, toward some other state of affairs not immediately present. Husserl defines the essence of the indicative sign as follows: "that certain objects or states of affairs *of whose reality someone has actual knowledge* indicate to him *the reality of certain other objects or states of affairs*, in the sense that *his belief in the reality of the one is experienced... as motivating a belief or surmise in the reality of the other.*"[19] Thus the meaning of an indicative sign is exhausted in its role as a detour or transfer.

To clarify the nature of the indicative sign, Husserl takes up the case of so-called facial expressions that often accompany speech, noting that such gestures must be classed as indications and not as expressions. When I am conversing with someone, I see their gestures and I can make inferences about their inner states and feelings, but those inner states and feelings are not directly presented. The gestures act as detours or relays in that they lead me from an actual state of affairs (these particular facial gestures I see before me) to a presumed state of affairs (my interlocutor's inner states and feelings, which I cannot see). Gestures and facial expressions fail to be expressions, in Husserl's sense of the term, because they are "not phenomenally one with the experiences made manifest in them in the consciousness of the man who manifests them, as is the case with speech." Through them, a speaker does not intend to "put certain 'thoughts' on the record expressively," and thus, gestures and facial expressions "have properly speaking, *no meaning.*"[20]

However, Husserl's definition of an expression might be too strict for its own good, in that his criterion would classify the words we speak to each other as indications and not expressions. When I am listening to someone speak, aren't the words that he or she utters vocal signs that lead me to infer, on the basis of one state of affairs (the words presently being spoken), another state of affairs (the inner states or intentions of my interlocutor)?

Husserl agrees with the characterization that discourse spoken to another is indeed indicative, but with qualifications. When a listener hears another's speech, "he takes the speaker to be a person, who is not merely uttering sound but *speaking to him*, who is accompanying those sounds with certain sense-giving acts, which the sounds reveal to the hearer, to whose sense they seek to communicate to him."[21] In this respect, spoken words qua *spoken* are indications. At the same time, the discourse itself is still expressive—there are sense-giving acts that animate the speaker's discourse, and thus meanings that are expressed by the speech. As Husserl puts it, "all expressions in *communicative* speech function as *indications*."[22] Although we might have originally assumed that expressions and indications are mutually exclusive, this is not the case. Rather, the criteria for distinguishing an indication from an expression are different: An indication is determined by whether or not a sign points beyond itself to some other state of affairs; an expression is determined by whether or not a sign is meaningful, that is, whether the expression is "phenomenally one" with the experiences manifested in it. To clarify the difference, it would be useful to isolate an expression from an indication. Yet, if "all expressions in *communicative* speech function as *indications*," then where is the proper place to locate an expression that is not *also* functioning as an indication? When is one being *expressive* but not at the same time *communicative*? When can expressions be isolated from indications *altogether*?

Husserl provides a clue at the beginning of the First Investigation. He notes that "*expressions* function meaningfully even in *isolated mental life, where they no longer serve to indicate anything.*"[23] Only in the internal soliloquy that accompanies one's mental life do we find an example of non-indicative expressions. There, in solitary mental life, expressions continue to function just as they do in communicative discourse, only without *also* operating as indications for an interlocutor. When I soliloquize, I have no need to communicate anything to myself. Thus the "vocal" signs I employ are not indicative: I am not led to infer the existence of one state of affairs (my own mental state) based on the presence of another state of affairs (my internal soliloquy).[24] Solitary mental life is wholly expressive, and thus distinguishes and separates the indicative and expressive strata of language.

With indication reduced, Husserl clarifies the nature of the expressive sign by differentiating the "physical phenomenon" of speech—that is, the sound of the spoken words or the look of written words—from the mental "acts" that give the spoken utterance its "meaning." These mental acts come in two different kinds, sense-giving acts and sense-fulfilling acts, or what Husserl calls meaning-intentions (*Beduetungsintentionen*) and their fulfillment. Every meaningful utterance is animated by a meaning-intention that is related to (or directed toward) an intentional object. Thus every meaning-intention is paired with an intuition of an intentional object that can, potentially, fulfill it. All intentional objects are presented to a subject as intuitions. A meaning-intention is fulfilled depending on whether the intuition of the intended object is present or not. Whether the intentional objects are actually perceived or simply imagined does not matter for Husserl; in either case, they are presented intuitively, albeit in different modes of presentation.

To illustrate how Husserl's theory works, imagine the following: You are having a conversation about the city of Juneau, Alaska, although you have never been there or even seen pictures of it. You might think of the city of Juneau without having a real or imagined intuition of the actual city in mind. In that case, you would simply have an empty meaning-intention of the city; it would still be meaningful, but "unfulfilled" because it is lacking an intuition of the object intended. However, you could visit the city or see a photo of it, and those new intuitions would, by degrees, begin fulfilling the empty intention. Husserl describes this process:

> A *name*, e.g., names its object whatever the circumstances, in so far as it *means* that object [insofar as it is animated by a meaning-intention aimed at that object].

> But if the object is not intuitively before one...mere meaning is all there is to it. If the originally *empty* meaning-intention is now fulfilled, the relation to an object is realized, the naming becomes an actual conscious relation between name and object named.[25]

The passage provides a description of the relationship between a meaning-intention and its intuitive fulfillment. There are many instances when I can speak (or understand speech) about things with which I am not personally acquainted. My speech still has a meaning even if I lack an intuition of the object that I intend. However, the meaning-intention must always be present, for there is no act of speech that lacks a meaning-intention or is not directed at an intentional object that could fulfill it. If that were to happen, then, for Husserl, my speech would not actually be meaningful—it would be nonsense.[26]

There is a further consequence to Husserl's theory. Since the essence of the expressive sign depends only on the relationship between meaning-intentions and the intuited intentional objects at which they aim, the actual existence of these objects—just like the ontological status of the sound object in Schaeffer—is irrelevant. In writings subsequent to the *Logical Investigations*, the inexistence of the intentional object is preserved. In §49 of *Ideas*, Husserl provides an astonishing illustration. In order to clarify why the phenomenological reduction brackets all positing of existence to objects, thus reducing the sphere of phenomenological investigation to the immanent contents of consciousness and their transcendental structures alone, he contemplates the destruction of the world.[27] "*While the being* [i.e., the actual existence] *of consciousness...would indeed be necessarily modified by an annihilation of the world of physical things its own existence would not be touched.*"[28] The immanent contents of consciousness, Husserl's privileged objects of study, do not require *any actual* worldly manifestation. They are absolutely indifferent to the existence of the exterior world.

The same argument holds for the First Investigation. As Derrida observes, Husserl's analysis of the meaning of speech does not require an actual sounding voice or external listener. The genuine meaningfulness of an expression resides "not in the sonorous substance or in the physical world, [nor] in the body of speech in the world." The actual sonorous voicing of an expression is irrelevant to the expression's meaning. Whether actually sounded or voiced in internal soliloquy, the only necessary condition is that a meaning-intention animates one's speech and directs it toward some potential intuitive fulfillment. Husserl's interest is not in the sonorous speaking voice but in "the voice phenomenologically taken, speech in its transcendental flesh, in the breath, in the intentional animation that transforms the body of the word into flesh.... The phenomenological voice would be this spiritual flesh that continues to speak and to be present to itself—*to hear [entendre] itself*—in the absence of the world."[29]

Inscription and Ideality

While "the annihilation of the world" reduces the materiality of the sign as inessential, Husserl is acutely aware of the pragmatic and communicative role of the actual physical signifier, whether in the form of sonorous speech or written letter.

Only through the use of an actual physical signifier can meaning-intentions be made available for others. I must inscribe my thoughts in some way—in speech, in writing, in a figure—in order for someone else to be able to encounter them. But it is not the actual inscription that matters; rather, my interlocutor approaches an inscription in such a way that he or she can reactivate the intention that animated it.

This is an issue that Husserl addressed at various times in his career—and it is a touchstone for Derrida's work on the problem of writing. In the *Logical Investigations*, Husserl analyzes the relationship of inscription and intention in §10 of the First Investigation. Say, for instance, I am handed a sheet of paper with a name written on it by a friend. When we regard the note, the written word "remains intuitively present, maintains its appearance," yet we do not attend to it *as a set of written marks*. Rather, "our interest, our intention, our thought... point exclusively to the thing meant in the sense-giving act," that is, to the person intended by our friend. For Husserl, this means, "phenomenologically speaking, that the intuitive presentation [i.e., the physical inscription]... undergoes an essential phenomenal modification when its object begins to count as an *expression*. While what constitutes the object's appearing remains unchanged, the intentional character of the experience alters."[30]

The utility of the physical inscription is due to its permanence; it allows for a meaning-intention to be exteriorized and preserved. There would be no archives without inscriptions. But, as Husserl notes in his later work on the *Origin of Geometry*, there is a danger involved in writing. Meaning-intentions, which animate all forms of inscription, might not be fully recovered when we encounter an act of inscription. Since all inscriptions are indications, they can only redirect a reader or listener back toward the meaning-intentions that animated them (and the ideal objects that are being intended), but they cannot *guarantee* the full recovery of prior meaning-intentions. Since there is no guarantee that my inscription will make the object I intend intuitively present to a reader or listener, there is always the threat of a loss of fidelity, a danger that my intentions will not be fully reactivated by my interlocutors. Yet, without forms of inscription, it would be impossible to communicate my meaning-intentions to another at all. There is no other route to communication other than (potentially) inadequate forms of inscription.[31]

Thus inscription is a necessary but insufficient condition for the communication of meaning. Contrastingly, in solitary mental life, we are not communicating anything to ourselves; there is no inscription in which we need to lodge and recover meaning. The danger of exteriority, of inscription, of indication is bracketed away. We remain within a purely internal space, where our expressions can be "phenomenally one with the experience [i.e., the meaning-intentions] made manifest in them...."[32] We can now see what Husserl was aiming at in his first, rough characterizations of indication and expression. Only in solitary mental life can an intuited intentional object be immediately and fully present with the meaning-intention directed toward it. The physical aspects of the inscription, the sonorous or written signifier, fall away completely. Derrida summarizes the situation nicely:

> While in real communication existing signs *indicate* other existences which are only probably and mediately invoked, in monologue, when expression is *full*,

non-existent signs *show* significations (*Bedeutungen*) that are ideal (and thus non-existent) and certain (for they are present to intuition). The certitude of inner existence, Husserl thinks, has no need to be signified. It is immediately present to itself. It is living consciousness.[33]

In everyday communicative acts, expressions must traverse the medium of the signifier, functioning as the middle term that links a speaker with a listener. The meaning-intentions transmitted must always undergo a passage to the exterior, through the signifier, in order to be recovered. Recovery is fraught with infidelity, because the meaning-intention may not be correctly transmitted through the signifier. However, in solitary mental life the signifier has no role to play. It is purposeless because I require no middle term or mediator, being simultaneously both speaker and listener.

Enter the *phenomenological voice*. In internal soliloquy, the inner voice—the phenomenological voice—operates like a medium that affords the connection of the speaking tongue and the receiving ear without distorting the signal. It is a fantastic medium that mediates perfectly, leaving no trace on the message it transmits. It binds together meaning-intention and fulfilling intuition without sacrificing fidelity. To use a recent coinage, one might say that the phenomenological voice is "lossless." This is in contrast to the sonic or written signifier, where its materiality makes it an imperfect medium, trading fidelity for inscriptive permanence. What grants the phenomenological voice this power? The answer is its inexistence, its lack of dependence upon any worldly fact or thing, a status that the phenomenological voice shares with meaning-intentions and fulfilling intuitions. As Derrida argues, an intentional object can be repeated infinitely because it is free from all worldly spatiality; "it is a pure noema that I can express without having...to pass through the world."[34] The same could be said for the phenomenological voice. Unlike spoken or written speech, where the material signifier and the ideal objects that are indicated are of different orders, the phenomenological voice is freed from all worldly spatiality or exteriority. It too does not "pass through the world." In solitary mental life, all parts of the transmission (sender, message, medium, and receiver) are ideal. There is simply nowhere in the system to lose fidelity.

The Phenomenological Voice

How does the phenomenological voice compare with the phonographic voice? First, the phonographic voice places a premium on the voice's sonorousness by capturing its sound in detail. The perfect recording is, ideally, a form of inscription without loss of sonic fidelity. It is a fidelity to the sonic signifier, not to its inexistent intention. Yet the fantasy of the phonograph often includes a transmutation: Extreme fidelity to transcribing the very materiality of the voice captures, at the same time, the very soul that animates the voice. In this fantasy, the phonographic voice is more than a recorded inscription; it inscribes something "real" about the voice. This trope—the transmutation of the inscription of the voice into the inscription of the soul—appears in "The Phonograph's Salutation" when the phonograph intones both sides of the equation: "I seize the palpitating air" becomes "In me are souls embalmed." In fact, a good materialist might claim that the soul *is* nothing more than the materiality of the voice.

In contrast, the phenomenological voice would dismiss this fantasy. No matter how faithful a recording might be, it can only be faithful to the "physical phenomenon" of the voice, not the meaning-intentions that animate its speech. One might argue that the phonographic voice is nothing more than a very detailed form of writing, different from the letter only in degree, not in kind. For Husserl, the true test of distinguishing the voice's meaning-intentions from its physical phenomenon requires reduction; only in the interiority of silently speaking-to-oneself is the expressive stratum of the speech act isolated. The phonographic voice is doomed to be a voice of communication since the only kinds of signs that it can produce are indicative signs, ones that transfer a listener from one state of affairs (the sounds emitted from the horn of the phonograph) to another (the supposed meaning-intentions that animate the discourse). This *necessarily* indicative function of the phonographic voice can be exploited to make it seem *as if* the phonograph has the power of spontaneity, as if the words it speaks indicate its power to intend. This occurs in "The Phonograph's Salutation," especially at the moments when the phonographic voice employs the shifter.

Second, considering the ease with which a listener can entertain claims about the phonograph's intentions (it greets, it flatters, it recites), one might view Husserl's appeal to the silent soliloquy of the subject—and the tremendous privilege given to the act of silent mental "speech" over communicative speech—as a way of holding at bay the uncanny simulation of intention found in the phonographic voice. There is in Husserl's project an attempt to locate something utterly resistant to the technological simulation of human activities. The lingering consequences of the philosophical joke, presented in Jacquet-Droz's Scribe and revisited in the "I" of the phonographic voice, could be dismissed with a single blow by revealing both cases as elaborate forms of inscription, nothing more than a bevy of indicative signs. All the media machines in the world could not reach the stratum where expression is to be found—expression, which grounds the possibility of all communication and thus the very condition of the possibility of all media machines. That expressive stratum can only be securely located in solitary mental life. In contrast to the actuality of a mechanically reproduced voice that could resound in the absence of a speaker, Husserl asserts that the meaning of the voice is found precisely elsewhere than its sound, in some necessarily un-inscribable mark of the human, untouched (and untouchable) by machinery. The spontaneity of intention perfectly fits this bill. Perhaps Husserl's investigation into the voice is simply a philosophically modern way of reasserting a very old idea: that the soul is the essence of the voice.

Third, if the essence of the voice is the soul, then Husserl can assert that its sound is ultimately irrelevant. Throughout the First Investigation, the emphasis is never on the sounding of the voice per se, but on the circuitry between the voice and the ear. As Derrida shows, Husserl privileges the voice as a medium over other sonic or written media because the voice permits one to speak and hear oneself at the same time. The subject who speaks and hears what he/she says is simultaneously expressing and affecting him/herself, a feedback loop that ideally transmits and receives all at once.[35] The phonograph is incapable of this feat, for it cannot both record and playback at the same instant. The inscription on the wax cylinder must precede its representation. Although "The Phonograph's Salutation" may claim that "I am an ear, flawless as truth, and truth's own tongue am I," it cannot be both ear and tongue simultaneously. That is a power that only a subject possesses. No phonographic feedback loop can simulate the loop of the subject, for it will always trace the same groove—first ear, then tongue.[36]

The feedback loops explored in the history of electronic music always involve a moment of temporal lag between the sound output and its input. This lag might be very small, simply the time required for an electronic signal to travel the length of a patch cord or for air compressed by the cone of the loudspeaker to move a microphone's transducer. That little temporal lag, when the closed circuitry opens if only for a moment to something exterior, is crucial in any feedback system. Derrida's deconstructive critique of Husserl's phonocentrism is explicable in these terms. In essence, Derrida argues that the closed system of speaking-to-oneself depends on a moment of temporal lag or delay that Husserl ignores. Derrida employs Husserl's own understanding of the "living present" from his lectures on time consciousness against Husserl, in order to deconstruct the instantaneity of speaking and hearing oneself in the First Investigation. For Derrida, there is always a delay between these two events; the living present always requires the presence of protention and retention, that is, something outside itself in order for it to be itself, an internal interval that it cannot include. Derrida describes this paradoxical interval under the term spacing (*espacement*).[37]

But rather than pursue that critique of Husserl further, another route is available that leads back to the problem of the shifter and the acousmatic voice. As we know, Husserl's First Investigation sets out to draw a strict distinction between the indicative and expressive sign. Yet the shifter ultimately poses a challenge to this project. In §26 of the First Investigation, Husserl addresses the shifter explicitly. He observes that "Each man has his own I-presentation (and with it his individual notion of I) and this is why the word's meaning differs from person to person." Each individual has a unique intuition of himself or herself, an "I-presentation" that would presumably act as the intuitive fulfillment of any utterance in which the speaker uses the word "I." According to Husserl, this I-presentation must be capable of unification with the meaning-intention that points toward it if it is to be a meaningful utterance. This is precisely what happens in the internal soliloquy: "In solitary speech the meaning of 'I' is essentially realized in the immediate idea of one's own personality, which is also the meaning of the word in communicated speech."[38]

But in communicative speech, as opposed to solitary speech, a problem emerges. "...Since each person, in speaking of himself, says 'I,' the word has the character of a universally operative indication of this fact."[39] That is, each person uses the word "I" to refer to himself or herself when speaking to others (or even speaking to oneself in solitary mental life), but Husserl notes that this word has the character of an indicative sign. Why? "Through such *indication* the hearer achieves understanding of the meaning, he takes the person who confronts him intuitively [that is, in person] not merely as the speaker, but also as the immediate object of the speaker's speech."[40] When I hear someone say "I" to me, I take this as a sign that the speaker is intending to refer to himself or herself. However, the intuition that could fulfill this intention is unavailable to me since I have no access to the speaker's I-presentation. "The word 'I' has not itself directly the power to arouse the specific I-presentation.... It does not work like the word 'lion' which can arouse the idea of a lion in and by itself. In its case, rather, an indicative function mediates, crying as it were, to the hearer, 'Your *vis-à-vis* intends himself.'"[41] The word "I" does not intend an ideal object, and this makes it different from the word "lion," whose intentional object is intuitively available to all. The sign "I" can only function as an indication; it is a clue for another that a speaker intends himself.

This leads Husserl to note that "we should not suppose that the immediate presentation of the speaker sums up the entire meaning of the word 'I.'... Undoubtedly the idea of self-reference, as well as an implied pointing to the individual idea of the speaker, also belong... to the word's meaning."[42] The meaning-intention of the word "I" should not be understood as being utterly fulfilled by an intuition of the speaker's I-presentation, but that part of its meaning-intention includes an aspect of self-reference. When I say "I," I am not only pointing out myself, but including that self-referentiality of the gesture as part of its meaning. Thus indication, relation, and detour are all built into the intention. Strangely, the strata of expression and indication are confounded, since the expressive sign "I" is an expression of indication. The same could be said about other shifters, like "here," "there," "above," "below," "now," "yesterday," and such. All of these words must be understood ultimately as expressions of indication, in that they cannot directly arouse the specific intuition or object that is being intended—rather, they can only operate as a way of saying that "your *vis-à-vis* intends" the surrounding environs, the area above or below, the temporal present, a duration that has passed, and so forth. To coin a paradoxical term, they are *indicative expressions* or *expressive indications* that tell another that a subject is intending something, yet they cannot explicitly say what is being intended—since they intend reference to an intuition that is absolutely unavailable to others.

Husserl is forced to acknowledge that "an essentially indicating character naturally spreads to all expressions which include these and similar presentation as parts: this includes all manifold speech-forms where the speaker gives normal expression to something concerning himself, or which is thought of in relation to himself. All expressions for percepts, beliefs, doubts, wishes, fears, commands belong here."[43] At the heart of speech, at the very moment where we articulate "I, now, here," we encounter a detour of unfulfillable indication. If we apply Husserl's argument to the statement "I think therefore I am," a dilemma arises. Speaking such a statement in solitary mental life results in a successful but purposeless act; the I-presentation is successfully paired with its intuitive fulfillment. However, the word "I" also includes an indicative component and, as Husserl argued earlier, in isolated mental life, indications serve no function.[44] For the indicative aspect of the word "I" to serve a purpose, the sentence must be understood as a communicative act and not an internal expression. However, when the sentence is uttered to another, its meaning will necessarily be empty because the speaker's I-presentation is a priori unavailable. If we grant the indicative aspect of the sentence a purpose, the expressive aspect of the sentence remains unfulfilled; yet, if we fulfill the expressive aspect of the sentence, the indicative aspect becomes purposeless.

This dilemma reveals just how significantly the phenomenological voice is affected by the acousmatic voice. The spacing of source, cause, and effect characteristic of acousmatic sound is made perspicuous by the fact that the sonic effect underdetermines attributions of the source or cause. When discussing the acousmatic voice, we might adjust the terms slightly and say that the underdetermination of the source by the voice reveals the structural spacing of the voice and its source. That spacing, rather than encouraging a reduction of the voice *either* to the status of an autonomous entity *or* to the physicality of its source, makes the voice into a site of endless detour or reference. The acousmatic voice directs the listener toward the absent presence or present absence of the source, without ever allowing the completion of that passage. Analogously, in Husserl's case, the shifter "I" functions as an *indicative expression* or *expressive indication*—indicative, in that it points the listener toward an I-presentation, and yet incapable of completing the passage of that indication, since the listener can have no access to the speaker's I-presentation. The most I can know with certainty is that someone is speaking about himself or herself, but lacking access to their I-presentation, I can never definitively find out who. This is the moment when the phenomenological voice is shaken by an insistent question from the acousmatic voice: "Who speaks?" This has strong consequences for Husserl's attempt to define the phenomenological voice. If we follow Husserl strictly, we are left with this unhappy result: The only voice that is not acousmatic is one's own.

Perhaps securing one voice at all is no small achievement. Yet the collateral damage is tremendous. Husserl is forced to take recourse to a drastic solution: a silent, wordless, solitary voice, existing only in the interior space (i.e., for Husserl, without spacing) between silent vocalization and silent hearing where the intended-I and the intuited-I would coincide. By the end of his philosophical career, Husserl would come to see that secure interior space put under intense scrutiny. Indeed, how certain can we be that even this silent inner voice is expressing *itself*? Can one be certain that the loop between tongue and ear, which survives even in the face of the annihilation of the world, is as closed and ideal as Husserl demands? Can one be certain from where this silent voice is emitted? Is it possible to hear the sound of the silent inner voice acousmatically? This is what Heidegger, Husserl's most famous pupil, will demonstrate—even against his own intentions.

§3. HEIDEGGER AND THE ONTOLOGICAL VOICE

A Phenomenology of the Natural Standpoint

Methodologically, Husserlian phenomenology begins when we suspend our naïve views of the world and our naturalistic ways of understanding mental life by undergoing the test of the *epoché*. Only when we bracket this "natural standpoint" do we locate a presuppositionless basis on which to begin philosophical inquiry. But many philosophers since Husserl have contested this beginning move, arguing that the attempt to understand lived experience is already sullied the moment the natural standpoint is bracketed. Heidegger, Husserl's most important student, is among this group. Rather than beginning with the reduction of the world, Heidegger, in *Being and Time*, uses Husserl's philosophical methods against his teacher to provide a phenomenological analysis of the natural standpoint. He describes and analyzes the ways in which we are always already involved with the world around us, in order to arrive at an undistorted phenomenology of our lived experience.

One of Heidegger's first discoveries is that in the natural standpoint, we do not by and large encounter "mere things"; rather we encounter things that matter with respect to our goals and projects. The things that we encounter, and the way that we encounter them, are experienced in relation to what we are doing or what we are concerned with. Heidegger designates these things as "equipment," intending the term in a sense that is broad enough to cover anything that is useful for some human agent—tools, materials, toys, clothing, dwellings, etc. When properly functioning, equipment requires a context that includes other pieces of equipment. Objects qua equipment are primarily experienced within a network of other objects related according to the purposes to which they are used, not as independent substances with properties. That discovery leads Heidegger down a certain path of thought: If what something *is* (i.e., its essence) depends on its role in some given context, then that thing can have no independent essence of its own. So, to take an example from Heidegger, if what it is to be a hammer depends on being related in appropriate ways to nails, boards, carpenters, furniture, and so on—that is, other pieces of equipment—then hammers not only have no essence independent of their functional role, they have no essence independent of the existence of actual hammers, nails, boards, etc.

To clarify Heidegger's position, we can recast it in relation to the analyses of chapter 1 concerning Schaeffer's sound object and his various modes of listening. Schaeffer's method, modeled on Husserlian phenomenology, was to undergo a series of reductions and imaginative variations in order to arrive at the essence of a sound, to hear it as a sound object. *Entendre* is the mode of listening most appropriate to hearing the sound object, since it strips away indexical and indicative significations. In contrast, Heidegger argues that "'initially' we never hear noises and complexes of sound, but the creaking of the wagon, the motorcycle. We hear the column on the march, the north wind, the woodpecker tapping, the crackling fire."[45] In the natural attitude, we attend to the sources of sounds. We listen in the mode of *écouter*. Detaching the sound from the source in order to discover its essential qualities is a very different act from our commonplace mode of listening to sounds. "It requires a very artificial and complicated attitude in order to 'hear' a 'pure noise.'"[46] The difference between hearing a motorcycle and hearing a pure noise is that of hearing a piece of equipment, which is guided by our routes of interest and projects, versus hearing a "thing" or "entity." This is often referred to as the difference between something that is, respectively, "ready-to-hand" (*Zuhandenheit*) and "present-to-hand" (*Vorhandenheit*).

Since we first hear the motorcycle and not "pure noise," this gives us "phenomenal proof" that in our everyday dealings with the world, we are always already "*together with* innerworldly things [ready] at hand and initially not at all with 'sensations' whose chaos would first have to be formed to provide the springboard from which the subject jumps off finally to land in a 'world.'"[47] It would be mistaken to think that our initial relation to a sound is to hear it as a sensation or an effect, and then work our way back to the cause or source by inference. That view would promote a distorted picture of how we relate to the world—for it would make our world appear secondary, as if it were something that was constructed by, first, encountering uninterpreted things, second, giving them interpretations and, finally, relating them all together to make a world. For Heidegger, our everyday dealings with the world reveal that the world is always already given; it comes first, and we only learn to abstract and reify it later.

This little sonic example can be expanded, even taken as exemplary for Heidegger's project in *Being and Time*. Heidegger's phenomenological discoveries about the "natural standpoint," or everyday naïve experience, cannot be reconciled with Husserl's claim that phenomenology is a rigorous science of essences that remains after the "annihilation of the world." Heidegger's research leads him to a decidedly different claim about the relationship of essence and existence, namely, that what a thing *is*—its essence—is dependent on its existence, on its position and operation within a whole network of use. Contrast this with Husserl's programmatic claim that "Pure phenomenology as science, so long as it is pure and makes no use of the existential position of nature, can only be an investigation into essences."[48] This is exactly what Heidegger cannot abide, because he has discovered that existence is an essential part of the essence to be investigated. Existence and essence are not separable because the essence of a thing is distorted when its existence is bracketed.[49]

The Introduction of *Dasein*

Heidegger's project in *Being and Time* is even more radical than simply offering a phenomenology of the natural standpoint. Just as we have a distorted picture concerning beings because we have thought about them as substantial (as present-at-hand) rather then understand them in terms of their everyday functioning (as ready-to-hand), Heidegger argues that we also have a distorted view about ourselves. For the traditional philosophical view of the subject—the view that is often referred to as the "Cartesian subject"—is that, like all other entities, it is a substance imbued with certain properties. For Descartes, the subject is both a "thinking thing," a substance with the power of thought, as well as an "extended thing," a substance with the property of physical extension, the occupation of space. Human beings are understood as an amalgam of these two substances. For Heidegger, this substantialist view about the subject is wholly determined by a traditional philosophical prejudice for understanding beings as present-at-hand. But if we were to consider ourselves as we are in our everyday actions, we would have a very different view. We would not understand ourselves primarily as substances imbued with properties, but as beings that are involved with our everyday world, pursuing various projects, concerned about the future, and so forth.

Instead of using the traditional philosophical term "subject" to name the kind of beings that we are, Heidegger chooses another term, *Dasein*. In German, *Dasein* means "everyday human existence."[50] For Heidegger, *Dasein* is used as a technical term to designate the kind of beings that we, as inquirers, are. Much of *Being and Time* is given over to exploring the various modes of *Dasein*'s being, that is, the various ways in which *Dasein* will encounter the world in its everydayness and what kinds of possibilities are available. It is important to note that *Dasein*, although not separated from the world of beings that it encounters, is a being of a special sort. For unlike the equipment that it uses, *Dasein* exhibits a unique characteristic, namely, that *Dasein* "is ontically distinguished by the fact that, in its very Being, that Being is an *issue* for it."[51] As *Dasein*, we are uniquely interested in the ways that we encounter the world, the ways we interact with others, and the ways in which the projects we pursue have meaning and value for us. Moreover, it must be emphasized that *Being and Time* is not a philosophical anthropology; it is not an account of the various ways that particular *Daseins* are involved in their worlds. Heidegger is not really interested in describing particular cases of a *Dasein*'s involvement with particular beings. He designates that level of investigation with the term *ontic*. Rather, Heidegger is interested in the ways that those individual possibilities are structured; that is, he seeks to disclose the necessary structures or modes of *Dasein*'s being. This level of investigation is designated as *ontological*. For instance, *Dasein* is always being-in-the-world, in the sense that no matter what individual projects individual *Dasein* might follow, that is, what ontical projects it may pursue, there is no escaping the fact that those projects are always already defined and pursued within a world that is given ahead of time. Being-in-the-world is an ontological structure of *Dasein*.

Being-in-the-world can be further specified. From the analysis of equipment, we have already seen that *Dasein* is always already absorbed in activity. The useful things (equipment) that it encounters are determined on the basis of *Dasein*'s involvement in various projects and goals, and its commerce in the world. *Dasein*'s commerce with the world is not simply reducible to its involvement with equipment, but also includes its involvement with others. We are never alone in the world; we are always in the world with others. Heidegger calls this being-with [*Mitsein*], an entailment of the ontological structure of *Dasein* as being-in-the-world. In our everyday commerce with others, we get involved in the projects that others have started or with situations that others have found captivating. Just as objects are never simply present-at-hand, but ready-to-hand in the sense that we are concerned in our dealings with them, we also never relate to others as if they were simply things present-at-hand. In our dealings with others, we exhibit solicitude [*Fürsorge*], meaning that we care about our relations with others. Unlike Descartes or Husserl who follow the traditional philosophical strategy of first describing *my* world and then moving outward, on that basis, to encompass *the* world, Heidegger's philosophy does not downplay the significance of shared social norms, values, and projects in shaping the world that *Dasein* experiences.[52] In *History of the Concept of Time*, Heidegger writes,

> In order to give a more accurate portrayal of the phenomenal structure of the world as it shows itself in everyday dealings, it must be noted that what matters in these dealings with the world is not so much anyone's own particular world, but that right in our natural dealings with the world we are moving in a common environmental whole.[53]

Yet, when considering shared values, norms, projects, and roles, one cannot definitively ascribe the existence of any of these shared features to some particular causal agent. I cannot say that the reason I should hold a door open for my neighbor derives from some particular individual in the world who invented the practice. For me, these kinds of shared practices have simply come to be. Heidegger's term for describing the "agent" responsible for them is *das Man*, which is translated as "the They" or "the One." For example, we might say that there is an appropriate way to use a hammer (i.e., hold it from the bottom to get the most leverage) and this is simply what "one" does. We might also say that it is polite to cover your mouth when you cough; again, that is simply what "one" does. Our everyday dealings with the world and with others are shaped and facilitated by a shared set of practices—simply, what "one" does. But it is also important to note that while Heidegger is aware of the constitutive *conformity* that allows us, as beings-in-the-world who are always being-with others, to have a world in common, he is also concerned about the way that this can produce a *Dasein* who is fundamentally conformist.[54]

We can find conformism in *Dasein*'s everyday discourse. By the term *Gerede*, Heidegger designates *Dasein*'s everyday chatting, its "idle talk." The phrase is "not to be used here in a disparaging sense,"[55] but rather to describe our everyday use of language—language in its "average intelligibility." Heidegger is of the view that language or discourse is a form of communication that involves both an object (a being) and a statement about it. The goal of discourse is to make the listener "participate in the disclosed being toward what is talked about in discourse." But in idle talk, the goal of disclosing an object (or being) to the listener is not entirely attained:

> In the language that is spoken when one expresses oneself [in idle talk], there already lies an average intelligibility; and in accordance with this intelligibility, the discourse communicated can be understood to a large extent without the listener coming to a being toward what is talked about in discourse so as to have a primordial understanding of it. One understands not so much the beings talked about, but one does listen to what is spoken about as such. This is understood, what is talked about is understood, only approximately and superficially.... And since this discoursing has lost the primary relation of being to the being talked about, or else never achieved it, it does not communicate in the mode of a primordial appropriation of this being, but communicates by *gossiping* and *passing the word along*.[56]

Heidegger's view about the function of discourse is reminiscent of Husserl, despite the great differences in their respective phenomenology. Just as Husserl tried to stave off the possible lack of fidelity that inhabits all written or spoken language by holding it up to a standard of a meaning-intention that can be entirely and adequately fulfilled by an intuition, Heidegger critiques the superficiality of idle talk by holding it up to a standard of "primordial understanding," where a word (or piece of discourse) is fulfilled by the degree to which it discloses the object (or being) under discussion. In both cases, the true function of language is to be adequate to something non-linguistic—an object, a being, an intuition. Heidegger will describe the potential loss of the object of discourse as one of discourse's very own possibilities. He writes, "Discourse...has the possibility of becoming idle talk."[57] This is because discourse can function inauthentically, when it idly and superficially proceeds without truly revealing its subject matter, or authentically, when it fully discloses the object about which it speaks. Discourse, in its inauthentic mode of idle talk, passes along what "one" says; in so doing, it turns "disclosing around into a closing off" in the sense that idle talk "*omits* going back to the foundation of what is being talked about."[58]

Because *Dasein* is ontologically being-in-the-world, it is always already in the midst of its dealings with the world. Heidegger describes this as the "entanglement of *Dasein*."[59] And just as one can get wrapped up in idle talk by passing along the gossip and chatter that are already circulating, *Dasein* can also lose itself in the projects and tasks that others give to it, or that it sees others doing. *Dasein* always has the possibility of "falling prey" to the world, by losing itself in its very involvement with the world. Thus *Dasein* might be living in a way that is inauthentic, in the sense that it unreflectively takes the projects and tasks of others as its own, rather than trying to discover how its own being is an issue for it. As Heidegger describes it, "In the self-certainty and decisiveness of the 'they,' it gets spread abroad increasingly that there is no need of authentic, attuned understanding. The supposition of the 'they' that one is leading and sustaining a full and genuine 'life' brings a *tranquilization* to *Dasein*, for which everything is in 'the best order' and for whom all doors are open. Entangled being-in-the-world ... is at the same time *tranquilizing*."[60] By handing over the task of finding projects and making values to the "they," *Dasein* tranquilizes itself in a way that Heidegger describes as inauthentic. Akin to inauthentic speech, which is one possible mode of discourse, Heidegger argues that we can live our lives inauthentically by avoiding the task of "primordially appropriating" the projects and values in which we participate.

The problem of *Dasein*'s authenticity is not simply a question of leading the good life or avoiding the discourse of the "they." In fact, Heidegger is suggesting that the entire philosophical tradition should be diagnosed as a case of inauthentic idle talk. The fundamental philosophical question is the question of being. But that question has always been "closed off" and "covered up" by posing the question not at the ontological level, but at the ontic level of beings. Treating objects as present-at-hand, as substances with properties, has led us to think of ourselves as similarly present-at-hand—as thinking and extended substances. Traditional philosophical discourse would be a case of idle talk in the sense that it does not reach the level of primordially understanding its subject matter—being. It, for the most part, continues to speak a language that has been passed down and is unable to reach the object of inquiry, being, by focusing only on beings.

The call of conscience as the ontological voice

If philosophical discourse has for the greater part of its history been inauthentic discourse, how can the philosopher ever come to discover that his projects, goals, and language are inadequate? From what position could the philosopher determine that she is living an inauthentic life? How can authenticity ever challenge the unbroken pervasiveness of inauthentic life?

Enter the *ontological voice*. In order to break the grip of inauthentic life, there must be some clue in our everyday actions that discloses an alternative to the conformism of idle chatter and falling prey to the "they." This clue is found in the "voice of conscience." Heidegger states, "We shall claim that this potentiality [for authenticity] is attested by that which, in the everyday interpretation of itself, *Dasein* is familiar to us as the 'voice of conscience.'"[61] This voice of conscience appears to *Dasein* in the form of a "call" (*ein Ruf*). However, it is important for Heidegger to note the specific way this call is manifested, in order to be able to contrast it with idle talk and other forms of inauthentic discourse. Heidegger enumerates the specific features of the call through a series of questions.

First, *what is summoned in the call of conscience*? The answer is that *Dasein* is summoned; it hears the call as directed toward itself. *Dasein* is the receiver of the call.

Second, *to what is one summoned in the call*? *Dasein* is called to *itself*, but in a very specific sense. The call is not directed toward *Dasein* in its everyday activity. In our everyday commerce with others, Heidegger claims that "I *myself* am not for the most part the who of *Dasein*, but the they-self is,"[62] meaning that when I am most entangled in being-in-the-world, I am least genuinely myself and more a "they-self," a self whose projects are determined by others. However, the call of conscience does not summon *Dasein*'s inauthentic they-self but only its authentic self. In the face of the call, the self is distinguished from the they-self. "Because only the *self* of the they-self is summoned and made to hear, the *they* collapses."[63] The call challenges my entanglements with others, in that it is wholly focused on the genuine self and not the parts of the self that are wrapped up with others.

Third, *what is spoken in the call of conscience*? Here, the answer is surprising. Heidegger writes, "Strictly speaking—nothing."[64] This is because "the call does not say anything, does not give any information about events of the world, has nothing to tell." This is not because the call is void of content. Rather, the fact that the call says nothing is itself important. For if the call were to speak of anything in particular, of any beings, then the content of the call would be ontical in nature. It would speak of things in the world, things with which we could be involved. But by refraining to speak of anything, that is, by refraining to speak on an ontical register akin to that of idle talk or even average everyday discourse, the call speaks on an ontological register. Just as one should not confuse the question of being with the totality of beings, one should not confuse the call with just another voice. While Heidegger notes that the call "is lacking any kind of utterance" and that conscience "*speaks solely and constantly in the mode of silence*," he is insistent that the call is "not at all obscure and indefinite," nor can its lack of utterance allow for *Dasein* to "shut this phenomenon [of the call] into the indefiniteness of a mysterious voice."[65] Rather, the ontic contentlessness of the call is precisely what allows the call to be heard in the proper ontological register. In Heidegger's marginalia to *Being and Time* (§55, "Existential and Ontological Foundations of Conscience"), he reasserts the ontological register of the call, noting that "We don't 'hear' it with the senses."[66] If we did, this would be another form of ontic activity. In another bit of marginalia, he writes, "Where does this listening [to the call] and being able to listen come from? Sensuous listening with the ears is a thrown mode of being affected."[67] When the call is heard in this ontologically appropriate way, then there is no possibility of mistaking the call as emerging from some ontic entity or as a piece of everyday discourse. For "deceptions" about the content and recipient of the call "occur not by an oversight of the call (a mis-calling) but only because the call is *heard* in such a way that, instead of being understood authentically, it is drawn by the they-self into a manipulative conversation with one's self and is distorted...."[68]

Fourth, *who calls*? Again, Heidegger's answer is surprising. He claims that the caller "remains in striking indefiniteness."[69] The caller cannot be identified in terms of a "worldly orientation," that is, there is no "name, status, origin, and repute" that we can attach to the caller. It prohibits "any kind of becoming familiar." Yet, at the same time, the voice of conscience seems to emerge from within the self, while not being identifiable as the self's own voice. "The call comes *from* me and yet *from beyond* me."[70] The paradoxical claim that the caller is somehow both absolutely distanced from me yet seems to emerge from me is intelligible if we recall that Heidegger is not simply positing the "voice of conscience," but describing the features of this voice as it manifests itself in everyday activity. In other words, Heidegger begins by providing a phenomenology of the voice of conscience, noting how its manifestation is experienced by *Dasein*, and using this phenomenal description as a basis or "clue" for explicating *Dasein*'s own essence, its way of being.

Before discussing Heidegger's ultimate answer to the question "Who calls?," I want to note that this ontological voice of conscience shares many features with the acousmatic sounds discussed in the previous chapter. Based on the manifestation of the call alone, *Dasein* cannot definitively answer the question "Who speaks?" The source of the voice is underdetermined by the phenomenal clues given. We could describe the phenomenological manifestation of the voice of conscience as a sonic effect spaced from its source or cause, one that is noteworthy because of the way that it has no specific content, says nothing, speaks silently, and yet functions as an address directed at *Dasein*. Because it silently refrains from saying *anything*, the effect underdetermines *Dasein*'s attributions of a source or cause to the voice. The source cannot be confidently pinned down. Like the sound in Kafka's burrow, the beating of the heart in Poe, or the anechoic sounds heard by Cage, the voice of conscience seems to come both from me and beyond me. It is chiastic in the sense that I have already addressed. This chiasm is noted as part of the call's phenomenal manifestation. Even though it seems to come from me (as well as beyond me), Heidegger explicitly refers to the call as an "it"—it calls, against my will, even as I recognize that it "without doubt [sic] does not come from someone else who is with me in the world."[71]

However, one must not be too hasty to simply classify the voice of conscience alongside the other examples of acousmatic chiasm found in Kafka, Poe, and Cage. In each of those cases, sounds under consideration were all sonorous—even if that sonorousness was only imagined. Regardless of the phenomenological "mode of presentation" of each sound, the real or imagined sonorous aspects of the sound offered clues about its source and effect. For example, although the *Pfeifen* remained unchanged in all parts of Kafka's burrow, that sonorous fact was used as the basis for making inferences about the source of the sound, its distance from the narrator, and the location from which it emerged. In contrast, Heidegger's ontological voice of conscience is silent. There are no specifically sonorous aspects of the ontological voice that can be used to support inferences concerning the source or cause of the sound.

At the same time, Heidegger notes that this voice is an address emitted from some location and directed toward a listener. This feature of the ontological voice is productively contrasted with the Schaefferian sound object. If one auditioned the ontological voice under Schaeffer's preferred mode of *entendre*, the silent effect of the ontological voice would be separated from its source and secured for a listener as an object in itself. The ontological voice would simply be a silent sound object, and there would be no reason for uncertainty or uncanniness in the experience of hearing it. This would not fit with Heidegger's phenomenological description of the call. The silent ontological voice is not simply silence as such but, according to Heidegger's description, "a keeping silent."[72]

Heidegger's claim is explicable in terms of acousmatic spacing. An absent source *restrains itself*, and this restraint is *audible* in the silence of the ontological voice. The silence of the voice of conscience, this sonic effect, is heard as being in the mode of keeping silent, so it implies that a source or cause is restraining itself or actively producing silence. It is a silence that demands attribution. In the silent effect, we can hear the present trace of an absent source. At the same time, the source of the voice is underdetermined by its silent effect—in no small part because the silent voice, with its utter lack of sonorous features, offers nothing to the listener upon which to make an attribution. The fact that the silent voice of conscience remains an address depends on the structure of acousmatic sound that I articulated in chapter 5; that is, the voice of conscience is constituted by spacing from its source. The ontological voice, the voice of conscience, reveals this structure in a form that is more austere than in Kafka, Poe, and Cage.

Who Speaks?

Heidegger is not content to remain at this level of analysis. There is a drive to dis-acousmatize the voice of conscience, to pin the ontological voice to a source and overcome the spacing heard in the call's phenomenal manifestation. Heidegger considers a few possible solutions before positing a definitive attribution. First, the voice of conscience could be "an alien power entering *Dasein*."[73] If this interpretation is pursued, one "supplies an owner for the power [i.e., the voice] thus localized,"— which one might do when identifying the voice of conscience with the superego or an introjected voice of authority—"or else one takes that power as a person (God) making himself known."[74] This interpretation would suffice for explaining the voices heard in mystical union or in religious visions, like Battier's "state of acousmate," when a god speaks directly by means of a voice that is both internal to *Dasein* yet beyond it. Another interpretation is to "explain [the call of] conscience away 'biologically,'" which would entail the reduction of the voice to naturalistic explanation, to an evolutionary adaptation, a byproduct of neural activations, or an auditory hallucination caused by physiological factors. Each of these explanations would locate a source for the voice within the sphere of ontic beings. Each would erase the spacing heard in the ontological voice by identifying the source of the voice with God, an alien power, or some physiological or biological mechanism. Heidegger dismisses these interpretations on grounds that are guided by an unflagging belief that "what is... must be *present-at-hand*; and what cannot be demonstrated as *present-at-hand* just is not at all."[75] This final characterization aligns with the thesis, put forth in the chapter 5, that an acousmatic sound is not itself an entity. Rather, a sound becomes acousmatic when a listener apprehends the spacing between source, cause, and effect. Spacing is not an entity; it is nothing that can be found in terms of objectively present (i.e., present-at-hand) things. The strange phenomenon (if one can still use that word) of acousmatic sound surges forth only with the spacing of the source, the cause, and the effect. Dis-acousmatization occurs at the moment when the spacing of source, cause, and effect is overcome or banished by locating an object that can occupy the position of the source and reestablish its plenitude. Thus, each of these attempts to locate the source of the call of conscience, to dis-acousmatize the alien voice and make it familiar, must cross over the threshold of ontological difference from the ontological register of spacing to the ontic register of entities. If this analysis is correct and the characteristics of the voice of conscience are indeed inexplicable in terms of ontic entities, one would be compelled to agree with Heidegger's rejection of these various interpretations on the grounds that they "hastily pass over the phenomenal findings."

After rejecting these interpretations, Heidegger puts forth a bold idea: The voice of conscience must come from a source that is not an entity present-at-hand. One candidate would be *Dasein* itself, whose "facticity...is essentially distinguished from the factuality of something objectively present [present-at-hand].... *What if Dasein...were the caller of the call of conscience?*"[76] According to Heidegger, nothing speaks against this interpretation since the phenomenal findings concerning the voice of conscience could be explicated in terms of *Dasein* being both the caller and the called. How so?

1. The caller is inexplicable in terms of worldly entities. Yet, *Dasein* is not a worldly entity like the things it encounters and concerns itself with. *Dasein* only inauthentically thinks of itself this way when it defines itself based on the influence of the "they." Thus the call does not come from everyday *Dasein*, but from *Dasein*'s more primordial self. The alien character of the call is not because it comes from somewhere outside *Dasein*; rather it emerges from a part of *Dasein* that is unfamiliar and alien only to the they-self, the self of *Dasein*'s everyday commerce. "The caller is unfamiliar to the everyday they-self, it is something like an *alien* voice. What could be more alien to the they, lost in the manifold 'world' of its heedfulness, than the self individualized to itself in uncanniness thrown into nothingness?"[77] By uncanniness (*unheimlichkeit*), Heidegger designates *Dasein*'s perpetual sense of never quite feeling at home (*heimlich*) in the world. *Dasein* is always already preceded by a situation that is not of its own making, always already entangled in the world, and always being projected forward. Uncanniness is an authentic trait of *Dasein* that, for Heidegger, counters the tranquilizing effect of being lost in the "they." "Uncanniness is the fundamental kind of being-in-the-world, although it is covered over in everydayness."[78] The authentic self is uncanny; it has taken on the projects of others as its own, but it has yet to make the world an authentic home for itself. The alien quality of the voice of conscience—alien only to inauthentic and tranquilized *Dasein*—is a manifestation of authentic *Dasein*'s primordial uncanniness.
2. The caller says nothing; it speaks of no facts and offers no commands or injunctions. "The call speaks in the uncanny mode of *silence*." It is silent because it does not command *Dasein* to participate in this or that project, or attain its authenticity by following this or that goal. The call's silence must be contrasted with idle chatter. Instead of passing language along without motivating the object of which it speaks, the call pulls *Dasein* away from its entanglements in the "they" and its everyday gossip. The voice of conscience, in saying nothing, prescribes nothing; but it is not without force, in the sense that the call proscribes *Dasein* and demands an end to its inauthenticity.

Based on these observations, Heidegger concludes that (i) the "caller is *Dasein*," (ii) "the one summoned is also *Dasein*," and (iii) "what is called forth by the summons is *Dasein*," called to end its entanglements with the "they" by being called to its "ownmost potentiality-of-being,"[79] that is, its potential to become authentic *Dasein*. Yet, at the moment Heidegger asserts this conclusion, the acousmaticity of the ontological voice of conscience is dissolved, and the source of the voice is revealed as *Dasein* addressing itself, about itself. Heidegger will argue that this solution fits the phenomenological findings better than the other interpretations. But we may not find ourselves very satisfied with this answer, for a couple of reasons.

First, Heidegger's solution asserts the auto-affectivity of the voice, replicating Husserl's central claim about phenomenological voice: *Dasein* affects itself by being both the sender and the receiver of its (contentless) message. Despite the critique of Husserlian phenomenology presented in *Being and Time*, Heidegger's ontological voice presents us with another version of speaking-to-oneself.[80] Second, given that the ontological voice says nothing, how can we ever determine the source of this voice? There is no sonorous evidence upon which to make our attribution. Anything that keeps silent could just as reasonably be the source as *Dasein* itself. Normally, one could appeal to other sense modalities to determine the source of the silence. But given the phenomenological description of the voice of conscience, to what other sense modalities could one appeal? Heidegger would no doubt argue that the voice of conscience could not be seen or felt, since it is not a present-at-hand entity. Only hearing is granted privileged access to the register of *Dasein's* being. This is why Heidegger, in his marginalia, is so emphatic about asserting that this mode of hearing is not a mode of being simply sensuously affected, that we cannot hear the ontological voice with the senses.[81] Yet, without any content to hear, there is nothing to appeal to in making our attributions. A double bind arises: By emphasizing the "nothing" heard in the voice of conscience as the clue that this voice is spoken by *Dasein* in the ontological register, Heidegger undercuts the possibility of making this claim secure. For how could we ever be certain?

In fact, there is a moment in *Being and Time* when Heidegger seems to undermine his own claim about the source of the ontological voice. In a passage from §34, on *Dasein's* relationship with discourse and language, Heidegger writes this very curious sentence: "Hearing even constitutes the primary and authentic openness of *Dasein* for its ownmost possibility of being, as in hearing the voice of the friend whom every *Dasein* carries with it."[82] The first half of the sentence is explicable in terms of Heidegger's analysis of the call of conscience. Hearing—ontological hearing, not sensuous hearing—is the mode that allows *Dasein* to encounter itself in its entanglements and summon itself back to its possibility for an authentic life. *Dasein* bootstraps itself from inauthenticity to authenticity by hearing itself address itself in the call. From that moment on, *Dasein* responds to the call's silent proscription by changing its way of life, by leaving its entanglements in the they, shedding its they-self, and discovering how to live authentically in its uncanniness. But who is this "friend whom every *Dasein* carries with it?" How can this be squared with Heidegger's ultimate position, where auto-affective *Dasein* transports itself away from inauthenticity? In this curious sentence, the ontological voice finds its source in another, in the voice of the friend. And if this were so, if the ontological voice were to be the voice of a friend—even a silent friend—would this not also fit with the "phenomenal findings"? The voice of the friend that *Dasein* carries would appear to both "come *from* me and yet *from beyond* me." The call of the friend would be silent, yet addressed to me, prescribing nothing, and so forth.

The possibility that *Dasein's* ontological voice is the voice of a friend would also resolve a persistently troubling feature of Heidegger's account, namely, how can authenticity ever challenge the unbroken pervasiveness of inauthentic life? Upon what basis would *Dasein* first get the inkling or suspicion that it was living inauthentically? In Heidegger's account, there is simply no motivation for the call of conscience to appear. *Dasein*'s ontological self just somehow suddenly breaks through and calls itself to itself. Would it not make more sense if the source of the call came from somewhere else, from the voice of a friend, someone involved in the situation with the capacity to point out *Dasein*'s inauthenticity? The interruption of the call would be better motivated if it did not emerge from within the closed circuitry of auto-affection but from without. This voice would not have to be a voice that prescribes some specific path to authenticity, but a voice that simply halts the flow of *Dasein*'s current state in order to help it discover its potential to live otherwise. It could be a voice of interruption, one that challenges *Dasein*'s absorption in inauthentic modalities. It could be a therapeutic voice, one that keeps silent while watching over *Dasein*'s progress toward finding its ownmost potentiality. Such voices were not unknown in Heidegger's day. Indeed, with a small modification, the ontological voice may indeed be better understood as a psychoanalytic voice.

§4. THE PSYCHOANALYTIC VOICE

Heidegger's curious sentence concerning "the voice of the friend whom every *Dasein* carries with it" initiates a radical change in post-Heideggerian theories of the voice. Figures like Emmanuel Levinas and Jacques Derrida challenge the privileged identification of the inner voice with the subject by placing emphasis on exteriority, alterity, and difference within the very constitution of the voice.[83] Similarly, Slavoj Žižek and Mladen Dolar, two cultural theorists and philosophers whose work is grounded primarily in the theories of French psychoanalyst Jacques Lacan, offer a post-Heideggerian critique of the voice. The nature of their critique differs from Levinas and Derrida, primarily over the epistemological status of the voice. Rather than treat the voice as a medium that discloses (or deconstructs) the auto-affective subject, Žižek notes that in Lacanian theory, the voice is "not on the side of the *subject* but on the side of the *object*." For Žižek and Dolar, the voice is not something that I can identify as my own, as something that is properly mine and buried deep within the self. Nor is it exactly the voice of another. The *object voice* is simply obstinate or inert. "... This voice—the superegoic voice, for example, addressing me without being attached to any particular bearer—functions as a stain, whose inert presence interferes like a strange body and prevents me from achieving my self-identity."[84]

I want to linger for a moment on Žižek's phrase: "addressing me without being attached to any particular bearer." This is nothing less than what I have been calling the acousmatic voice—the voice whose source remains constitutively underdetermined. In the analyses of the voice presented thus far, I have tried to show how the acousmatic voice comes to affect a range of other voices: the phonographic voice, the phenomenological voice, and the ontological voice. However, in each of the authors read (Husserl, Heidegger, Edison, Powers, and Gouraud), the term "acousmatic" was not explicitly used, given its previous rarity. I have argued that one should still be authorized to use this term, even if the authors did not, since we are interested in a history of the phenomenon of unseen sound—of moments when the separation of seeing from hearing is privileged or central—and not simply a history of the word acousmatic.

With Žižek and Dolar, however, we explicitly encounter the *voix acousmatique*.[85] They borrow the term from Pierre Schaeffer's student, Michel Chion, who developed a theory of the acousmatic voice in his groundbreaking book, *The Voice in Cinema*. This voice appears in cinema in the guise of the *acousmêtre*. In Chion's theory, the voice is less the vehicle for the transmission of linguistic propositions or a character's psychological state than it is a special kind of cinematic object whose transparency or obscurity in relation to the source becomes meaningful in the development of a film's narrative. The source of the voice can be veiled or unveiled, often at crucial moments, like those found in Alfred Hitchcock's *Psycho* or Fritz Lang's *The Testament of Dr. Mabuse*—to name two of Chion's central examples. In developing his theory of the acousmatic voice as a cinematic object, Chion was influenced by Lacan's writing on the voice and its development by psychoanalysts like Denis Vasse.[86]

Given the Lacanian background that undergirds Chion's theory, Žižek and Dolar press the voice into action in ways that Chion or Lacan did not imagine. In particular, they develop it in order to challenge Derrida's deconstruction of the voice in *Speech and Phenomena*. Žižek is keen to challenge the alleged one-sidedness of Derrida's critique of "phonocentrism," namely, that the voice (in the long tradition of Western metaphysics) is the site of living presence and, as such, establishes the self-identity of the subject, immunizing it from all threats of exteriority, alterity, and non-identity. In Žižek's view, it was a mistake to ever treat the voice as the guarantor of the subject's self-identity because the voice, as "object voice," is also the site where the subject's self-identity is most stringently challenged. Žižek writes:

> Derrida proposed the idea that the metaphysics of presence is ultimately founded upon the illusion of "hearing-oneself-speaking" [*s'entendre-parler*], upon the illusory experience of the Voice as the transparent medium that enables and guarantees the speaker's immediate self-presence.... True, the experience of *s'entendre-parler* serves to ground the illusion of the transparent self-presence of the speaking subject. However, is not the voice *at the same time* that which undermines most radically the subject's self-presence and self-transparency?... I hear myself speaking, yet what I hear is never fully myself but a parasite, a foreign body in my very heart.... This stranger *in myself* acquires positive existence in different guises, from the voice of conscience to the voice of the persecutor in paranoia.[87]

If every moment of the subject's speech is, at the same time, the speech of someone other than the subject, the *acousmaticity* of the Voice can never be overcome. We are always hearing the obstinate voice of another in our heads, obeying its commands, speaking and being spoken by it. In Lacanian theory, this is because, in order to have access to language at all, we must learn to speak the language into which we are born but which we did not invent—a situation that Lacan calls *alienation in language*.[88] Alienation operates as the condition of the possibility of all speech; in order to speak, we are always speaking a language that is not our own. The acousmatic voice is our permanent condition.[89]

However, this condition can devastate the subject, stultified and hectored by the voice of the other. One way to cope with this omnipresent acousmatic voice is through the use of psychoanalysis. As I said, Lacan never really developed his thinking about the object voice beyond a few passing suggestions. Dolar's book, *A Voice and Nothing More*, is perhaps the most thoroughgoing and comprehensive attempt to develop a Lacanian theory of the object voice to date.[90] Of particular concern to Dolar is the exchange of voices in the psychoanalytic session. Now, the Freudian and Lacanian psychoanalytic situation, in which the analyst sits out of view of the analysand, is also, broadly speaking, an acousmatic situation—a situation of hearing without seeing. But, in Dolar's account, psychoanalytic exchange engenders a new voice, what I will call the *psychoanalytic voice*, a voice that counters and quells alienation in language. Acousmatic sound plays a seminal and structural role in Dolar's account of the object voice, and one must pay close attention to it in order to understand his psychoanalytic alternative.

The success or failure of Dolar's endeavor hinges on his treatment of acousmatic sound. Dolar's strategy for exposing the object voice is to subject it to a series of reductions. These reductions reveal what the object voice is by showing what it is not. This process has three stages: First, the voice is differentiated from the meaningfulness of linguistic statements; second, the voice is differentiated from the source from which it is emitted; third, the voice is differentiated from the sound that it makes. After clarifying the nature of the object voice, I turn to Dolar's treatment of the voice in terms of ethics and finally draw the contrast between the acousmatic voice and the psychoanalytic voice.

The voice as object: What the voice is not

FIRST REDUCTION: THE VOICE AND MEANING (*PHONÉ* VERSUS LOGOS)

Unlike Husserl, who believes the voice to be the ideal medium for the expression of meaning, Dolar argues that the voice is distinct from the meaning of its utterances. In a chapter on the "Linguistics of the Voice," Dolar appeals to structuralist linguists, such as Saussure and Jakobsen, who distinguish the signifier from the voice that speaks. The signifier is simply a sign. It has two conditions: It requires other signs from which it is differentiated and a medium in which to present these differences. The medium, however, need not necessarily be sonorous, for any medium capable of articulating differential signs would do. According to Saussure, "It is impossible that sound, as the material element, should in itself be a part of the language. Sound is merely something ancillary that language uses...."[91] For Dolar, "the inaugural gesture of phonology [as opposed to phonetics, which is interested in the sonorous aspects of language] was thus the total reduction of the voice as the substance of language." Thus, in structuralist linguistics, the difference between the voice, *phoné*, and the chain of signification, logos, is asserted for the sake of the latter.

Dolar, while affirming this difference, inverts the valuation. He, following Žižek, argues that the voice is always an "excessive voice," an "eclipse of meaning."[92] As evidence, Žižek and Dolar make a surprising appeal to the history of music, mining it for examples where the intelligibility of the text is challenged by the sound of the singing voice. The history of Western music repeatedly offers us examples of voices that threaten the "established order" and thus must be "brought under control, subordinated to the rational articulation of spoken and written word, fixed into writing."[93] Overtaking the text that it is supposed to articulate, the voice is a threat to order, becoming an aesthetic object in its own right. Attempts to control the voice threatened to subordinate it to the rational order of the spoken and written word. Žižek's examples, which are the same that Dolar mentions in "The Object Voice" and *A Voice and Nothing More*, range from Hildegard of Bingen to Elvis Presley, but anyone familiar with the repeated attempts by religious authorities to curtail music that challenges the intelligibility of the sung text could insert their favorite examples. The point remains the same: The various attempts to reduce the singing voice to logos, to subordinate the extravagance and autonomy of the voice for the sake of the order and intelligibility of the text's meaning always fail. There is something excessive about *phoné* that resists the conceptual reduction to logos.

SECOND REDUCTION: THE VOICE AND THE BODY (*PHONÉ* VERSUS *TOPOS*)

After distinguishing the voice from its meaning, Dolar challenges the view that the voice is fundamentally tied to a speaker. This is precisely where the acousmatic voice comes into his argument.

In a chapter on the "The 'Physics' of the Voice," Dolar introduces the reader to the acousmatic voice in the traditional manner, by retelling the myth of the Pythagorean veil. He preserves its basic features (as I have described them in chapter 2), citing Diogenes as his source. But unlike the Schaefferians, Dolar reads the myth with a subtle focus on the voice and its powers. "The point of this device [the veil] was ultimately to separate the spirit from the body. It was not only that the disciples could follow the meaning better with no visual distractions, it was the voice itself which acquired authority and surplus-meaning by virtue of the fact its source was concealed; it seemed to become omnipresent and omnipotent."[94] There are two strands of thought here. First, Dolar notes that the Pythagorean veil allows the listener better access to the meaning of the discourse by removing all visual distractions. Attention is wholly given over to logos. This aspect of Pythagorean practice supports the idealization of meaning associated with the long tradition of Western philosophy. Pythagoras is the first to "describe himself as a 'philosopher,'" and the use of the curtain is a "stroke of genius which stands at the very origin of philosophy."[95] The occultation of the body allows for the transcendental truth of the discourse to be more clearly articulated.

But Dolar's aim here is the voice, not meaning. That is why he also notes that, despite the disappearance of the voice into the meaning of the discourse, the Pythagorean veil also grants the voice an "authority and surplus-meaning." The very act of hiding the voice is also a technique for giving the voice certain powers—namely, omnipresence and omnipotence.[96] These powers of the voice are irreducible to the meaning of the statement; they are the product of a surplus-meaning that has nothing to do with logos. Surplus- meaning emerges from the difference between the voice and its bearer or source, that is, between *phoné* and *topos*. Thus it is no coincidence that Pythagoras enjoyed cultic status in the ancient world. By means of the veil, he imbued his utterances with special powers. Dolar's reading of the Pythagorean veil is emblematic for philosophy in general; the power of philosophy itself does not wholly rely on the meaning or logic of its statements; rather, it resides in a hidden kernel of meaninglessness in excess of the statement. The voice becomes the voice of authority the moment it detaches from the speaker. The statement (logos) is enriched by this extra (alogical) power, a magic trick for compelling conviction.

But is it simply a trick? Dolar wonders, "Could we go so far as to say that the hidden voice structurally produces 'divine effects'?"[97] If the divine effect is simply a result of veiling the voice, then we should be able to simply diffuse it by unveiling its source. Dolar mentions the famous scene from the end of *The Wizard of Oz*, where the powerful wizard is revealed as "a ludicrous and powerless old man" the instant that Toto pulls back the curtain.[98] It is a scene of dis-acousmatization—a scene where the source of the acousmatic voice is exposed and its powers suddenly vanish. In it, "the aura crumbles, the voice, once located, loses its fascination and power, it has something like castrating effects on its bearer, who could wield and brandish his or her phonic phallus as long as its attachment to a body remained hidden."[99]

Dolar considers a possible scene of dis-acousmatization concerning the Pythagorean disciples. The Pythagoreans were of two classes—the *akousmatikoi* and the *mathematikoi*. According to legend, the former were positioned outside the veil and the latter within. Those first initiated into the school would begin as *akousmatikoi*, as exoteric disciples, keeping a vow of silence for five years. After that initiatory period, they could be promoted to *mathematikoi* and brought inside the veil, becoming esoteric disciples. Embellishing on the legendary account, Dolar speculates that the effect of seeing Pythagoras for the first time would have been "not unlike the scene in the *Wizard of Oz*." One would expect the scene of dis-acousmatization to strip the philosophical voice of the master of all its powers, exposing it as the fragile emission of an aged and powerless figure. It should have a "castrating effect" on the philosopher, cutting his discourse down to size. Yet, in Dolar's analysis, just the opposite occurs. The castrating effect takes its toll on the newly initiated *mathematikoi*. Dolar imagines that the disciples, horrified by what they saw, would do their best to deny that the event ever happened. In Lacanian terms, they would maintain their illusions for the sake of the symbolic order, the big Other, "the agency for which one has to maintain appearances."[100] Dolar writes, "It may well be that, once the lifted screen uncovered a pitiable old man [Pythagoras], the disciples' main concern was to maintain the illusion, so that the disillusionment which they must have experienced did not affect the big Other. Another screen had to be raised to prevent the big Other from seeing what they saw."[101] For Dolar, the screen is always a site of fantasy. After the first screen is lifted and the fantasy of the master's power is exposed in the scene of dis-acousmatization, a new screen is installed in its place. That new fantasy can be cast in Lacan's famous formulation: "*Je sais bien, mais quand même...*" or "I know very well, but nevertheless...."[102] The disciples know very well that Pythagoras' powers are false, but nevertheless...they believe him to be the master.

This Lacanian formulation appears quite often in Žižek and Dolar's work. In its fetishistic form, the phrase is completed as "I know very well that Mother hasn't got a phallus, but nevertheless...I believe she has one." According to Freud's theory of fetishism, "one stops at the last-but-one stage, just before the void becomes apparent, thus turning this penultimate stage into a fetish, erecting it as a dam against castration, a rampart against the void."[103] The formulation captures the defensive aspect of the subject's belief, whether pathological or normal, a strategy to hold onto what one believed (and wants to believe) was the case, despite knowing better. Žižek uses it to describe the subject under the thrall of ideology, one whose very behavior perpetuates the current social order, even when they ostensibly know that this is the case. Speaking of the pursuit of money, one might say, "I know that money is a material object like others, but nevertheless... it is as if it were made of a special substance over which time has no power."[104] The behavior of the Pythagorean disciples is akin to the ideologue: After encountering the castrating scene of dis-acousmatization, the *mathematikoi* stop at the penultimate stage, at the very moment when they could still believe that the acousmatic voice possessed powers of omnipotence and omnipresence. Their fetishistic belief in the voice could be formulated as "I know very well that the voice must have some natural and explicable cause, but nevertheless... I believe it is endowed with secret powers."[105] Thus, after the scene of dis-acousmatization, the screen of fetishism follows.

In Lacan's writings, the veil is emblematic of desire, figuring in one of his most famous passages. In *The Four Fundamental Concepts of Psychoanalysis*, Lacan recounts Pliny's parable of the contest between Zeuxis and Parrhasios, two artists engaged in a challenge to see who can paint the most lifelike picture. At first, Zeuxis appears to be the winner by painting a picture of grapes so real that the birds peck at it:

> In the classical tale of Zeuxis and Parrhasios, Zeuxis had the advantage of having made grapes that attracted the birds. The stress is placed on the fact that these grapes were not in any way perfect grapes, but on the fact the even the eye of the birds was taken in by them. This is proved by the fact that his friend Parrhasios triumphs over him for having painted on a wall a veil, a veil so lifelike that Zeuxis, turning toward him said, *Well, and now show us what you have painted behind it.*[106]

This passage articulates two "opposed strategies of deception," one where birds are deceived by the imitation of reality, the other where humans are deceived by their own desire. Conventionally, we might think of Zeuxis' grapes as "objects of desire," that is, something that could satisfy one's hunger, thus providing a reason or source for one's desire. However, Parrhasios' veil is a better illustration of Lacan's notion of desire. Desire has *no* object that can satisfy it, for it, like the veil, is a kind of lure. Desire constantly seeks an object that can satisfy it while disavowing the fact that its satisfaction is structurally impossible. Desire is lured by its own desire for satisfaction, by the endless sliding of various signifiers or objects into the position of the desired thing. The figure of the veil illustrates desire's own infernal operation; we always want what is behind the veil, even though there is nothing there.

According to Jacques-Alain Miller, the structural function of veiling is its ability to make the subject believe that there is an object of desire when there is in fact nothing: "The veil that hides *causes* what cannot be seen to exist... the veil creates something ex nihilo.... Thanks to the veil, the lack of object is transformed into object."[107] Dolar and Žižek augment this claim by appealing to a famous passage from Hegel's *Phenomenology*, also concerning veils: "It is manifest that behind the so-called curtain which is supposed to conceal the inner world, there is nothing to be seen unless *we* go behind it ourselves, as much in order that we may see, so that there may be something behind there which can be seen."[108] Their gloss on this passage is simple and unequivocal. For Žižek, "there is nothing behind the curtain except the subject who has already gone beyond it...this 'nothing' behind the curtain is the subject."[109] This is because the desired object behind the curtain is nothing but the void of desire itself. The object is not itself desirable, since any object would do. It is desire itself that makes this, or any object, desirable in the first place. There is in reality nothing behind the veil, yet the structure of the veil allows us to catch desire in operation. For Dolar, "with the acousmatic voice we have 'always already' stepped behind the screen and encircled the enigmatic object with fantasy."[110]

This fantasy operates at two levels in Dolar's account of the Pythagorean school. First, there is the moment when the *akousmatikoi* are filled with desire by the voice screened by the Pythagorean veil. This fantasy is manifest in the power of the acousmatic voice as omnipotent and omnipresent. Yet, even after the scene of dis-acousmatization, there is another veiling of the voice. "The source of the voice can never be seen, it stems from an undisclosed and structurally concealed interior, it cannot possibly match what we can see."[111] The voice is simply incommensurable with anything visible. For any visual thing must be *some-thing*; but the voice is not something, it is simply the *no-thing* of our desire. For Dolar, "there is always something totally incongruous in the relation between the appearance, the aspect, of a person and his or her voice...the fact that we see the aperture [the mouth] does not demystify the voice; on the contrary, it enhances the enigma."[112] When read in conjunction with Hegel, Dolar's point is clear. There is no interior to be seen—for there is nothing behind the curtain except ourselves—that is, the nothingness of our own desire. The speaker's body functions like another veil, causing what cannot be seen, the voice, to become an object of desire. Because the voice is always of a different structural order than the visible, it necessarily remains an acousmatic voice, even after the scene of dis-acousmatization. Instead of being ideological in nature, the second veil, the screen of fetishism, is propped up by the structure of desire. Even when peering behind the veil, the *mathematikoi* will never see the voice they desire to find.

Whether we focus on the Pythagorean veil or the veil that is the speaker's body, the voice is always an emblematic object of desire—in Lacan's terms, an *objet a*.[113] The voice, as *objet a*, structurally occupies the impossible position of the object imagined behind Parrhasios' (or Hegel's or Pythagoras') veil. There is nothing there; there is nothing to be disclosed, no source to be uncovered. *Phoné* can never be identified with topos. This is the reason for Dolar's emphatic statement that "*there is no such thing as disacousmatization....*The voice as the object appears precisely with the impossibility of disacousmatization."[114] Although Dolar entertains the possibility of the *mathematikoi's* scene of dis-acousmatization, this is never actually viable. Allow me to underscore that Dolar, through his reading of the Pythagorean veil, establishes the impossibility of dis-acousmatizing the voice. The scene of dis-acousmatization will always founder because the voice is an *objet a*. As such, the voice can never be unveiled; it is, pardon the pun, structurally un-a-*veil*-able. The separation of *phoné* and topos tantalizes us with the possibility of a source for the voice while structurally disallowing it to be located.

Dolar's reading of the Pythagorean veil, with its emphasis on the installation of the screen of fetishism, grounds a different kind of originary experience than found in the Schaefferian tradition. Instead of inscribing the Pythagorean veil at the origin of a practice of *musique concrète*, it establishes the impossibility of dis-acousmatizing the voice. It stands on the same horizon as the veils of Parrhasios and Hegel, as an emblem of the *objet a*, a monument to the voice as lure. Dolar crystalizes the point in his analysis of Francis Barraud's iconic painting, *His Master's Voice*. This image is best known as the logo of the HMV label, where Nipper the dog sits, head cocked, listening to the gramophone. Dolar notes that the "dog exhibits the emblematic posture of listening," of displaying an "exemplary attitude of dog-like obedience," where "listening entails obeying...."[115] Like the *akousmatikoi*, "the dog doesn't see the source of the voice, he is puzzled and staring into the mysterious orifice, but he believes—he believes all the more for not seeing the source; the acousmatic master is more of a master than his banal visible versions."[116] Here is, perhaps, the paradigmatic case of the acousmatic voice as the voice of obedience and belief. All the features are present; the omnipotence and omnipresence of the voice and the submission to its power are effects generated by the disavowal that organizes our beliefs, our submissiveness to our own desire. We are always lured in by the voice without a source, regarding it—like Nipper—as our master's voice.

THIRD REDUCTION: THE SILENT VOICE (*PHONÉ* VERSUS *PHONÉ*)

In Lacanian theory, there is an important distinction between desire and drive. Desire appears as an endless chain of substitutions, where the subject roves from one object to another in the pursuit of impossible satisfaction. The concatenation of unsatisfactory objects functions as a screen that veils the *objet a*—the name for the structural gap, lack, or *no-thing* that the desiring subject confounds for *some-thing* behind the veil. The chain of desirable objects is structurally akin to the endless chain of signifiers, the endless production of meaning, statement, and interpretation that constitutes Lacan's symbolic order. As Dolar notes, "the dimension of signification...concurs with the dimension of desire," in that desire follows the signifying logic of substitution.[117]

When the voice is taken as an object of desire, one is lured into an impossible hunt for its source; in contrast, Dolar directs his investigation toward the voice as an "object of the drive." The Lacanian drive, modeled on the Freudian death drive, is the propulsive force that endlessly, repetitively circles around the *objet a*; but, unlike desire, the drive derives its enjoyment not from any supposed satisfaction but only from the very act of perpetually going on. According to Dolar, "the dimension of the drive...does not follow the signifying logic [of substitution] but, rather, turns around the object, the object voice, as something evasive and not conducive to signification." It is not easy to separate the voice as an object of desire from the voice as an object of drive. But that is precisely the project of the latter half of *A Voice and Nothing More*. "In every spoken utterance one could see a miniature drama, a contest, a diminished model of what psychoanalysis has tried to conceive as the rival dimensions of desire and the drive."[118]

How can one isolate the voice as an object of drive? "In order to conceive the voice as the object of the drive, we must divorce it from the empirical voices that can be heard. Inside heard voices is an unheard voice, an aphonic voice, as it were."[119] Dolar is keen to differentiate the voice from anything sonorous or "phonic" that might lure the subject into substantializing and desiring the object voice. Sonority functions like the Pythagorean veil; it "both evokes and conceals the voice; the voice is not somewhere else, but it does not coincide with voices that are heard."[120] Just as the voice must be distinguished from the meaningfulness of its statement and the source from which it is emitted—that is, *phoné* must be grasped as alogical and atopical—to disclose the voice as an object of drive, it must also be differentiated from its sonorousness. The object voice is not the sound of the voice. Paradoxically, *phoné* as an object of drive is aphonic.

Ethics, Statement, and Enunciation

Once the voice has been fully reduced—once it has been grasped as the alogical, atopic, aphonic voice—Dolar is in position to expose its nature:

> If the voice does not coincide with any material modality of its presence in speech, then we could perhaps come closer to our goal if we conceive of it as coinciding with the *very process of enunciation: it epitomizes something that cannot be found anywhere in the statement*, in the spoken speech and its string of signifiers, nor can it be identified with their material support. In this sense the voice as the agent of enunciation sustains the signifiers and constitutes the string, as it were, that holds them together, although it is invisible because of the beads concealing it.[121]

The act of enunciation is distinct from the chain of signifiers that constitutes the statement, the linguistic meaning of the utterance. Because every subject is alienated in language, because the language spoken is always the language of the Other, for Lacan, the statement is never the place where the subject appears. Rather, it is only in the inverse of spoken language, in the other of the statement (the other of the Other), that one can detect the fleeting presence of a subject. According to Bruce Fink, the subject of enunciation "is *not* something which or someone who has some sort of permanent existence: it only appears when a propitious occasion presents itself," classically as a slip of the tongue, a gaff, an aphasia, and such.[122] By breaking with the alienation in language, such propitious moments reveal, if only in passing, the subject of enunciation as the place from where the act of address really emerges.

The difference between statement and enunciation also defines two modes of ethical discourse. "A certain opposition has persisted between the voice, its pure injunction, its imperative resonance, on the one hand, and on the other discursivity, argument, particular prescriptions or prohibitions or moral judgments, a wide variety of ethical theories."[123] The latter is an ethics of *statement plus enunciation*, that is, an ethics of prescriptions, orders, and commands, where the statement steals its force from the power of enunciation. This is an ethics of "His Master's Voice," where one dutifully obeys the orders that the Other hands down. In this mode, we behave like Nipper—a good dog, but hardly an attractive model of human ethical action. The former is an ethics of *enunciation without statement*, where as subjects, we have to supply the statement ourselves. The pure injunction of the silent voice "is like a suspended sentence...but a sentence demanding continuation, a sentence to be completed by the subject, by his or her moral decision, by the act. The enunciation is there but the subject has to deliver the statement and thus assume the enunciation, respond to it and take it on his or her shoulders."[124]

These are two very different possibilities: a choice between obeying the commands of the Other or assuming the responsibility for one's actions. Clearly, the two choices are not equally attractive, for who would willingly choose the former? Dolar claims that "the dividing line is very thin," that it is not so easy to differentiate responsibility from obedience. "This voice is utterly ambiguous: if it is at the very core of the ethical, as the voice of the pure injunction [enunciation] without positive content [statement], it is also at the core of straying away from the ethical, evading the call, albeit in the name of ethics itself."[125] Here, an astute reader cannot miss the Heideggerian script that underwrites this claim. The ethical "ambiguity" follows the Hölderlinian logic much beloved by Heidegger—*where the danger is, so grows the saving power*. The aphonic voice of enunciation is a Lacanian version of the ontological voice. One could substitute the silent call of conscience for the enunciation without statement, and *Dasein*'s inauthentic escape into "idle chatter" for the evasion of the ethical voice of enunciation. The moment Dolar introduces ethics, Heidegger's ontological voice resounds. The encounter with the silent voice, whether addressing the Lacanian subject or *Dasein*, demands an authentic act—an act of taking responsibility for one's actions and doing so only by absenting oneself as much as possible from the Other. The Lacanian subject is just as heroic as authentic *Dasein*; the ethics of the silent voice is an ethics of authenticity.

Dolar deploys the same ethical alternatives when analyzing the political aspects of the voice. "We have to disentangle, from the sonorous and shrill voices [of fascism and totalitarianism], the non-sonorous voice of pure enunciation, the enunciation without a statement: the enunciation to which one has to supply the statement, the political statement in response to that voice—not by listening/obeying, not by merely performing social rituals, but by engaging in a political stance."[126] More emphatically, Dolar presents a schematic description of the two alternatives in terms of the interpellation of the subject into the social order. This is the inauthentic option:

On the one hand there is the process of becoming a subject by recognizing oneself as the addressee of that call, which would then be a version of His Master's Voice issuing positive prescriptions.... [In this case], one turns into a subject precisely by assuming the form of the autonomous "I" [which Lacan diagnoses as an imaginary construction], disavowing its heteronomic origin, so that ideological domination and autonomous subjectivity work hand in hand....[127]

And there is the authentic option:

On the other there is at the same time a voice that interpellates without any positive content—something one would perhaps rather escape by obeying the sonorous voice of statements and commands; nevertheless this pure excess of the voice is compelling, although it does not tell us what to do and does not offer a handle for recognition and identification.... [In this case] one becomes a subject only by fidelity to the "foreign kernel" of the voice which cannot be appropriated by the self, thus by following precisely the heteronomic break in which one cannot recognize oneself. The ideological interpellation can never quite silence this other voice....[128]

The two alternatives are asserted repeatedly: The ethics and politics of inauthenticity always involve sonorousness, listening as a form of obedience, the disavowal of enunciation in the name of the statement, the perpetuation of the social rituals of the big Other, and the false belief in the integrity of the (imaginary) ego; in contrast, the ethics and politics of authenticity always involve silence, responsibility, a resistance to escapism and the big Other, a fidelity to the heteronomy of the self, a lack of prescriptive demands, and so forth.

The psychoanalytic voice

Heidegger's ontological voice is echoed in Dolar's voice of enunciation without statement, but with a difference: psychoanalysis. The Heideggerian struggle for authenticity occurs without any worldly interlocutor; the call of conscience calls *Dasein* to itself, but with no expectation of where or when the call may appear. Heidegger suggests no technique for bringing the call into audibility, nor can he motivate reasons why the call should suddenly sound. The fugitive suggestion that the ontological voice comes from the *Dasein*'s "friend" is not developed in *Being and Time*. However, it opens up the possibility that the ontological voice requires another as a condition for its manifestation.

Enter the *psychoanalytic voice*. For Dolar, the psychoanalytic session is a special situation in which the voice of enunciation without statement can be heard. The analyst allows this silent voice to be heard and to affect the discourse of the analysand. Thus I refer to this ontological voice, now placed in the context of the psychoanalytic session, as the psychoanalytic voice. In Lacanian analysis, following the Freudian model, the analyst is typically sitting out of the analysand's line of sight, which not only allows for the patient to address their discourse toward various imaginary egos projected upon the analyst, but also places a premium on the vocal interaction.[129] Instead of putting forth interpretations, the Lacanian analyst is for the most part silent, letting the analysand's speech flow—an endless chain of signifiers. The invisibility of the analyst seems analogous to the invisibility of Pythagoras behind the veil; yet Dolar puts forth a fascinating comparison between the psychoanalytic session and the Pythagorean legend, intended to show the radical difference between these two situations, a difference that is of great interest for the theory and practice of acousmatic sound:

> We have already seen that the voice is the very medium of analysis, and that the only tie between analyst and patient is the vocal tie. The analyst is hidden, like Pythagoras, outside the patient's field of vision, adding another turn of the screw to Pythagoras' device: if with Pythagoras the lever was the acousmatic voice, then here we have an *acousmatic silence*, a silence whose source cannot be seen but which has to be supported by the presence of the analyst.[130]

On one side, there is the acousmatic voice of Pythagoras, the voice that hides itself behind a veil in order to enjoy the power of omnipotence and omniscience. It is a voice that incites our desire by veiling itself and supplements the statements it makes with forcefulness and conviction drawn from its unlocatability. The acousmatic voice is "His Master's Voice," a voice that not only addresses a subject—interpellates a subject, as Dolar likes to say—but also prescribes or commands a subject to obey. It is a voice of statement plus enunciation. On the other side, there is the psychoanalytic voice, the voice of acousmatic silence. Unlike the master, the psychoanalytic voice is a silent voice—a voice that enunciates without a statement. The difference between these two voices can be articulated in terms of the difference between desire and drive, between a voice that endlessly chatters on and the paradoxical aphonic voice. Silence, according to Dolar, "never appears as such, it always functions as the negative of the voice, its shadow, its reverse, and thus something which can evoke the voice in its pure form. We could use a rough analogy to start with: that silence is the reverse of the voice just as the drive is the reverse side of desire, its shadow and its 'negative.'"[131] The psychoanalytic voice remains silent like the drive. It never appears because there is no place in the imaginary and symbolic orders where it can be manifested. Yet, somehow, at the level of the real, it addresses the subject. Acousmatic silence operates as the other, opposite side of the acousmatic voice.

Although the presence of the analyst supports this voice, Dolar is at pains to argue that the psychoanalytic voice is not the analyst's voice. "Psychoanalysis, in its elementary form, places side by side an analysand who speaks...and an analyst who keeps silent. The analyst's stance, in a different register, consists in turning himself into the agent of a voice which coincides with the silence of the drives...thus turning the silence into an act."[132] The term "act" is a constant in Žižek and Dolar's discourse that contrasts with the term "activity." If an activity is simply the perpetuation of the symbolic order, an act performed in the name of the big Other is "the moment when the subject who is its bearer *suspends* the network of symbolic fictions which serves as a support to his daily life and confronts again the radical negativity on which they are focused...."[133] The analyst, by acting as a phenomenal manifestation of the silence of the drives, creates a situation that suspends the symbolic fictions of everyday discourse and forces the analysand to do the same.[134]

The psychoanalytic voice recapitulates the three reductions that allowed the object voice to emerge. It is an alogical voice in that it makes no statement; it is an aphonic voice in that it is silent; and it is an atopical voice in that it has no definitive source. The analyst's job is to become an agent for the silent psychoanalytic voice, to turn the silence of the drive into an act. This forces the analysand's voice to undergo the same three reductions. How so?

First, the analyst's silence disturbs the analysand's flow of language. The analysand talks and talks, sliding along the chain of signifiers and gathering surplus pleasure from the endless flow of words. For Dolar, "the function of the analyst's silence is to interrupt this process, to bring it to a halt, to introduce a break, a gap in that flow, in this production of meaning....The poetry of the unconscious [its endless production of speech] falls on the deaf ears of the analyst...."[135] By halting the endless sliding of the signifier, the analyst's silence reveals the analysand's speech to be a stream of nonsense. It exposes the lack of logos in the analysand's discourse, disclosing it as an alogical production.

Second, the analyst's silence dispossesses the analysand of his or her voice. "In the presence of this silent other, a simple and striking effect sets in: words are suddenly transposed into a dimension where they start to sound strange and hollow; the moment the analysand hears his or her own voice against the backdrop of that silence, there is a structural effect which we could call the *dispossession of the voice*, its expropriation.... It ceases to be the asset of self-presence and auto-affection."[136] The analysand begins to hear that their voice is not their own. It speaks the very things that it has been ordered to speak by the big Other. The voice that emerges from the mouth of the analysand is a ventriloquial voice, one that has reproduced the network of symbolic fictions that supports daily life and commerce. This is the moment when the analysand discovers that the language and desire of the Other is lodged in them—that they are alienated in language. The subject learns that they are split between statement and enunciation; that they are not in the place where they speak. The voice becomes anonymous and thus atopical.

Third, the voice returns to the analysand but in inverted form. In the final stage of analysis, "the voice comes back to us through the loop of the Other, and what comes back to us from this Other is the pure alterity of what is said, that is, the voice."[137] This is the paradoxical aphonic *phoné*, the silent voice of the subject of enunciation; it must be silent since enunciation is never itself phenomenally manifested in sound. Thus silence is "the other of speech, not just of sound, it is inscribed inside the register of speech...."[138] Since alienation in language is the condition of every speech act, sounding speech is always the speech of the big Other. The only way to break this alienation is through othering the Other, through the arrival at the alterity of the statement—that is, silent enunciation. In everyday discourse, "we expect a response from the Other, we address it in the hope of a response."[139] This expectation is broken in the discourse between analyst and analysand when the analyst does not offer a response. Instead, "all we get is the voice," that is, acousmatic silence.[140] The words of the analysand are returned, but with their logos, topos, and *phoné* undercut. The analysand can now hear the silent strand within their speech that marks the other of the big Other (which is language). They no longer hear their discourse as voicing unsatisfied desires. "*The message of desire is returned as the voice of the drive.*"[141] In returning, the endless, insatiable dialectic of desire is avoided. The analysand hears their speech differently, as a split subject who recognizes both their inextricable alienation in language and the unsatisfiability of their desires; yet, by learning to hear themselves as the other of this language, the other of desire, and to take responsibility for this otherness, they learn, like the drive, how to go on. "The voice," in this final phase, "is what is said turned into its alterity, but the responsibility is the subject's own, not the Other's, which means that the subject is responsible not only for what he or she said, but must at the same time respond for, and respond to, the alterity of his or her own speech. He or she said something more than he or she intended, and this surplus is the voice which is merely produced by being passed through the loop of the Other."[142]

Phantasmagoria and fidelity

In the final stage of the psychoanalytic treatment, the subject has traversed all three stages in the reduction of the voice (alogical, atopical, aphonic) to encounter the psychoanalytic voice proper—the acousmatic silence that comes from within me yet from beyond me in the form of an address without prescription. At the beginning of the treatment, the analyst assumed the role of supporting this voice; but at the end, the analysand has learned to take responsibility for this psychoanalytic voice by hearing it anew. "The last stage of this [therapeutic] trajectory," writes Dolar, "would be the passage from the position of the analysand to that of the analyst: it is a way of remaining faithful to this experience, to this event, to this voice, by assuming its position, by representing the very object voice."

Remaining *faithful*? The word should give us pause. Faithfulness, fidelity—these were the terms previously used to describe the subject enthralled by the acousmatic voice of the master. In Dolar's bestiary, there is no animal more faithful or possessing more fidelity to its master than the dog.[143] Nipper, who listens faithfully to "His Master's Voice," functioned as an icon for the ideological subject—obedient to interpellation by the big Other. The dog is Dolar's figure of the subject subjected to "His Master's Voice."

> The sound [of the gramophone] is so realistic that even animals are taken in. The high *fidelity* of the sound finds it perfect match in the high *fidelity* of the dog. The dog doesn't see the source of the voice, he is puzzled and staring into the mysterious orifice, but he believes—he believes all the more for not seeing the source; the acousmatic master is more of a master than his banal visible versions.[144]

Fidelity is precisely what keeps the ideological subject in the grip of the big Other. It is a form of behavior well-suited to Lacan's famous formulation: "I know very well that…but nevertheless…." Dolar champions psychoanalysis as an alternative to prescriptive interpellation and its demands for fidelity. The analyst guides the analysand away from the acousmatic master toward recognition of the silent voice of enunciation, where the subject discovers that they are split; the analysand hears the silent voice as a voice that interpellates without prescriptions, thus allowing them the possibility of becoming responsible for the fullness of their own speech. Yet, at the termination of the therapy, Dolar tells us to remain "faithful to this experience, to this event, to this voice." Fidelity switches sides, now appearing as a non-prescriptive prescription for the split subject, not the dominating tool of the ideologue. Where should we place fickle fidelity?

It is hard to tell. We remain faithful to either the intoning acousmatic voice of the master (as Nipper did) or the silent acousmatic voice heard in the analytic session. Dolar clearly contrasts the two alternatives:

> On the one hand, there is the [ideological] process of becoming a subject by recognizing oneself as the address of that call, which would then be a version of His Master's Voice issuing positive prescriptions; on the other hand there is at the same time a [silent] voice which interpellates without any positive content—something one would perhaps rather escape by obeying the sonorous voice of statements and commands….[145]

Faithfulness to "His Master's Voice" invokes one form of fidelity—the canine fidelity of the ideological subject who remains obedient to the prescriptive interpellation of the big Other. Faithfulness to the silent voice invokes a contrasting form of fidelity—a therapeutic fidelity of the split subject who remains attached to the epiphany of the analytic session. These two forms of fidelity delineate two forms of the subject.

> In the first case one turns into a subject precisely by assuming the form of the autonomous "I," disavowing its heteronomic origin, so that the ideological

domination and autonomous subjectivity work hand in hand…in the second case one becomes a subject only by *fidelity* to the "foreign kernel" of the voice which cannot be appropriated by the self, thus by following precisely the heteronomic break in which one cannot recognize oneself.[146]

In both cases, the subject's fidelity is to a particular kind of acousmatic voice: the acousmatic voice of the master in contrast to the acousmatic silence of the psychoanalytic voice.[147]

Perhaps this is no coincidence. I would like to suggest that there is a reason why the acousmatic voice demands fidelity or faithfulness from a subject. In chapter 4, I argued that acousmatic sound (a classification under which we can include the acousmatic voice) requires the use of *technê*. However, the contribution that *technê* makes to the production of acousmatic sound is, for the most part, dismissed or bracketed out of consideration. The simultaneous necessity and expulsion of *technê* aligns the tradition of acousmatic sound with a tradition of phantasmagoria, or the occultation of the means of production. By expelling *technê*, acousmatic sounds are often experienced as autonomous, transcendent, or otherworldly.

Dolar's use of the acousmatic voice is no exception to this tradition of phantasmagoria, of the simultaneous employment and dismissal of *technê*. Take, for instance, his reading of the Pythagorean veil. First, we encounter the *akousmatikoi* in the thrall of the philosopher's voice—obedient, silent disciples who cannot see the master. Dolar describes the shock that the newly initiated *mathematikoi* would have experienced as they cross the threshold of Pythagoras' veil. The moment when they see the powerlessness of the master, the *mathematikoi* would also come to realize the constitutive, technical role of the veil in producing the effect of the acousmatic voice. The technical trick would be exposed in a scene of dis-acousmatization. But, as we know, in Dolar's reading, the scene of dis-acousmatization is not sustained; instead, the disciples retroactively deny that the trauma ever happened by fetishistically maintaining their belief in the power of the acousmatic voice. They install a screen of fetishism, motivated by a need to repress (and exscribe) the castrating scene. At the same instant that the disciples install this fetishistic screen, the *technê* of the Pythagorean veil would be expelled. The acousmatic voice of the master returns when *technê* is disregarded.

But Dolar goes even further than the Pythagorean disciples. In his own retroactive revision, Dolar, as you will recall, argues that "*there is no such thing as disacousmatization*," because "the source of the voice can never be seen, it stems from an undisclosed and structurally concealed interior, it cannot possibly match what we can see."[148] When Dolar utters these words, he draws an impassable border between the two sensory registers, an incongruity that makes the voice's acousmaticity into an a priori effect. At that moment, simply because of a primordial sensory mismatch, the acousmatic voice becomes *permanently* phantasmagoric—no longer affected by its source, context, or means of production. *Technê* is permanently expelled. And yet, if the voice is structurally and permanently acousmatic, then how could the initiates have experienced anything traumatic when they stepped behind Pythagoras' veil? If "there is no such thing as disacousmatization," then it trivially follows that there is no such thing as traumatic dis-acousmatization. The revelation of (or even the desire to reveal) the source of the acousmatic voice would be simply irrelevant. Moreover, if "there is no such thing as disacousmatization," then why would one ever need to employ the technology of the Pythagorean veil in the first place? The reasons to employ it become unmotivated, and Dolar's account becomes incoherent.

Like Heidegger, writing about the "friend," Dolar occasionally comes close to acknowledging the phantasmagoric nature of his claims. "Voices," he writes in a fugitive passage, "may be all in the head, without an external source, because we always hear the voice inside the head, and the nature of its external source is always uncertain the moment we close our eyes."[149] Again, if "there is no such thing as disacousmatization," then why does closing our eyes make any difference? The irreconcilability of the voice with the visible order would hold regardless of whether our eyes are open or closed. Closing our eyes would be redundant; it would produce no effect. Yet, by claiming that the source of the voice becomes uncertain the moment that one closes one's eyes, Dolar replicates the most common bodily technique for the production of phantasmagoric acousmatic sound, while simultaneously dismissing the role of *technê* in its production. *Technê* is implicitly acknowledged and explicitly ignored.

In every context where Dolar discusses the acousmatic voice, *technê* plays a significant, albeit unacknowledged, role. Why should we require the Pythagorean veil to grant philosophical discourse the power of sourceless enunciation, omnipotence, and omnipresence? Why should the horn of the gramophone make the obedient subject believe all the more in the power of the voice? Why would the source of the voice become uncertain upon closing one's eyes? Why should the analyst sit outside the analysand's field of vision?

Dolar's emphasis on fidelity to the acousmatic voice is phantasmagoric because phantasmagoria is a form of fidelity. Phantasmagoria is a mode of belief precisely captured by the Lacanian formulation "I know very well...but nevertheless...." Fidelity to the acousmatic voice only operates when *technê* is bracketed. We can modify Lacan's formula to capture this bracketing of *technê*: "I know very well that the voice has a source, but nevertheless [when I close my eyes], I believe that it doesn't." We can also apply this formula to other cases of acousmatic phantasmagoria: "I, Joseph Berglinger, know very well the mundane origin of music, but nevertheless [when I avert my eyes], I believe that my soul is detached from my body, that the tones become words." "I, anonymous author of the *Musikalische Eilpost*, know the hideous and distracting expressions of the musical performer, but nevertheless [when the concert hall is correctly lit and the musicians are veiled], I believe the spring-like sounds to come from the heavenly regions of a more beautiful world." "I, Richard Wagner, know very well that the orchestra is a machine, but nevertheless [when I construct a theater to hide the machinery of musical production], I believe that I enter a state akin to hypnotic clairvoyance." "I, Pierre Schaeffer, know very well the sources of the sounds I recorded, but nevertheless [when I remove their attack or lock them into a groove], I believe that they are only intentional objects, and that I myself have constituted them." In order for the acousmatic effect to be produced, the technical clause—here, literally bracketed—is typically suppressed or taken out of consideration.[150] The linguistic formulation helps to capture the suppression and bracketing of *technê*, the phantasmagoric occultation of the means of production required for the fantasy to operate.

Dolar goes too far when he claims that there is no such thing as dis-acousmatization. His claim is also easily recast in Lacan's formula: "I, Mladen Dolar, know very well that the voice is not acousmatic, but nevertheless [when I close my eyes, listen on the far side of the veil, attend to the speaker's mouth or the phonograph's horn], I believe that there is no such thing as disacousmatization." "*Je sais bien, mais quand même....*" This is the structure of fidelity. It too is preserved in the psychoanalytic session. "I, analysand, know very well that the analyst is in the room speaking with me, but nevertheless [when I lie on the couch and cannot see the analyst's face], I believe the voice is someone else's, the acousmatic voice of the Other, the acousmatic silence of the drive...." Moreover, I remain faithful to its therapeutic value. The fantasy of the Lacanian psychoanalytic session is that the analyst, functioning as a support for the silence of the drive, can create an impasse in the analysand's speech and eventually force them to recognize their responsibility over their own language. But one must not forget that Lacanian psychoanalysis is a technique, involving much more *technê* than simply repeating back the words that the patient speaks or giving presence to silence. (If it were just repetition or silence, then the phonographic voice would do just as well. For the sense of hollowness and otherness that inhabits our speech, the alogical and atopical aspects of speech, could just as easily be experienced by listening to a recording of our voice.) The moment we forget that psychoanalysis is a form of *technê*, we begin to take the analyst's silence for the silence of the drives; we confound the psychoanalyst's voice with the psychoanalytic voice.

Is the analyst ever able to occupy the place of the drives? Only through the application of a little technical trick. To say so is not to dispute the therapeutic value of psychoanalysis. It is simply to acknowledge that when we hear the psychoanalyst's voice as the psychoanalytic voice, when we permit the effects of this little technical trick to do its work—we do so just as we do with any other kind of acousmatic sound. *Technê* is the supplement that allows the acousmatic effect to emerge. To ignore that is to remain faithful to phantasmagoria.

Conclusion

We have come a long way from Moodus. By way of Schaeffer's studio, ancient Athens and Alexandria, Ansacq, Rome, Bayreuth, the Dalmatian seaside, Kafka's impenetrable passageways, the set of *Singin' in the Rain, Romeo and Juliet*'s balcony, Les Paul's in-house recording studio, and Freiburg's forest paths, the pursuit of acousmatic sound delivered us, unexpectedly, to the analysand's couch.

All trips must conclude somewhere, and a line from one of Bertrand Russell's lectures strikes me as germane to the conclusions I would like to draw. "The point of philosophy," Russell writes, "is to start with something so simple as not to seem worth stating, and to end with something so paradoxical that no one will believe it."[1] Although I do not think that the point of philosophy is to produce incredible paradoxes *for their own sake* (which I do not think is Russell's point, either), it is surely the case that philosophical arguments often travel the path from the simple to the paradoxical, and rarely the other direction. But there is a good reason; the obvious is never as obvious as it seems. When scrutinizing a belief, claim, or practice—especially one that concerns the unruly, inconsistent, and often contradictory domains of listening and sound—the thing scrutinized tends to shatter under the pressure of inquiry, perpetually pointing beyond itself to a whole network of supporting beliefs, claims, techniques, and practices that, in turn, are supported in precisely the same manner.

While this situation can lead to a kind of relativistic despair, there is another way to interpret it. Given a web of practices of a certain size and complexity, of perpetual relay and return, its tensile properties begin to emerge. From it comes a "form of life," as Wittgenstein famously called it. Stanley Cavell once described Wittgenstein's notion as a "whirl of organism," which gathers its strength from the fact that we share "routes of interest and feeling, modes of response, senses of humor and of significance and of fulfillment, of what is outrageous, of what is similar to what else, what a rebuke, what forgiveness, of when an utterance is an assertion, when an appeal, when an explanation...."[2] Of course, these routes of interest are shaped by the practices that came before, connected with the biological norms of human beings (that it has *these* sense organs, that its body is shaped in *these* ways, that it possesses *these* sensory ratios and thresholds...). It is in the coordination and agreement, in the overlapping and sharing of forms of life, that the relativist worry is set aside. All of our cultural practices "rest upon nothing more, but nothing less" than forms of life.

Acousmatic listening is one such practice. More than simply a static mode of hearing without seeing, when scrutinized, it relies on the support of other practices. (The fact that it points to other practices is perhaps what makes it a "route," while the value we place on it makes it a "route of interest.") It depends on historically cultivated ways of employing the senses and their ratios; it points us to a conceptualization of sound on the basis of a tripartite ontology of source, cause, and effect; it employs a panoply of techniques (bodily, architectural, and otherwise) to direct attention onto transcendent effects; it imbues transcendent effects with the values it affirms in its cultural surroundings; and it cleaves to historical and cultural forms of fidelity. The philosophical structure of acousmatic listening is nothing more, and nothing less, than the articulation of its genesis and location in a form of life.

If we consider acousmatic listening as a node in a network of cultural practices, perhaps now is the moment when, after defining and exemplifying the various routes of interest to which it points and on which it depends, we can imagine ourselves standing back and inhabiting the impossible position of overlooking the whole form of life. Imagine reaching an index finger into this web to pluck the node of acousmatic listening as one would pluck an instrument's taut string. As we release this string and let it sound, perhaps I can direct your attention to some of its higher partials, to some of the overtones that spill out, as all sounds invariably do, into the surrounding environment.

An acousmatic sound is often defined as a sound that one hears without seeing its cause, a sound heard in the absence of any visual information. However, I have tried to argue that acousmatic sound is not best characterized in terms of a division between two sensory registers—registers of the kind that Marshall McLuhan used to invoke in his comparisons of visual and auditory space. Rather, the experience of acousmatic sound is epistemological in character, articulated in terms of knowledge, certainty, and uncertainty. Even when described in terms of seeing and hearing, the experience of acousmatic sound is not fundamentally *about* seeing and hearing. The difference between the eye and the ear is really a synecdoche for the different ways the mind apprehends the exterior world, modulated through the sense organs. While the tripartite ontology of sound (with its three moments of source, cause, and effect) is specific to the way the world is accessed via the sense of hearing, hearing does not give the listener a world distinct from the world encountered in seeing. It simply shapes the way that world is constituted in a form of life, just as the other senses do the same. In contrast to our ubiquitous, multi-modal experience of the world, where knowledge is gathered and correlated from multiple sources, the experience of acousmatic sound is an exception. Far from simply isolating the ear and offering a privileged glimpse into the essence of listening, the acousmatic listener continually attempts to use the knowledge he or she has garnered from fellow senses to make sense of his or her auditory experience. If the experience of acousmatic sound relied on the ear alone, there would be no accounting for the imaginative supplementation that nearly always occurs—the projection of the vocalic body onto sonic effects, the belief in the omniscience and omnipresence of the *acousmêtre*, or the frantic hypothesis about source and cause—nor could one account for the acousmaticity of a sound, which can happen even in the full light of day.

If we make acousmaticity the criterion for identifying a sound as acousmatic, it follows that not every sound heard from a loudspeaker is de jure acousmatic.[3] Even in cases of *musique acousmatique*, when listening to sounds coming from loudspeakers, one is often quite certain about the sound's source, cause, and effect.[4] By placing the emphasis on a sound's acousmaticity, one discovers grounds for distinguishing between an *acousmatic* sound and a *schizophonic* sound—two terms that are often treated synonymously.[5] Schizophonic sounds, in R. Murray Schafer's description, are sonic copies that have been separated from their original context, usually as the result of mechanical reproduction. The criterion for schizophonic sounds requires both a copy and an original. Acousmatic sound, when defined in terms of acousmaticity, does not require this; it only requires spacing of the source, cause, and effect. Dolar's incongruous voice that seems incommensurable with the body that emits it and the final scene of *Singin' in the Rain* are instances of acousmatic sounds that are not schizophonic. In those instances, the issue of original and copy is simply not pertinent.

Acousmaticity, the determination or degree of spacing between source, cause, and effect, depends on the cognitive state of the listener and the knowledge they possess about the sound heard, its environmental situation, and its means of production, among other factors. To put it bluntly, acousmaticity is ultimately a judgment of the listener, not an intrinsic quality of the sound itself. Two listeners may experience different degrees of acousmaticity (or lack thereof) in the same sound. The acousmaticity of the sounds heard on Les Paul's recording of "Lover" would be very different for Les Paul and for his listeners. Similarly, in the vocal interactions between analyst and analysand, the psychoanalyst's voice is not acousmatic, but it becomes so when it is taken for the psychoanalytic voice. Frequently, knowing the means of production is an effective way of reducing acousmaticity. Thus my theory of acousmatic sound could be described as subject-oriented; to hear an acousmatic sound is simply to hear a sound acousmatically.[6]

This subject-oriented aspect of acousmatic sound should be differentiated from a current thesis in sound studies, one that promotes a wholly affective approach to sound. Inspired by the work of Deleuze, this thesis starts from a view about the material properties of sound but moves quickly in the direction of the subject's affective states. The claim is that sound is a material, vibrational force; when it encounters a body, this force makes a direct impact on the nervous system of the listener, one that bypasses his or her cognitive categories and forms of representation (to use Kantian language); the impact produces immediate somatic, affective states. This has become a common view in theories of noise, immersive sound, and sound art.[7] In lieu of a longer exposition and critique of this vibrational-affective thesis—something for another occasion—we might pose this question: How can it account for the experience of acousmatic sound? Take, for example, the case when an acousmatic sound becomes dis-acousmatized. To experience a change in the acousmaticity of a sound, there need not be *any* alteration in the vibrational properties of the sound.[8] A change in acousmaticity can only be articulated in cognitive terms. The vibrational-affective theorist has no interest in the cognitive, epistemological dimension of listening, which, I take it, is precisely their point. Yet that leaves them unable to account for acousmatic sounds. Although it is uncontroversial that acoustical vibrations produce sounds, the information that vibrations carry—information about the source, cause, and effect of a sound, and on which perceptual qualities like pitch and timbre depend—is crucial for nearly all our dealings with sound. The tripartite sonic ontology employed in this book can accommodate the acoustical fact that vibrations produce sounds, but it is pitched at a slightly higher register, one better suited to both capture the different ways that we ordinarily talk about (and distinguish) sounds and vibrations, and articulate central and salient aspects of sonic experience.[9] This register does not exclude the existence of affect, either—for what is the anxiety of Kafka's burrower if not an affective state?—but it situates affect in relation to the conceptual determination (and underdetermination) of sounds. To put it concisely: While my argument is subject-oriented and mediately affective, theirs is object-oriented and immediately affective.

Technê, in this context, is often deployed for the sake of producing acousmatic sounds. The latter are often taken as transcendent, where the value of transcendence is inflected through connection to the various practices—religious, secular, aesthetic, ethical, and so forth—available in the cultural situation. While neither *technê* nor *physis* are reducible to individual moments in the tripartite ontology of sound, they can be described in terms of the relationship that holds among all three moments. Techniques intervene in the spacing of source, cause, and effect, creating conditions whereby the effect can be experienced apart from its source or cause and taken as a product of the sound's *physis*. Bodily techniques can prepare the listener for attentive focus on the effect while bracketing the source or cause; physical barriers can obscure the source or cause; technologies can create conditions whereby a presumed source (the black box on Les Paul's guitar) masks a real source (the playback device) in order to produce extraordinary effects that seem in excess of the source. *Physis* shines through the effect in those extraordinary situations (usually prepared by *technê*) when the source and cause seem wholly irrelevant, impotent, or are imaginatively replaced with sources and causes from another realm entirely.

In the trip from Moodus to the psychoanalyst's couch, I have considered many cases of acousmatic sound. They cover a territory that is broader than music alone, dissipating into many kinds of sonic practice. Acousmatic sounds do not respect that unfortunate line that is often drawn between music studies and sound studies—a line that I hope this book goes some distance to erase. Although it seems a truism to even say it, music studies *is* a species of sound studies and auditory culture. I suppose that, to some, this proposition is disconcerting, as if some alien discipline has come to territorialize music.

I see it differently. Music and the practices concerned with it represent one of the most highly developed parts of our shared auditory culture. While certain domains of sound or music are neatly bound within the web of practices that make up an auditory culture, others—like acousmatic sound—are not so sharply delimited. Operating as a node in the tensile mesh of a form of life, acousmatic sound is a point where disparate auditory and cultural practices intersect. It is the secret vibration that runs through their knotted fibers. Acousmatic sound gathers them together—music studies, sound studies, philosophy, literature, film, and psychoanalysis—with a sublime indifference to disciplinary propriety. Sound respects no boundaries, and neither does sound unseen.

Notes

Introduction
1. Modeled on the parallel term "landmark," a soundmark, according to R. Murray Schafer, "refers to a community sound which is unique or possess qualities which make it specially regarded or noticed by the people in that community." Schafer, *The Soundscape: Our Sonic Environment and the Tuning of the World* (Rochester, NY: Destiny Books, 1994), 10.
2. Odell Shepard, *Connecticut Past and Present* (New York: Alfred A. Knopf, 1939), 104.
3. Stephen Hosmore to Thomas Prince, August 13, 1729, in *Collections of the Connecticut Historical Society*, Vol. 3 (1890), 280–281.
4. Ibid., 281.
5. Richard Cullen Rath, *How Early America Sounded* (Ithaca, NY: Cornell University Press, 2003), 29.
6. Ibid., 41.
7. Anonymous, *Connecticut Gazette* (August 20, 1790), 3.
8. John Warner Barber, *Connecticut Historical Collections Containing a General Collection of Interesting Facts, Traditions, Biographical Sketches, Anecdotes, &c., Relating to the History and Antiquities of Every Town in Connecticut, with Geographical Descriptions* (New Haven, CT: J. W. Barber, 1836), 526.
9. For a sampling of legends concerning Dr. Steel, see Samuel Adams Drake, *A Book of New England Legends and Folk Lore* (Rutland, VT: C. E. Tuttle, 1971), 427–431; Federal Writer's Project for the State of Connecticut, *Connecticut; A Guide to Its Roads, Lore, and People* (Boston: Houghton Mifflin Co., 1938), 404–405; Hosford B. Niles, *The Old Chimney Stacks of East Haddam* (New York: Lowe & Co., 1887), 28–31; David E. Philips, *Legendary Connecticut* (Hartford, CT: Spoonwood Press, 1984), 199–203; Carl F. Price, *Yankee Township* (East Hampton, NY: Citizens' Welfare Club, 1941), 167–181; Charles Burr Todd, *In Olde Connecticut; Being a Record of Quaint, Curious and Romantic Happenings There in Colonial Times and Later* (New York: Grafton Press, 1906), 142–152; Clarence M. Webster, *Town Meeting Country* (New York: Duell, Sloan & Pearce, 1945), 94; and Glenn White, *Folk Tales of Connecticut* (Meriden, CT: The Journal Press, 1977), 23–25.
10. Barber, *Connecticut Historical Collections*, 527.
11. Jelle Zeilinga de Boer, *Stories in Stone* (Hanover, CT: Wesleyan University Press, 2009), 143.
12. Barber, *Connecticut Historical Collections*, 527.
13. de Boer, *Stories in Stone*, 144.
14. Todd, *In Olde Connecticut*, 151–152.
15. de Boer, *Stories in Stone*, 148.
16. Ibid., 149.

17. Price, *Yankee Township*, 181.
18. Erik Hesselberg, "'Moodus Noises' Strike Again," *Hartford Courant*, March 24, 2011 [accessed 12/27/2012: http://www.courant.com/news/connecticut/hc-moodus-quake-0325-20110324,0,6968586.story].
19. Pierre Schaeffer, *Traité des objets musicaux* (Paris: Éditions du Seuil, 1966), 91. All translations from the *Traité* are mine, except for passages from chapter 4, "L'acousmatique." An abridged translation of that chapter is available as "Acousmatics," in *Audio Culture: Readings in Modern Music*, eds. Christoph Cox and Daniel Warner (New York: Continuum Books, 2004), 76–81.
20. See Michel Chion, "Acousmatique," in *Guide des objets sonores* (Paris: Buchet/Chastel, 1983), §1, 18. John Dack's translation (which I use here, slightly modified) is published online by EARS (the ElectroAcoustic Resourse Site) as *Guide to the Sound Object*, trans. John Dack [accessed 1/9/2013: http://www.ears.dmu.ac.uk/spip.php?page=articleEars&id_article=3597].
21. Ibid., "Acousmatique," §2c, 19.
22. Ibid., "Acousmatique," §2b, 19.
23. Roger Scruton, *The Aesthetics of Music* (New York: Oxford University Press, 1997), 2–3.
24. On Freud's office, see Edmund Engelman, *Berggasse 19: Sigmund Freud's Home and Offices, Vienna, 1938* (New York: Basic Books, 1976), especially plates 12 and 13; on the late topography, see Sigmund Freud, *The Ego and the Id*, trans. James Strachey (New York: Norton, 1989); on technique and role of modern audio technologies, see Sigmund Freud, "Recommendations for Physicians on the Psycho-Analytic Method of Treatment," in *Collected Papers, Vol. 2*, trans. under the supervision of Joan Riviere (New York: Basic Books, 1959); on the sonorous envelope, see Philippe Lacoue-Labarthe, "The Echo of the Subject," in *Typography*, trans. Christopher Fynsk (Stanford, CA: Stanford University Press, 1989); and Kaja Silverman, *The Acoustic Mirror: The Female Voice in Psychoanalysis and Cinema* (Bloomington, IN: Indiana University Press, 1988).
25. On *clausura*, see Craig Monson, "Putting Bolognese Nun Musicians in their Place," in *Women's Voices Across Musical Worlds*, ed. Jane Bernstein (Boston: Northeastern University Press, 2004); on *Dunkelkonzerte*, see Bryan Gilliam, "The Annexation of Anton Bruckner: Nazi Revisionism and the Politics of Appropriation," *The Musical Quarterly*, Vol. 78, No. 3 (Autumn 1994), 584–604; and Friedrich C. Heller, "Von der Arbeiterkultur zur Theatersperre," in *Das Wiener Konzerthaus*, eds. Friedrich C. Heller and Peter Revers (Vienna: Wiener Konzerthausgesellscaft, 1983), 101; on Bayreuth's architecture, see *Bayreuth, the Early Years*, ed. Robert Hartford (Cambridge, UK: Cambridge University Press, 1980); on other hidden orchestras after Bayreuth, see Heinrich W. Schwab, *Konzert: öffentliche Musikdarbietung vom 17. bis 19. Jahrhundert* (Leipzig, DE: Deutscher Verlag für Musik, 1971), 186–189.
26. Michel Chion, *The Voice in Cinema*, trans. Claudia Gorbman (New York: Columbia University Press, 1999); Carolyn Abbate, *In Search of Opera* (Princeton, NJ: Princeton University Press, 2001); Slavoj Žižek, *Looking Awry: An Introduction to Jacques Lacan through Popular Culture* (Cambridge, MA: MIT Press, 1991), 124–130; Mladen Dolar, *A Voice and Nothing More* (Cambridge, MA: MIT Press, 2006); on Ellison and acousmatic technology, see Mark Goble, *Beautiful Circuits: Modernism and the Mediated Life* (New York: Columbia University Press, 2010); on "acousmatic

blackness," see Mendi Obadike, "Low Fidelity: Stereotyped Blackness in the Field of Sound" (doctoral dissertation, Duke University, 2005); and Nina Sun Eidsheim, "Voice as a Technology of Selfhood: Towards an Analysis of Racialized Timbre and Vocal Performance" (doctoral disseration, University of California, San Diego, 2008). While mentioning work on acousmatic sound, I should give special note to Mandy-Suzanne Wong's "Sound Objects: Speculative Perspectives" (doctoral dissertation, University of California, Los Angeles, 2012), which treats the various ways that the term "sound object" has been used in musical discourse in the wake of Schaeffer's coinage, with different degrees of fidelity to his intentions.

27. I am aware of conflating acousmatic *sound* and acousmatic *listening* in this introduction. However, the conflation is not pernicious. Later, I argue that there is no acousmatic sound *as such*; its acousmaticity depends on the conditions under which a sound is heard.

28. Appealing to Occam's razor, I would argue that despite the simplicity of this model, it is sufficiently robust for analyzing the cases of acousmatic sound discussed in this book.

29. For instance, auditory scene analysis and other ecological approaches to listening would support this claim, committed as they are to the primacy of the environmental situation of a listener in all discussions of auditory perception. See Albert S. Bregman, *Auditory Scene Analysis: The Perceptual Organization of Sound* (Cambridge, MA: MIT Press, 1990); and Eric F. Clarke, *Ways of Listening: An Ecological Approach to the Perception of Musical Meaning* (Oxford, UK: Oxford University Press, 2005).

30. I had a similar experience while attending the SuperCollider Symposium, a computer music conference, at Wesleyan. As I walked under a tree, I heard an odd-sounding bird chirping a strange but familiar tune. I spotted an enclosed loudspeaker and circuit hanging from the tree. The sound was nearly indistinguishable from a real bird, except that the bird was chirping select strains of Madonna's "Like a Virgin." The artist is Dan St. Clair, and the work is entitled *Call Notes*. For documentation, see http://www.hearingthings.net/projects/callnotes/.

31. To his credit, Chion is well aware of this fact. He writes, "Schaeffer thought the acousmatic situation could encourage reduced listening, in that it provokes one to separate oneself from causes ... in favor of consciously attending to the sonic textures, masses, and velocities. But, on the contrary, the opposite often occurs, at least at first, since the acousmatic situation intensifies causal listening [i.e., listening for the source] in taking away the aid of sight. Confronted with a sound from a loudspeaker without a visual calling card, the listener is led all the more to ask, 'What's that?' (i.e., 'What is causing the sound?') and to be attuned to the minutest clues (often interpreted wrong anyway) that might help to identify the cause." Michel Chion, *Audio-Vision*, trans. Claudia Gorbman (New York: Columbia University Press, 1994), 32. Yet, in the following paragraph, Chion wipes all of this away as merely preparatory for the discipline of reduced listening, aided by the repeated playback of sounds. "When we listen acousmatically to recorded sound it takes repeated hearings of a single sound to allow us gradually to stop attending to its cause and to more accurately perceive its own inherent traits." The premium remains on reduced listening (see chapter 1).

32. Steven Connor, *Dumbstruck: A Cultural History of Ventriloquism* (Oxford, UK: Oxford University Press, 2000), 35.

33. For a suggestive analogy, see Whitney Davis, *A General Theory of Visual Culture* (Princeton, NJ: Princeton University Press, 2011) and his use of the term "visuality."
34. In chapter 3, I discuss a moment in 18th-century France when a historical agent explicitly attempted to coin a word to describe an extraordinary audible phenomenon that was heard but not seen, what I would call an acousmatic sound.

Chapter 1

1. Journal entry of March 1948. Pierre Schaeffer, *A la recherche d'une musique concrète* (Paris: Éditions du Seuil, 1952). All further citations are from *In Search of a Concrete Music*, trans. Christine North and John Dack (Berkeley, CA: University of California Press, 2012). I have occasionally modified North and Dack's rendering when necessary.
2. March. Ibid., 4.
3. April 18. Ibid., 6.
4. May 5. Ibid., 12.
5. May 10. Ibid., 13.
6. April 19. Ibid., 7.
7. April 21. Ibid.
8. May 15. Ibid., 14.
9. Ibid.
10. April 21. Ibid., 7.
11. Schaeffer, *Traité des objets musicaux*, 20.
12. However, Schaeffer's method for theorizing the sound object would change dramatically. Throughout *In Search of a Concrete Music*, Schaeffer relies on information theory to indentify the sound object. This is in contrast to the phenomenological method of the *Traité*.
13. *Traité*, 95.
14. My emphasis on Schaeffer's Husserlianism contrasts with scholars who claim Merleau-Ponty as the central phenomenological influence on Schaeffer. See Michel Chion, *Guide des objets sonores*, 32; and Makis Solomos, "Schaeffer phénoménologue," in *Ouïr, entendre, écouter, comprendre après Schaeffer* (Paris: Buchet/Chastel, 1999), 53–67.
15. Marc Pierret, *Entretiens avec Pierre Schaeffer* (Paris: P. Belfond, 1969), 97.
16. *Traité*, 262.
17. Solomos, "Schaeffer phénoménologue," 57–58.
18. Maurice Merleau-Ponty, *Phenomenology of Perception*, trans. Colin Smith (London: Routledge & Kegan Paul, 1962), ix.
19. Merleau-Ponty, ibid., viii; *Traité*, 662.
20. Merleau-Ponty, ibid.
21. By the time *Traité* was published, Schaeffer would have had access to French translations of many of Husserl's central texts. The *Cartesian Meditations* were published in France in a translation by Gabrielle Peiffer and Emmauel Levinas in 1931. See *Méditations cartésiennes: introduction à la phénoménologie* (Paris: Librairie philosophique J. Vrin, 2001). Paul Ricoeur published a translation of *Ideas* in 1950 with an extensive commentary. See *Idées directrices pour une phénoménologie*, trans. Paul Ricoeur (Paris: Gallimard, 1950); an English translation of Ricoeur's commentary is available as Paul Ricoeur, *A Key to Edmund Husserl's Ideas I*, trans. Bond Harris

and Jacqueline Bouchard Spurlock (Milwaukee, WI: Marquette University Press, 1996). *Formal and Transcendental Logic* was available in 1957; see *Logique formelle et logique transcendantale: essai d'une critique de la raison logique*, trans. Suzanne Bachelard (Paris: Presses universitaires de France, 1957). Jean Wahl's lectures on the *Crisis of the European Sciences*, with copious quotations, appeared the same year; see *L'ouvrage posthume de Husserl: la "Krisis": la crise des sciences Européennes et la phénoménologie transcendantale* (Paris: Centre de Documentation Universitaire, 1957). From the material related to Husserl's *Crisis*, Husserl's important essay on the "Origin of Geometry" was available in a translation with extensive commentary by Jacques Derrida in 1962; see *L'origine de la geometrie*, trans. and ed. Jacques Derrida (Paris: Presses Universitaires de France, 1962); an English translation of Derrida's commentary is available as Jacques Derrida, *Edmund Husserl's Origin of Geometry: An Introduction*, trans. John. P. Leavey, Jr. (Lincoln, NE: University of Nebraska Press, 1982).
22. *Traité*, 262.
23. Schaeffer agrees, claiming that "a [sound] object is an object *only* for our listening, it is relative to it." *Traité*, 95.
24. A. D. Smith, *Husserl and the Cartesian Meditations* (New York: Routledge, 2003), 34. I acknowledge a debt to Smith's clear description of Husserl's theory of the object.
25. *Traité*, 263.
26. Ibid., 263–264.
27. Edmund Husserl, *Ideas Pertaining to a Pure Phenomenology and to a Phenomenological Philosophy*, First Book, trans. F. Kersten (The Hague, NL: Martinus Nijhjoff, 1982), §41. Although I am relying on Kersten's translation, I have also consulted W. R. Boyce Gibson's translation of *Ideas* (New York: Collier-Macmillan, 1962) and the original German edition, occasionally modifying Kersten's translation for the sake of clarity.
28. *Traité*, 263.
29. Ibid., 264.
30. The classic example of the transposable melody first appears in von Ehrenfels before it receives differing interpretations in the neo-Helmholtzian Graz school and the gestalt psychologists. For an overview of this history, see Jean Piaget, *The Psychology of Intelligence*, trans. Malcolm Piercy and D. E. Berlyne (London: Routledge, 2003), 61ff. Although Husserl talks about the temporal perception of a melody in his lectures on time consciousness, he discusses the transposition of a melody to make a different point concerning the productivity of fantasy. See Edmund Husserl, *On the Phenomenology of the Consciousness of Internal Time (1893–1917)*, trans. John B. Brough (Dordrecht, NL: Kluwer Academic Publishers, 1991), §4.
31. *Traité*, 267 (Schaeffer's emphasis).
32. Ibid., 264–265.
33. This is the closest Schaeffer comes to arguing for the intersubjectivity of the sound object: "We must therefore stress emphatically that [a sound] object is something real [i.e., objective], in other words that something in it endures through these changes and enables different listeners (or the same listener several times) to bring out as many aspects of it as there have been ways of focusing the ear, at the various levels of 'attention' or 'intention' of listening [*d'entendre*]." Pierre Schaeffer, *Solfège de l'objet sonore* (Paris: INA-GRM, 1998), 59. I have more to say about this passage below.

34. A. D. Smith, *Husserl and the Cartesian Meditations*, 230. Edmund Husserl, *Cartesian Meditations*, trans. Dorion Cairns (The Hague, NL: Martinus Nijhoff, 1970), §54, 149.
35. *Traité*, 265.
36. Husserl, *Ideas*, §27.
37. Ibid.
38. Dan Zahavi, *Husserl's Phenomenology* (Stanford, CA: Stanford University Press, 2003), 44.
39. Husserl, *Cartesian Meditations*, §2.
40. Aristotle, "Metaphysics," 982b, in *The Basic Works of Aristotle*, ed. Richard McKeon (New York: Random House, 1941).
41. *Traité*, 266.
42. Husserl, *Ideas*, 96.
43. Merleau-Ponty, *Phenomenology of Perception*, 21–22.
44. *Traité*, 92.
45. Husserl, *Ideas*, §32.
46. *Traité*, 91.
47. I offer an extensive treatment of Pythagorean origins of this term in chapter 2.
48. *Traité*, 91.
49. Ibid., 93.
50. Ibid.
51. Edmund Husserl, *Logical Investigations*, Vol. 2, trans. J. N. Findlay (New York: Humanities Press, 1970), §6, 503–505.
52. Schaeffer presents evidence for the non-correlation of frequency and pitch perception in his *Solfège*, CD 1, tracks 26–28.
53. *Traité*, 91.
54. Ibid., 268.
55. Ibid., 93.
56. Timothy Taylor, *Strange Sounds* (New York: Routledge, 2001), 46.
57. *Traité*, 268.
58. Ibid., 93–94.
59. Ibid., 33.
60. Ibid., 270.
61. May 5. Schaeffer, *In Search of a Concrete Music*, 12.
62. *Traité*, 103–128.
63. Ibid., 104.
64. Ibid.
65. Ibid. John Dack's decision to translate *ouïr* as *perceiving* captures the phenomenological aspects of Schaeffer's usage by allowing for a closer relation between Schaeffer's theories and Merleau-Ponty's. However, it does lead to some translation issues because of the presence of both *ouïr* and *percevoir* in French. For example, when Schaeffer writes, "*Ouïr, c'est percevoir par l'oreille...*" (104), we end up with the strange tautology "Perceiving, is to perceive by the ear," or even "Perception is auditory perception." See Chion, *Guide*, §6; and Dack's translation.
66. Chion, *Guide*, §6.
67. *Traité*, 106.

68. Denis Smalley, "The Listening Imagination: Listening in the Electroacoustic Era," *Contemporary Music Review*, Vol. 13, No. 2 (1996): 79.
69. *Traité*, 107. "Dirigé vers l'événement, j'adhérais à ma perception, je l'utilisais à mon insu."
70. May 7. Schaeffer, *In Search of a Concrete Music*, 12.
71. May 10. Ibid., 13.
72. This claim is substantiated by appeal to the rest of the *Cinq études de bruits*. Other than the *Étude aux tourniquets*, all other titles from the set do not make reference to their original sound sources. Schaeffer's later studies, such as the *Étude aux objets*, continue in this manner.
73. *Traité*, 104.
74. Ibid., 270–272.
75. Derrida, in his critique of Husserl, argues that "indicative signification," that is, the indexical sign, is always connected to "empirical existents in the world" and "covers everything that falls subject to [phenomenological] 'reductions': factuality, worldly existence, essential non-necessity, nonevidence, etc." See Jacques Derrida, *Speech and Phenomena*, trans. David B. Allison (Evanston, IL: Northwestern University Press, 1973), 30. An analogous case can be made for Schaeffer's proscription of *écouter*. I revisit Husserl's theory of the indicative sign in chapter 7.
76. *Traité*, 268.
77. John Dack, "Ear-Training Using the Computer and PROGREMU," 4 [accessed 12/1/2010: http://www.cea.mdx.ac.uk/local/media/downloads/Dack/Ear_training.pdf].
78. Pierre Schaeffer, *La musique concrète* (Paris: Presses Universitaires de France, 1967), 29.
79. Chion, *Guide*, §38.
80. Dack, "Ear-Training," 5. Schaeffer's *Tableau récapitulatif de la typologie* (TARTYP) is an attempt to organize just such a taxonomy. See *Traité*, 459.
81. Pierre Schaeffer, *L'œuvre musicale*, ed. François Bayle (Paris: INA-GRM, 1990), 85.
82. Ibid. A reliance on classical form runs throughout Schaeffer's compositions. The fugal-like layout of parts of the *Étude aux objets* could be traced back to the *Bidule en ut*, composed with Pierre Henry.
83. Pierre Henry, *Variations pour une porte et un soupir* (1963) from the CD *Mix 03.1* (France: Philips Music Group, 2001).
84. Schaeffer establishes these negative claims in a section of the *Traité* entitled "On the sound object: what it is not." See *Traité*, 95–98.
85. *Traité*, 95–96.
86. Ibid., 97.
87. Ibid.
88. Edmund Husserl and Ludwig Landgrebe, *Experience and Judgment*, trans. James S. Churchill and Karl Ameriks (Evanston, IL: Northwestern University Press, 1973), §87.
89. Ibid.
90. Husserl, *Cartesian Meditations*, §34.
91. Edmund Husserl, *The Idea of Phenomenology*, trans. William P. Alston and George Nakhnikian (The Hague, NL: Martinus Nijhoff, 1964), 53–54.
92. *Traité*, 92.

93. On "ideal objectivites," see Husserl's essay "The Origin of Geometry," in Husserl, *The Crisis of the European Sciences and Transcendental Phenomenology*, trans. David Carr (Evanston, IL: Northwestern University Press, 1970). For a more thorough analysis, see Jacques Derrida, *Edmund Husserl's Origin of Geometry: An Introduction*, which reprints Carr's translation of the essay.
94. Schaeffer, *Solfège de l'objet sonore*, 59–61.
95. Ibid., 59, "...les multiple attentions ou intentions d'entendre." The English translation misses the close association between attention, intention, and reduced listening.
96. Ibid., 61.
97. Husserl, *Ideas*, §89–90.
98. *Traité*, 269.
99. Martin Heidegger, *The Basic Problems of Phenomenology*, trans. Albert Hofstadter (Bloomington and Indianapolis, IN: Indiana University Press, 1982), 21.
100. *Traité*, 270.
101. Ibid.
102. Ibid.
103. Husserl, "The Origin of Geometry," 354. An interesting parallel is found in René Leibowitz, who apes this passage nearly phrase by phrase in the introduction to *Schoenberg and His School*: "Let us begin at the beginning by questioning the very origin of polyphony. It goes without saying that the question, as we put it here, has no historical or philological interest. It is not a matter of investigating the first polyphonists, or of hunting down the earliest extant essays in polyphony. What matters is to find the *original meaning* with which polyphony must have made its first historic appearance, with which it has developed through the centuries (thanks to a living tradition), with which it still appears to us today, even though we know almost nothing about its first creators." See René Leibowitz, *Schoenberg and his School*, trans. Dika Newlin (New York: Da Capo Press, 1979), xx. Schaeffer speaks disparagingly of Leibowitz and his advocacy for dodecaphony in *In Search of a Concrete Music*, chapter 15, 126ff., apparently without realizing their shared methodological commitment.
104. The phrase "regressive inquiry," or *Rückfrage*, appears in Husserl, "Origin of Geometry." The ahistoricism of Husserl's project has often been noted by other scholars of phenomenology, most powerfully by Paul Ricoeur in an essay for the *Revue de métaphysique et de morale* from 1949. He argues that Husserl's "philosophy of essence," is mistrustful of "genetic explanations." "The 'eidetic reduction,' which parenthesizes the individual case and retains only the sense (and the conceptual significance which it expresses), is in itself a reduction of history." In Husserl's work, "the notion of origin...no longer signifies historico-causal genesis but rather grounding." See Paul Ricoeur, "Husserl and the Sense of History," in *Husserl: An Analysis of His Phenomenology*, trans. Edward G. Ballard and Lester E. Embree (Evanston, IL: Northwestern University Press, 1967), 146.
105. Husserl, *Ideas*, §7.
106. Ibid.
107. Frances Dyson, *Sounding New Media: Immersion and Embodiment in the Arts and Culture* (Berkeley, CA: University of California Press, 2009), 10.
108. For instance, the converse does not hold. One cannot say that "the acousmatic experience is never clearly revealed except in the sound object." Chapters 4, 5, 6,

and 7 explore this possibility—a theory of acousmatic sound that does not entail Schaeffer's ontology of the sound object.
109. Theodor Adorno and Ernst Krenek, *Briefwechsel* (Frankfurt, DE: Suhrkamp Verlag, 1974), 12.
110. Ibid., 174–175.
111. Theodor Adorno, *Against Epistemology: A Metacritique*, trans. Willis Domingo (Cambridge, MA: MIT Press, 1982), 49.
112. Ibid., 6.
113. Michel Chion, *The Voice in Cinema*, 17–18.
114. Ibid., 23.
115. *Traité*, 93 (emphasis added).
116. See Carlos Palombini, "Technology and Pierre Schaeffer: Walter Benjamin's *technische Reproduzierbarkeit*, Martin Heidegger's *Ge-Stell* and Pierre Schaeffer's *Arts-Relais*," *Organised Sound*, Vol. 3, No. 1 (1998): 35–43.
117. Martin Heidegger, *The Question Concerning Technology*, trans. William Lovitt (New York: Harper Colophon Books, 1977), 4.
118. Ibid., 12.
119. Pierre Schaeffer, *Machines à communiquer* (Paris: Éditions du Seuil, 1970), 92.
120. Roland Barthes, *Mythologies*, trans. Annette Lavers (New York: Hill and Wang, 1972), 151.
121. François Bayle, *Musique acousmatique: propositions—positions* (Paris: INA, 1993), 181.
122. Quoted in Barthes, *Mythologies*, 151.

CHAPTER 2

1. Lynn Hasher, David Goldstein, and Thomas Toppino, "Frequency and the Conference of Referential Validity," *Journal of Verbal Learning and Verbal Behavior*, Vol. 16, No. 1 (1977): 107–112. Perhaps the truth effect offers the proof for the old saying "A lie told often enough becomes the truth." Surprisingly, this statement is often attributed to Lenin, although there is no solid evidence to support the attribution—an example of precisely the kind of thing the saying asserts.
2. Claude Levi-Strauss, *The Raw and The Cooked: Introduction to a Science of Mythology I*, trans. John and Doreen Weightman (New York: Harper & Row, 1969), 35.
3. Chion, *The Voice in Cinema*, 19 and fn. 5. This passage does not appear in the original French edition; see Michel Chion, *La voix au cinema* (Paris: Cahiers du Cinéma, 1982). According to Claudia Gorbman (personal correspondence), it was added by Chion to the English translation.
4. Bayle, *Musique acousmatique*, 179.
5. Ibid., 180.
6. Ibid., 181.
7. Ibid.
8. "New Media Dictionary," *Leonardo* Vol. 43, No. 3 (2001): 261. The dictionary was edited by Louise Poissant and appeared in a number of issues of *Leonardo*. According to her introduction, the entries were prepared by Francis Dhomont, Robert Normandeau, and Claire Piché. Although this entry is anonymous, I think it is safe to say that Dhomont is its author. The entry is nearly identical to Dhomont's liner notes to *Cycle D'errance*. See Francis Dhomont, "Acousmatic,

what is it?" in liner notes to *Cycle D'errance*, sound recording (Montréal: Diffusion i Média, 1996).
9. Francis Dhomont, "Is There a Québec Sound?" *Organised Sound*, Vol. 1, No. 1 (1996): 24.
10. Denis Diderot, "Acousmatiques," *Encyclopédie ou dictionnaire raisonné des sciences, des arts et des métiers*, Vol. 1 (Paris: chez Briasson, 1751), 111.
11. Mladen Dolar, *A Voice and Nothing More*, 61.
12. Jérôme Peignot, "De la musique concrete à l'acousmatique," *Esprit*, No. 280 (Jan. 1960): 116.
13. Beatriz Ferreyra, "CD Program Notes," *Computer Music Journal*, Vol. 25, No. 4 (Winter 2001): 118.
14. Kim S. Courchene, "A Conversation with Beatriz Ferreyra," *Computer Music Journal*, Vol. 25, No. 3 (Fall 2001): 17.
15. Carolyn Abbate, "Debussy's Phantom Sounds," *Cambridge Opera Journal*, Vol. 10, No. 1 (1988): 69.
16. Schaeffer, *Traité*, 91.
17. Marc Battier, "What the GRM Brought to Music: From Musique Concrète to Acousmatic Music," *Organised Sound*, Vol. 12, No. 3 (2007): 196.
18. Ibid.
19. In the text from which M8 is taken, Ferreyra does not speak about the modern use of the term "acousmate." Thus, there is not enough evidence to determine if this change of name is intended to justify some particular reading of the modern reception of acousmate.
20. As I will show below, the name "acousmatics" comes from the division of the Pythagorean school into two camps, the *mathematikoi* and the *akousmatikoi*. The term "acousmate," as I show in chapter 3, has a non-Pythagorean origin. While picking nits, M1–11 contain a few other small mistakes: (1) Contra Chion (M1), Apollinaire wrote *two* poems entitled "Acousmate," both written before 1900. See Apollinaire, *Œuvres Poétiques* (Paris: Gallimard, 1965), 513, 671. Also, Chion botches the date of Clement of Alexandria's *Stromateis* by placing it at 250 B.C.E. That would be a remarkable date for one of the Church fathers! The work was likely written in the beginning of the third century C.E. (2) Dhomont[?] (M3) says that acousmate derives from the Greek word *akouma* when it derives from the word *akousma* (ἄκουσμα), a noun meaning "the thing heard." (3) Dolar (M6) says that Larousse's entry on the acousmatics, which speaks of Pythagoras being concealed by a curtain, is based on Diogenes, VIII.10. See Diogenes Laertius, *Lives of Eminent Philosophers*, trans. R. Drew Hicks (Cambridge, MA: Harvard University Press, 1950). This cannot be the case since Diogenes, VIII.10, mentions no veil—a point to which I will return below. Iamblichus is the more likely source of Larousse's entry.
21. "When in 1948, I proposed the term 'concrete music,' I intended by this adjective, to mark an *inversion* in the sense of musical work. In place of notating musical ideas by solfège symbols, and to confine their concrete realization to known instruments, I started to collect concrete sounds…and abstract the musical values that they potentially contained." *Traité*, 23. For Schaeffer's overview, see *La musique concrète*, 16–17. For a philosophical reception of Schaeffer's concrete/abstract divide, see Peter Szendy, "Abstraite musique concrète ou tout l'art d'entendre," *Le comentaire*

et l'art abstrait, eds. Murielle Gagnebin and Christine Savinel (Paris: Presses de la Sorbonne Nouvelle, 1999), 73–96.
22. This statement comes from a 15-minute radio broadcast by Philippe Arthuys and Jérôme Peignot, entitled "Musique animée," Chaîne nationale, no. LUR301 (1955). It is cited by Évelyne Gayou, *Le GRM: le groupe de recherches musicales* (France, Éditions Fayard, 2007), 345.
23. The "cover of darkness" derives from Diogenes, VIII.15. The relationship between the cover of darkness and the Pythagorean veil will be treated below.
24. Bayle, *Musique acousmatique*, 180
25. Dhomont (M4) ends with a quotation from Bayle: "More recently—1974—, in proposing the term *acousmatic music* I wanted to designate work in the studio by distinguishing it from work with electroacoustic instruments...(ondes Martenot, electric guitar, synthesizers, digital audio systems in real time) and furnish an appropriate term for a music which is...developed in the studio, projected in the hall, like cinema." Bayle, *Musique acousmatique*, 181. Dhomont follows Bayle's own self-inscription into the history of acousmatic sound.
26. For more on Bayle's images or "i-sounds," see François Bayle, "Image-of-Sound, or i-Sound: Metaphor/Metaform," *Contemporary Music Review*, Vol. 4 (1989): 165–170.
27. Claude Levi-Strauss, *The Raw and The Cooked*, 16.
28. Levi-Strauss was no foreigner to *musique concrète*. The "overture" to *The Raw and The Cooked*, in comparing music and myth, explicitly criticizes *musique concrète* for its inability to operate with the two requisite levels of signification, and thus "flounders in non-significance" (23). For an informed response to Levi-Strauss, see John Dack, "Acoulogie: An Answer to Lévi-Strauss," *Proceedings of EMS07* (2007) [accessed 12/28/2010: http://www.ems-network.org/IMG/pdf_DackEMS07.pdf]. Although music and myth are akin for Lévi-Strauss, what about the myth of a music's origin?
29. Jean-Luc Nancy, *The Inoperative Community*, trans. Peter Collins, Lisa Garbus, Michael Holland, and Simon Sawhney (Minneapolis, MN: University of Minnesota Press, 1991), 53.
30. Ibid., 46.
31. Ibid., 45.
32. Ibid.
33. Ibid.
34. Iamblichus, *De vita pythagorica*, Chap 18, §§80–89. I worked with two editions: (1) Iamblichus, *On the Pythagorean Way of Life*, text, translation, and notes by John Dillon and Jackson Hershbell (Atlanta, GA: Scholars Press, 1991). (2) Iamblichus, "The Life of Pythagoras," in *The Pythagorean Sourcebook and Library: An Anthology of Ancient Writings Which Relate to Pythagoras and Pythagorean Philosophy*, compiled and translated by Kenneth Sylvan Guthrie (Grand Rapids, MI: Phanes Press, 1987), 57–122. The first edition is vastly preferable, and all translations and citations come from Dillon and Hershbell, unless otherwise noted. Iamblichus' *De vita pythagorica* is the first of 10 books, referred to collectively by Dominic O'Meara as *On Pythagoreanism*. For an overview and reconstruction of the whole 10-book sequence, see O'Meara, *Pythagoras Revived: Mathematics and Philosophy in Late Antiquity* (Oxford, UK: Clarendon Press, 1989), chaps. 2–3. According to Dillon, "Very little in these books is original to Iamblichus. They are chiefly centos of

passages from earlier writers, including Plato and Aristotle, and they have been found useful mainly on that account" (876). John Dillon, "Iamblichus of Chalcis," *Aufstieg und Niedergang der Römischen Welt*, Band II.36.2 (1987): 862–909.
35. *De vita pythagorica*, chap. 17, §§71–72.
36. Ibid., §72. It is important to note that love of learning is not *philosophia* here, but *philo-matheias*—a clear nod toward Iamblichus' allegiance to the *mathematikoi*. I will come back to this theme below.
37. Ibid. (Guthrie translation).
38. Ibid. (Dillon-Hershbell translation, slightly modified).
39. Ibid.
40. The breadth of this word allows some accounts of the Pythagorean school to be described as divided by a "curtain," while others use the term "veil." In this book, I often use the terms veil and curtain interchangeably.
41. There is a similar effect in the French reception of Iamblichus' *sindon*. Larousse calls it a *rideau*. Schaeffer, *Traité*, 98, calls it a *tenture*. Thomas Stanley, in his influential *History of Philosophy* (1655–1662), refers to this *sindon* as a screen in a passage that repeats Iamblichus, *De vita pythagorica*, §72. See Thomas Stanley, *The History of Philosophy: Containing the Lives, Opinions, Actions and Discourses of the Philosophers of Every Sect* (London: printed for A. Millar, A. Ward, S. Birt, D. Browne, T. Longman, J. Oswald, H. Whitridge, and the executors of J. Darby and S. Burrows, 1743), chap. VII of "The Discipline and Doctrine of Pythagoras," 424.
42. *De vita pythagorica*, chap. 18, §80; Walter Burkert, *Lore and Science in Ancient Pythagoreanism*, trans. Edwin L. Minar, Jr. (Cambridge, MA: Harvard University Press, 1972), 192–208.
43. See Burkert, *Lore and Science*, 192–207; Carl Huffman, *Philolaus of Croton: Pythagorean and Presocratic: A Commentary on the Fragments and Testimonia with Interpretive Essays* (Cambridge, UK: Cambridge University Press, 1993), 11–12; Charles H. Kahn, *Pythagoras and the Pythagoreans: A Brief History* (Indianapolis, IN: Hackett, 2001), 15–16; Richard D. McKirahan, *Philosophy Before Socrates: An Introduction with Texts and Commentary* (Indianapolis, IN: Hackett, 1994), 89–93; Christoph Riedweg, *Pythagoras: His Life, Teaching, and Influence*, trans. Steven Rendall (Ithaca, NY: Cornell University Press, 2005), 106–108.
44. *De vita pythagorica*, chap. 18, §82.
45. Burkert, *Lore and Science*, 195.
46. Riedweg, *Pythagoras*, 107.
47. Alexis, quoted in Carl Huffman, "Pythagoreanism," in *Stanford Encyclopedia of Philosophy* [accessed 12/28/2010: http://plato.stanford.edu/entries/pythagoreanism]. The fragment can be found in W. Geoffrey Arnott, *Alexis: The Fragments. A Commentary* (Cambridge, UK: Cambridge University Press, 1996), fr. 223; and also cited by Athenaeus in *The Deipnosophists*, trans. C. B. Gulick (Cambridge, MA: Harvard University Press, 1927), Vol. 4, 161b. On the use of middle comedy as a source of information on the acousmatics, see Burkert, *Lore and Science*, 198–201; and Riedweg, *Pythagoras*, 108.
48. Burkert, *Lore and Science*, 195.
49. Nicomachus, who writes about Pythagoras from within this Academic tradition, describes many of the distinguishing features that still persist in the lay imagination: the brilliant mathematician, the discoverer of harmonic theory, the first

exponent of the distinction between the intelligible and the sensible, and so forth. This Academic view is not simply the invention of later Platonists, but has some precedent in Plato's own writings. For example, in Book VII of *The Republic*, Socrates subscribes to Pythagorean thinking about the relation of astronomy and musical harmony: "[We admit] that as the eyes are framed for astronomy so the ears are framed for the movements of harmony, and these are in some sort kindred sciences, as the Pythagoreans affirm…". Plato, *Republic*, Book VII, 530c, in *The Collected Dialogues of Plato, Including the Letters*, eds. Edith Hamilton and Huntington Cairns (Princeton, NJ: Princeton University Press, 1961).

50. Aristotle, *Metaphysics*, 984a7.
51. Aristoxenus, fr. 90, in *Die Schule des Aristoteles: texte und kommentar. Heft II. Aristoxenus*, ed. Fritz Wehrli (Basel, CH: Schwabe, 1967), 32–33.
52. Riedweg, *Pythagoras*, 109, citing Giovanni Comotti, "Pitgora, Ippaso, Laso e il metodo sperimentale," in *Harmonica mundi. Musical e philosophic nell'antichità*, eds. R. W. Wallace and B. MacLauchlan (Rome: Biblioteca di Quarderni Urbinati di Cultura Classica 5, 1991), 20–29.
53. Hippasus, 18.A.15, in Herman Diels and Walther Kranz, *Die Fragmente der Vorsokratiker* (Berlin: Weidmann, 1972).
54. *De vita pythagorica*, chap. 18, §88.
55. Riedweg, *Pythagoras*, 26.
56. On the reception of Hippasus within the Pythagorean school, see Burkert, *Lore and Science*, 194–196; Riedweg, *Pythagoras*, 106–108; Kahn, *Pythagoras*, 15.
57. *De vita pythagorica*, chap. 18, §81.
58. Ibid., §80.
59. Ibid., §87.
60. Ibid., §89.
61. Burkert, *Lore and Science*, 195–197.
62. Iamblichus, *De communi mathematica scientia*, ed. Nicolas Festa (Stutgardiae, DE: Teubner, 1975), §25, 76.16–78.8, quoted in Burkert, *Lore and Science*, 193–194. A more thorough analysis of this passage is given in Philip Sidney Horky, "Aristotle's Description of Mathematical Pythagoreanism in the 4th Century B.C.E.," 16 [accessed 12/27/2010: http://www.princeton.edu/~pswpc/pdfs/horky/051002.pdf].
63. Burkert supports his claim on the basis of philological evidence. He argues that Iamblichus' use of the word *akousmata* to designate the sayings of Pythagoras evinces his reliance on Aristotle. The term appears in Aristotle's writing but was uncommon; the more common, non-Aristotelian tradition uses the word *symbola* instead. Moreover, Burkert argues that Iamblichus' language in *De communi mathematica* "shows close kinship, without being a quotation, with expressions in the *Metaphysics*" (196). Such use of ancient sources is "quite consistent with [Iamblichus'] usual compilatory method of writing." Regarding Iamblichus' contradiction, Burkert claims that, for Iamblichus, "the *mathemata*…belong irrevocably to doctrine of Pythagoras. It seemed to him unthinkable that anyone could contest this, to say nothing of these doubters being acknowledged by their opponents as genuine Pythagoreans." Thus, Iamblichus' transposition is not a "slip of the pen" but rather an "alteration, whose motive is transparent," that is, to claim the *mathematikoi* as the genuine disciples (194–195).
64. Iamblichus, *De communi mathematica scientia*, §25, 77.4–77.18, quoted in Horky, "Aristotle's Description of Mathematical Pythagoreanism," 16.

65. Horky, "Aristotle's Description," 16
66. Horky cites *Posterior Analytics*, I.13, 78b34–79a8, as an example of Aristotle's use of this distinction.
67. Ibid., 34
68. Aristotle, *Metaphysics*, I.5, 985b.
69. Ibid., I.5, 986a.
70. Burkert, *Lore and Science*, 197.
71. Schaeffer, *Traité*, 657.
72. Two exceptions are Hugues Dufourt and Carlos Palombini. Dufourt has noted the association between Schaeffer's penchant for the spiritual and his perennial Pythagoreanism; see Dufourt, "Pierre Schaeffer: le son comme phénomène de civilisation," in *Ouïr, entendre, écouter, comprendre après Schaeffer*, 69–82. Palombini quickly mentions the division of the Pythagorean school into acousmatics and mathematics, which he characterizes as "practitioners of the mystic doctrine" and "remarkable scientists," respectively. Schaeffer sits firmly in the first group since, "for Schaeffer, music had not sprung from the numeric proportions of intervals." Palombini, "Pierre Schaeffer and Pierre Henry," in *Music of the Twentieth-Century Avant-Garde: A Biocritical Sourcebook*, ed. Larry Sitsky (Westport, CT: Greenwood Press, 2002), 438.
73. *De vita pythagorica*, chap. 18, §82.
74. Burkert, *Lore and Science*, 173.
75. Riedweg, *Pythagoras*, 68.
76. *De vita pythagorica*, chap. 18, §82.
77. Riedweg, *Pythagoras*, 77.
78. Ibid., 74.
79. Ibid.
80. Burkert, *Lore and Science*, 176–177. Yet, Burkert's interpretation is not universally accepted. Leonid Zhmud counters Burkert's emphasis on the *akousmata* as ritual commandments by dismissing their significance within the Pythagorean tradition, claiming that the acousmatics "were not taken seriously" by the Pythagoreans. See Zhmud, *Wissenschaft, Philosophie and Religion im frühen Pythagoreismus* (Berlin: Akademie Verlag, 1997), 99; Zhmud's view is dismissed by Riedweg, *Pythagoras*, 73.
81. Ibid., 174.
82. For a small sampling, see Diogenes, *Lives*, VIII.17–18; Porphyry, *Life of Pythagoras*, in Moses Hadas and Morton Smith, *Heroes and Gods: Spiritual Biographies in Antiquity* (New York: Harper and Row, 1965), §§41–44 (another translation of Porphyry is available in Guthrie, *The Pythagorean Sourcebook*, but the section numbers do not always correspond, e.g., §§41–44 in Hadas is §42 in Guthrie); and Clement, *Stromateis*, Book 5, chap. 5, §§27–31 ("On the symbols of Pythagoras"). For Clement in Greek, I worked with Clément d'Alexandrie, *Les Stromates*, ed. Alain Le Boulluec (Paris: Éditions de Cerf, 1981). English translations of Book 1 are from Clement of Alexandria, *Stromateis: Books 1–3*, trans. John Ferguson (Washington, DC: The Catholic University of America Press, 1991); those from Book 5 are from Alexander Roberts and James Donaldson, eds., *Ante-Nicene Fathers: Translations of the Writings of the Fathers Down to A.D. 325*, American reprint of the Edinburg edition, revised and arranged by A. Cleveland Coxe (Grand Rapids, MI: W. B. Eerdmans,

1978–1981), Vol. II. Since Roberts and Donaldson lacks section numbers corresponding to the critical text, I have provided the page numbers in their edition. Since Ferguson's translation is based on a critical edition, section headings are clearly marked.
83. Porphyry, *Life of Pythagoras* (§53 in Hadas, §53 in Guthrie). In a related passage (§§11–12 in Hadas, §12 in Guthrie), Porphyry also describes this manner of "enigmatical" speaking as "symbolic," where one "expresses the sense of allegory and parable," claiming that Pythagoras learned this manner of speaking from the Egyptians.
84. Ibid.
85. Burkert, *Lore and Science*, 174.
86. Ibid., 175.
87. *De vita pythagorica*, chap. 23, §§104–105.
88. Ibid.
89. Ibid., chap. 18, §87.
90. Ibid., §89.
91. Iamblichus' defense of the mathematic Pythagoreanism can also be placed in the context of his debates with Porphyry over the Pythagorean legacy. O'Meara characterizes the difference between Porphyry and Iamblichus along the lines of two kinds of Platonism: Pythagoreanizing Platonism and universalizing Platonism. Iamblichus, a Pythagoreanizing Platonist, "singles out Pythagoras as the fountainhead of all true (Platonic) philosophy," while Porphyry, a universalizing Platonist, "finds his Platonism both in Pythagoras and in very many other quarters" (27). Porphyry culls from many sources in Greek philosophy to arrive at a Neoplatonist philosophy centering on the problems of the plight of the soul imprisoned in the body, and reworking it in the wake of Plotinus' thought. As O'Meara notes, Porphyry feels no need "to give particular stress to the Pythagorean mathematical sciences" (28), focusing more on ethical and religious questions. In these respects, he appears to be more of an acousmatic Pythagorean. In Iamblichus' brand of Pythagoreanizing Platonism, which advocates for the importance of theurgy as the integration of Platonic wisdom with ancient pagan religious rites, Pythagoras is central in the transmission of ancient knowledge. In his *On Pythagoreanism*, Iamblichus argues for the significant role played by Pythagorean mathematics, music, and astronomy in the contemplation of the highest principles, relying substantially on Nichomachus' work on Pythagorean mathematics. Grafting this Pythagorean schema onto his own contemporary situation, Iamblichus could be seen to imply that Porphyry, in neglecting the *mathemata*, was just as superficial as the acousmatics in their understanding of the ancient Pythagorean truth. Thus, the division of the Pythagorean school into two classes by means of the veil metaphorizes the difference between the two camps, those who possess the proper key to unlock the ancient wisdom and those who simply follow the letter.
92. If there were something specifically visual about the demonstrations of Pythagoras, then one could imagine a reason why the split between seeing and hearing, and thus the veil, would be significant. This could be the case if the demonstrations involved geometric drawings or proofs. But this possibility is not developed in the ancient sources. Moreover, it would only apply to the *mathemata*, not the *akousmata*.
93. Chion (M1) is the only writer in the Schaefferian school who mentions Clement as a possible source of the Pythagorean veil. However, as noted above, he places Clement's *Stromateis* in the third century B.C.E.

94. Clement, *Stromateis*, Book 5, §59.1; *Ante-Nicene Fathers*, 458.
95. Ibid., §58.1–4. See *Ante-Nicene Fathers*, 458.
96. *A Greek-English Lexicon*, eds. Henry George Liddell and Robert Scott (Oxford, UK: Clarendon Press, 1979).
97. Clement, *Stromateis*, Book 5, §§58.6–59.1; *Ante-Nicene Fathers*, 458 (emphasis added).
98. Liddell and Scott, *Greek-English Lexicon*.
99. *De vita pythagorica*, chap. 23, §105.
100. For the fragments of Timaeus, see Felix Jacoby, *Die Fragmente der Griechischen Historiker* (Leiden, NL: Brill, 1950), 566; also available online with English translations at http://www.brill.nl/brillsnewjacoby [accessed 12/29/2010].
101. According to Burkert, the distinction "between a lower and higher degree of Pythagorean wisdom…goes back at least as far as Timaeus." Burkert, *Lore and Science*, 192.
102. Diogenes, *Lives* (VIII.10 in Hicks, §8 in Guthrie).
103. Ibid., VII.15.
104. Ibid., 328, fn. a.
105. Hicks's choice for the evening lecture should be differentiated from Bayle's introduction of both the Pythagorean veil and the cover of darkness in M2b. Where Bayle seeks a mimetic identification—deploying both the veil and the cover of darkness as grounds for his own practice—Hicks's note registers the strangeness of the topos of seeing and hearing Pythagoras and attempts to dispel one cryptic passage with the causal explanation introduced in another.
106. Clement also mentions Timaeus in the *Stromateis*, although not as a source of information about Pythagoras or his school. For a full account of the philological evidence, see Kurt von Fritz, *Pythagorean Politics in Southern Italy* (New York: Columbia University Press, 1940), 39. Von Fritz presents a series of parallel passages to show the pervasiveness of Timaeus as a source. The passages in question are Iamblichus, *De vita pythagorica*, §§71–72, 81; Diogenes, VIII.10; and Jacoby, 566, fn. 13.
107. Burkert and Dillon-Hershbell both skirt the oddity of the veil or *sindon* when it appears. Burkert implies that it comes from Timaeus, noting (after von Fritz) that Diogenes, VIII.10, and *De vita pythagorica*, §72 (the location of S2) are very similar, and that "[Iamblichus'] exposition as a whole must be based ultimately on Timaeus." To conjecture that the passage *as a whole* comes from Timaeus lacks evidence; nor can it account for the sudden discrepant appearance of the *sindon* in Iamblichus or the *parakalummati* of Clement. Moreover, the whole question is obscured by Burkert's odd, tangential addendum that, according to Athenaeus, 4.145b–c, "the king of Persia eats behind a curtain." The passage from Athenaeus, while interesting, does not help to locate the source of Iamblichus' veil. In fact, it complicates the philological evidence since Athenaeus uses the word *parakalummatos*—Clement's word, not Iamblichus'—to designate the veil. See Burkert, *Lore and Science*, 192n1. Dillon and Hershbell, in their translation of Iamblichus' *De vita pythagorica* repeat Burkert's reference to Athenaeus. See Dillon-Hershbell, *De vita pythagorica*, 99, n. 4.
108. On Clement's "method of concealment," see Henny Fiska Hägg, *Clement of Alexandria and the Beginning of Christian Apophaticism* (New York: Oxford University Press, 2006), chap. 4. Hägg draws a distinction between esotericism and concealment, arguing that "[e]soteric or secret doctrines are held to belong to an

elite, to an initiated, select group of people, an inner circle, and are never meant to be disclosed to all and sundry" (140–141). Concealment differs from esotericism in that it designates the practice whereby a message or teaching is adapted to different levels or kinds of students. Thomas Szlezák, in contrast to Hägg, describes esotericism as a "requirement of reason," in that it implies the conveyance of philosophical wisdom in tune with the level of preparation by the student. Plato's esoteric method of delivering just enough wisdom as is appropriate to the interlocutor is contrasted with the "secrecy" of the Pythagoreans. Secrecy, unlike esotericism, "is based on compulsion." See Thomas A. Szlezák, *Reading Plato*, trans. Graham Zanker (London: Routledge, 1999), 114. Hägg and Szlezák might be harmonized by synonymizing Christian-Platonic esotericism-concealment and contrasting it with Pythagorean secrecy.

109. Hägg, *Clement of Alexandria and the Beginning of Christian Apophaticism*, 141.
110. Although Derrida's central analysis of logocentrism is developed in a reading of Rousseau's writings on the origin of language, logocentrism is a classical inheritance: "Self-presence, transparent proximity in the face-to-face countenance and the immediate range of the voice, this determination of social authenticity is therefore classic: Rousseauistic but already the inheritor of Platonism." Jacques Derrida, *Of Grammatology*, trans. Gayatri Chakravorty Spivak (Baltimore, MD: Johns Hopkins University Press, 1976), 138. The Platonic critique of writing is presented in the *Phaedrus*. See Derrida, "Plato's Pharmacy," in *Dissemination*, trans. Barbara Johnson (Chicago: University of Chicago Press, 1981).
111. Clement, *Stromateis*, Book 1, §9.1.
112. Ibid., §14.2.
113. Ibid., §14.4
114. Ibid., Book 5, §19.3–4, 20.1; *Ante-Nicene Fathers*, 449.
115. Frank Kermode, *The Genesis of Secrecy: On the Interpretation of Narrative* (Cambridge, MA: Harvard University Press, 1979); and *The Art of Telling: Essays on Fiction* (Cambridge, MA: Harvard University Press, 1983), chap. 8.
116. Kermode, *The Genesis of Secrecy*, 2
117. Mark 4:11–12.
118. Kermode, *The Genesis of Secrecy*, 1. Kermode associates this history of hermeneutics with an organization of the sensorium that is in line with the acousmatic emphasis on hearing over seeing. Specifically, the ear becomes the mode in which hermeneutical interpretation occurs, often at the expense of the eye. The true interpreter is willingly placed behind the veil in order to blot out the evidence of the eye and appropriate the voice anew. Kermode cites the apocryphal epistle of Barnabas, which "distinguished between those within and those without by saying that the former has circumcised ears and the latter not" (3). "Yet all narratives are capable of darkness; the oracular is always, or thereabouts, accessible if only by a sensory failure" (14). The modality privileged by this sensory failure of vision—which forms the vehicle upon which the metaphors of dark narratives are constructed—is hearing. Moreover, Kermode characterizes the selection of textual portions by which allegorical reading operates in sensory terms. Cherry-picking is "just what the early exegetes did to the Old Testament, preserving it only because it cast shadows important to the perceiving eye, emitted signals intelligible to the understanding ear" (20).

119. Andrew C. Itter, *Esoteric Teaching in the Stromateis of Clement of Alexandria* (Leiden, NL: Brill, 2009), 14.
120. David Dawson, *Allegorical Readers and Cultural Revision in Ancient Alexandria* (Berkeley, CA: University of California Press, 1992), 206.
121. Clement, *Stromateis*, Book 5, §21.4; *Ante-Nicene Fathers*, 449.
122. Ibid., Book 5, §41.1–2; *Ante-Nicene Fathers*, 450.
123. Ibid., Book 5, §§32.2–33.3; *Ante-Nicene Fathers*, 452. The entirety of *Stromateis*, Book 5, chap. 6, is on the tabernacle.
124. Iamblichus, *De Mysteriis*, trans. Emma C. Clarke, John M. Dillon, and Jackson P. Hershbell (Leiden, NL: Brill, 2004), Book 8.3, §265.
125. Gregory Shaw, *Theurgy and the Soul: The Neoplatonism of Iamblichus* (University Park, PA: Pennsylvania State University Press, 1995), 175.
126. Iamblichus, *De Mysteriis*, Book 3.9, §120. See Shaw, *Theurgy and the Soul*, 175, for more.
127. O'Meara would dispute this claim. Although "the pagan philosophers of late antiquity were very aware of the threat that Christianity represented for them...it would be difficult to show, on the basis of the extant remains of *On Pythagoreanism*, that Iamblichus had Christianity specifically in mind as a target against which his Pythagoreanizing programme was to be directed." Rather, O'Meara selects Porphyry as the target. See O'Meara, *Pythagoras Revived*, 214–215. Riedweg claims just the opposite, stating that Iamblichus' *De vita pythagorica* can be described "as a pagan alternative to flourishing monasticism and the Christian way of life in general." Riedweg, *Pythagoras*, 31.
128. Peter Brown, *The Making of Late Antiquity* (Cambridge, MA: Harvard University Press, 1978), 59.
129. Ibid.
130. Riedweg, *Pythagoras*, 3. For example, both Porphyry and Iamblichus tell a story of Pythagoras' miraculous encounter with the fishermen of Croton. While the fisherman dragged their nets from the water, Pythagoras wagered that he could determine the exact number of fish, and if he was correct, the fish had to be released back into the water. After making his prediction, the fish were counted, and Pythagoras was found to be correct. But the miraculous thing about the story was that the fish remained alive although they were out of water for a considerable time while being counted. In Iamblichus' version, Pythagoras leaves while the fishermen speak of the miraculous event to others, creating a considerable interest in Pythagoras and his gifts. As Riedweg notes, such a tale bears some resemblance to the story of the exceptional catch, told in the Gospels of Luke (5:1–11) and John (21:1–14), when Jesus orders Simon to cast his nets after a fruitless evening of fishing and, miraculously, Simon catches so many fish that the boats nearly sink under the load.
131. For instance, Iamblichus' *De Mysteriis* was translated by Ficino, who also gave it its Latinized title.
132. Christiane L. Joost-Gaugier, *Pythagoras and Renaissance Europe: Finding Heaven* (Cambridge, UK: Cambridge University Press, 2009), 25.
133. Ibid., 29.
134. Johannes Reuchlin, *On the Art of the Kabbalah*, trans. Martin and Sarah Goodman (New York: Abaris Books, 1983), 233.

135. This account depends upon Reuchlin's view that Pythagoras derived his knowledge from the Jews and transmitted it to the Greeks. In this regard, he is differentiated from a host of other Greek philosophers and selected as the "most important one because he had access to the divine revelation of the Kabbalah." Joost-Gaugier, *Pythagoras and Renaissance Europe*, 43.
136. Giovanni Pico della Mirandola, *De Hominis Dignitate, Heptaplus, De Ente Et Uno, E Scritti Vari*, ed. Eugenio Garin (Florence, IT: Vallecchi, 1942), §45.286. The translation is from Pico della Mirandola, *On the Dignity of Man*, trans. Charles Glenn Wallis, Paul J. W. Miller, and Douglas Carmichael (Indianapolis, IN: Bobbs-Merrill, 1965), 33.

Chapter 3

1. Pierre Schaeffer, "Lettre à Albert Richard," in *La revue musicale*, No. 236 (Paris: Richard-Masse, 1957), iv.
2. Ibid.
3. For an exposition of the founding of the GRM, see Gayou, *Le GRM*, 107–110.
4. Battier, "What the GRM Brought to Music," 196.
5. Peignot, "De la musique concrète," 116.
6. Arthuys and Peignot, *Musique animée*. Cited by Gayou, *Le GRM*, 354; and Bayle, *Musique acousmatique*, 181. This passage is included in M2c.
7. For example, Jonathan Sterne, in *The Audible Past*, describes "acousmatic understandings of sound reproduction... as splitting copies of sound from their ontologically separate sources." The term becomes synonymous with R. Murray Schafer's negative term, "schizophonia." Sterne assimilates the two terms in order to critically interrogate them. "Acousmatic or schizophonic definitions of sound reproduction carry with them a questionable set of prior assumptions about the fundamental nature of sound, communication, and experience." Similarly, Jamie Sexton, in his *Music, Sound, and Multimedia*, claims that "Like the term 'acousmatic,' the concept of 'schizophonia'... refers to the separation of sound from source." Although I am sympathetic to Sterne's critique of the audiovisual litany, I ultimately think that his conflation of acousmatic sound and schizophonia is incorrect. Schizophonia, for Schafer, is a nervous term designed to describe the loss of the sound's unique referent in the age of recording, storage, and broadcast of sound. Acousmatic sound, as I will argue, has a history that predates recorded sound (a history that aligns with the one Sterne discusses in *The Audible Past* concerning 'audile techniques') and involves a much more complicated relationship between the source, cause, and effect of a sound. It cannot be summed up as a "split" between originals (or sources) and copies. See Sterne, *The Audible Past: Cultural Origins of Sound Reproduction* (Durham, NC: Duke University Press, 2003), 20, 25; Jamie Sexton, *Music, Sound and Multimedia: From the Live to the Virtual* (Edinburgh, UK: Edinburgh University Press, 2007), 122; and R. Murray Schafer, *The Soundscape*, 90–91.
8. The first of the two poems, written sometime around the summer of 1899, appears in the posthumous collection *Stavelot*. The other, collected in the *Poèmes retrouvés*, was first published in 1915, although it too is a youthful work that dates back to earlier in Apollinaire's career, likely from the same time as the first "Acousmate" poem. See Guillaume Apollinaire, *Œuvres poétiques*, 513, 671.

9. Stephen Eskilson, *Graphic Design: A New History* (New Haven, CT: Yale University Press, 2007), 177–178.
10. Margaret Davies, *Apollinaire* (New York: St. Martin's Press, 1964), 29.
11. Ibid., 30; and James R. Lawler, "Apollinaire inédit: Le Séjour à Stavelot," *Mercure de France*, No. 1098 (Feb. 1, 1955): 296–309.
12. *Dictionnaire de l'Académie française, nouvelle édition*, Vol. 1, A-K (Paris: Brunet, 1762).
13. Pierre Larousse, *Grand dictionnaire universel du XIXe siècle: Français, historique géographique, biographique, mythologique, bibliographique, etc.*, in 15 volumes (Paris: Administration du Grand dictionnaire universel, 1900–).
14. Bayle, *Musique acousmatique*, 180. Chion (M1) also mentions only one of Apollinaire's "Acousmate" poems and similarly dates it incorrectly to 1913. This suggests that perhaps Bayle's book was the source for Chion's footnote.
15. Battier, "What the GRM Brought to Music," 196.
16. Ibid.
17. Ibid.; Guillaume Apollinaire, "A propos de la poésie nouvelle," in *Œuvres en prose complètes* (Paris: Gallimard, 1977), 982.
18. Battier, "What the GRM Brought to Music," 196. See Guillaume Apollinaire, "Le Roi-Lune," in *Œuvres en prose completes*, 313–316, translated as "The Moon King," in *The Poet Assassinated*, trans. Ron Padgett (San Francisco: North Point Press, 1984). For another treatment of this material, see Douglas Kahn, "Introduction: Histories of Sound Once Removed," in *Wireless Imagination*, eds. Douglas Kahn and Gregory Whitehead (Cambridge, MA: MIT Press, 1994).
19. March 1948. Pierre Schaeffer, *In Search of a Concrete Music*, 4.
20. Battier, "What the GRM Brought to Music," 191; Apollinaire, "*Les Archives de la parole*," *Œuvres en prose complètes*, 213.
21. Battier, ibid.
22. Ibid.
23. Ibid., 196.
24. Ibid.
25. Battier's claim could also find support from the editors of the *Œuvres poétiques*, whose critical notes to the first "Acousmate" poem argue for a slightly different but related kind of prolepis. The editors suggest that Apollinaire's interest in the concept of an acousmate is a "prefiguration" of his later use of snatches of conversations and bits of speech, utilized so effectively in the later "conversation poems" of the *Calligrammes*. Apollinaire, *Œuvres poétiques*, 1130.
26. Battier, "What the GRM Brought to Music," 196; originally from Gabriel Feydel, *Remarques Morales, Philosophiques et Grammaticales, sur le Dictionnaire de L'académie Françoise* (Paris: Antoine-Augustin Renouard, 1807), 9.
27. See Wilhelm Heinrich Wackenroder, "The Remarkable Death of the Old Artist Francesco Francia" and "The Strange Musical Life of the Musical Artist Joseph Berglinger," in *Confessions and Fantasies*, ed. and trans. Mary Hurst Schubert (University Park, PA: Penn State University Press, 1971), 89–90, 152–3; Heinrich von Kleist, "Holy Cecilia or the Power of Music," in *Music in German Romantic Literature*, ed. and trans. Linda Siegel (Novato, CA: Elra Publications, 1983); Arthur Schopenhauer, *The World as Will and Representation*, trans. E. F. J. Payne, two vols. (New York: Dover Publications, 1969), 267; and Friedrich Nietzsche, "On Music and Words," in Carl Dahlhaus, *Between Romanticism and Modernism*,

trans. Mary Whittall (Berkeley, CA: University of California Press, 1980), 103–119.
28. All the instances appear in Battier, 196.
29. Ibid.
30. Feydel, *Remarques*, 9.
31. André Morellet, *Observations sur un ouvrage anonyme intitulé: Remarques morales, philosophique et grammaticales, sur le Dictionnaire de l'Académie française* (Paris: 1807), 13–14.
32. This is not to say that Feydel should not be understood as registering the moment of an *event* or *rupture* in the history of the term. But if that is the argument to be made, then Battier is the one who needs to make it against the evidence of the historical record.
33. Treüillot de Ptoncour, "Lettre de M. Treüillot de Ptoncour," *Mercure de France* (Dec. 1730, Vol. 2), 2807.
34. Ibid., 2807–2808.
35. Ibid., 2808.
36. Ibid., 2805–2806.
37. Ibid., 2812–2816.
38. Ibid., 2812.
39. Ibid., 2814.
40. Ibid., 2818–2819.
41. Patricia Fara, "Lord Derwentwater's Lights: Prediction and the Aurora Polaris Astronomers," *Journal for the History of Astronomy*, Vol. 27 (1996): 246; J.-J. Dortous de Mairan, *Traité physique et historique de l'aurore boréale* (Paris: n.p., 1754).
42. B. Langwith, "An Account of the Aurora Borealis That Appear'd Oct. 8, 1726. In a Letter to the Publisher from the Reverend Dr. Langwith, Rector of Petworth in Sussex," *Philosophical Transactions of the Royal Society*, Vol. 34 (1726–1727), 132–136.
43. de Ptoncour, "Lettre," 2823.
44. Fara, "Lord Derwentwater's Lights," 247.
45. Ibid., 244.
46. de Ptoncour, "Lettre," 2824–2825.
47. The problem still persists in contemporary epistemology of the senses when auditory theorists try to overcome ocularcentric metaphysics by coining alternative terms. For instance, F. Joseph Smith prefers the term "akumenon" to "phenomenon." See F. Joseph Smith, *The Experiencing of Musical Sound: Prelude to a Phenomenology of Music* (New York: Gordon and Breach, 1979), 30.
48. Eudoxus' text is preserved by Hipparchus, *Hipparchou Tōn Aratou kai Eudoxou Phainomenōn exēgēseōs vivlia tria = Hipparchi in Arati et Eudoxi Phaenomena commentariorum libri tres*, ed. and trans. Karl Mantiius (Lipsiae, DE: In aedibus B. G. Teubneri, 1894).
49. de Ptoncour, "Lettre," 2824–2825.
50. Ibid., 2826.
51. Ibid., 2827.
52. Ibid.
53. In his final rhetorical flourish, the Curé wonders, "Was it possible that all ears had been enchanted [*enchantées*], if one can put it this way, to believe they heard that

which they didn't hear? I cannot imagine that" (2828). It is noteworthy that the Curé mentions the possibility of enchantment in the final statement of his reflections. Was this perhaps the source for Feydel's objection concerning the synonymy of acousmate and incantation? If so, Feydel was a poor reader of the *Mercure*, for the Curé is doing everything in his power to disambiguate an acousmate from any state of enchantment, incantation, hallucination, and such. Moreover, Feydel gives no indication that he is aware of the context in which the word was baptized, or its Greek etymology. On the other hand, Morellet understands that the word is intended to designate an auditory event (one could even use it to designate a sound object, perhaps) by borrowing from the Greek *akousmata*, although there is no appeal to the baptism of the term in order to defend the Académie's definition.

54. John Locke, "An Essay on Human Understanding," in *The works of John Locke, in nine volumes* (London: C. and J. Rivington, 1823), Book II, chap. iii, §1, and chap. iv.
55. Ibid., Book II, chap. vii, §15.
56. Jonathan Rée, *I See a Voice: Deafness, Language, and the Senses* (New York: H. Holt and Co., 1999), 334.
57. William Cheselden's work on removing cataracts is discussed in Denis Diderot, "Letter on the Blind," in *Diderot's Early Philosophical Works*, trans. and ed. Margaret Jourdain (Chicago: Open Court Press, 1916). For critical and historical commentary, see Rée, *I See a Voice*, 337–341; Marjolein Degenaar, *Molyneux's Problem* (Boston: Kluwer, 1996); and Michael J. Morgan, *Molyneux's Question: Vision, Touch, and the Philosophy of Perception* (Cambridge, UK: Cambridge University Press, 1977).
58. The aesthetic development of this kind of thinking appears most famously in Gotthold Ephraim Lessing's *Laocoön*, where the eye and the ear are divided around the sister arts of painting and poetry, each developing a logic of space and time that is native to each modality, and only mixed at the peril of the art object. But the claim can also be found in a French source of the period in question, the Abbé DuBos's *Critical Reflections on Poetry, Painting and Music*. See Lessing, *Laocoön*, trans. Robert J. Phillimore (London: Routledge, 1905); and DuBos, *Critical Reflections*, trans. Thomas Nugent (London: Printed for J. Nourse, 1748).
59. Rée, *I See a Voice*, 336.
60. René Descartes, "Treatise on Light," in *The World and Other Writings*, ed. and trans. Stephen Gaukroger (Cambridge, UK: Cambridge University Press, 1998), 3.
61. On the role of the arbitrary sign in the classical age of France, see Michel Foucault, *The Order of Things: An Archaeology of the Human Sciences* (New York: Pantheon Books, 1971).
62. Descartes, "Treatise on Light," 4–5.
63. Fontenelle's enthusiasm for Descartes's scientific writings can be gleaned, as Aram Vartanian notes, from a famous passage in his "Eloge de Malebranche," where Malebranche's sudden awakening upon reading the *Traité de l'Homme* matches his own reaction. Fontenelle, through his election to the Académie Française and his natural talents as a writer, disseminated much of the era's scientific thought. See Aram Vartanian, *Science and Humanism in the French Enlightenment* (Charlottesville, VA: Rookwood Press, 1999), in particular, chap. 1, 41, n. 3.
64. Fontenelle, "Histoire des oracles" and "De l'origine des fables" *in Œuvres complètes*, Vol. 2, ed. G.-B. Depping (Geneva, CH: Slatkine Reprints, 1968).
65. Vartanian, *Science and Humanism*, 8.

66. The acousmate d'Ansacq is mentioned in the following issues of the *Mercure de France*: Dec. 1730, Vol. 2: 2804–2833; Feb. 1731: 333–338; March 1731: 447–457; May 1731: 1028–1034; June 1731, Vol. 1: 1254–1256; June 1731, Vol. 2: 1516–1531; July 1731: 1637–1649 (note, the pagination is faulty in this issue); Aug. 1731: 1841–1854; Nov. 1731: 2572–2580; March 1732: 416–426.
67. Anonymous, "Extrait d'une Lettre écrite de Bourgogne," *Mercure de France* (Feb. 1731): 336.
68. Ibid.
69. Anonymous, "Lettre sur les bruits Aeriens," *Mercure de France* (March 1731): 449.
70. "Account of the Performances of different Ventriloquists, with Observations on the Art of Ventriloquism," *Edinburgh Journal of Science* 9 (1928): 257–258; quoted in Connor, *Dumbstruck*, 255.
71. Anonymous, "Lettre sur les bruits Aeriens," 452–453.
72. Connor, *Dumbstruck*, 244.
73. Anonymous, "Lettre sur les bruits Aeriens," 451–452.
74. Anonymous, "Lettre écrite à M. de le R.," *Mercure de France* (May 1731): 1028–1029.
75. Ibid., 1033.
76. Ibid., 1034.
77. Laloüat de Soulaines, "Lettre de M. Laloüat de Soulaines," *Mercure de France* (June 1731), Vol. 2: 1527. N.B. The pagination is faulty; it is printed as 1427, but it should be 1527.
78. Anonymous, "Extrait d'une Lettre de M. l'Abbé," *Mercure de France* (Sept. 1738): 1986–1987.
79. "Compte-rendu des séances," paper read before the *Société académique d'archéologie, sciences et arts du département de l'Oise* (Beauvais, FR: Moniteur de L'Oise, 1901), 69–72. The basis for Thiot's suggestion comes from the original article of the Curé. In the final section on the topography of Ansacq, the Curé mentions that "considerable echoes" can be heard in the small valley near Ansacq, although not much is made of the suggestion. See de Ptoncour, *Mercure de France* (Dec. 1730): 2831.
80. Abbé Prévost, *Manuel lexique ou dictionnaire portatif des mots françois dont la signification n'est pas familière a tout le monde*, Vol. 1 (Paris: Chez Didot, 1755).
81. For instance, "Acousmate, a new term to designate a phenomenon which makes a great noise in the air," in P. Charles Le Roy, *Traité de l'orthographie Françoise en forme de dictionnaire* (Poitiers, FR: chez J. Félix Faulcon, 1764); or "Acousmate, or Akousmate. Phenomenon of a noise like that of several human voices, and different instruments, heard in the air," in Pierre-Charles Berthelin, *Abrégé du dictionnaire universel françois et latin, vulgairement appelé dictionnaire de Trevoux: contenant la signification, la définition & l'explication de tous les termes de sciences et arts* (Paris: Libraires Associes, 1762).
82. To be more precise, an acousmate, as defined by the Académie, is an intentional object in the mode of presentation of fantasy. In regard to Feydel's debate, the second vector of signification, which changes the ontological status of the object, does not turn it into a mental state. Feydel's reading, where one can be in a state of acousmate, remains idiosyncratic.
83. Charles Le Gendre, Marquis de Saint-Aubin-sur-Loire, *Traité historique et critique de l'opinion*, Vol. 6, corrected and augmented edition (Paris: Briason, 1758), 290ff.

84. A. Béclard, et al., *Nouveau dictionnaire de médecine, chirurgie, pharmacie, physique, chimie, histoire naturelle, etc.*, Vol. 1 (Paris: L.-T. Cellot, 1821).
85. Henry Power and Leonard W. Sedgwick, *The New Sydenham Society's Lexicon of Medicine and the Allied Sciences* (London: New Sydenham Society, 1881).
86. W. A. Newman Dorland, *The American Illustrated Medical Dictionary* (Philadelphia: Saunders, 1907).
87. [Louis Scipion Jean Baptiste Urbain de Merle] La Gorse, *Souvenirs d'un homme de cour; ou, Mémoires d'un ancien page* (Paris: Dentu, 1805), 339–340.
88. Petrus Borel, "Les Pressentiments," in *Album de la mode: chroniques du monde fashionable, ou, Choix de morceaux de littérature contemporaine* (Paris: Louis Janet, 1833), 334.
89. See Eustache-Hyacinthe Langlois, *Essai historique, philosophique et pittoresque sur les danses des morts: Accompagné de cinquante-quatre planches et de nombreuses vignettes* (Rouen, FR: A. Lebrument, 1852); and E. Henri Carnoy, "Les Acousmates et les chasses fantastiques," *Revue de l'histoire des religions*, Tome IX (1884): 375ff. Langlois writes, "Several scientists have given the name acousmate to a physical phenomenon that has not yet been explained in a satisfactory manner. It consists in extraordinary noises, lacking an appreciable cause, heard in the air..." (157).

Chapter 4

1. Jonathan Crary, *Techniques of the Observer* (Cambridge, MA: MIT Press, 1990), 132.
2. Ibid.
3. Karl Marx, *Capital*, trans. Ben Fowkes (New York: Vintage Books, 1976), 163 ff. The phantasmagoric quality of the commodity is often mistakenly attributed to the usurpation of use value by exchange value. Marx denies this: "The mystical character of the commodity does not therefore arise from its use-value. Just as little does it proceed from the nature of the determinants of value" (164), that is, exchange-value determined by the quantity and quality of labor required to produce the commodity.
4. Ibid., 165.
5. Ibid.
6. Theodor Adorno, *In Search of Wagner*, trans. Rodney Livingstone (London: Verso, 1985), 85.
7. Ibid., 86.
8. On the history of absolute music, see Carl Dahlhaus, *The Idea of Absolute Music*, trans. Roger Lustig (Chicago: University of Chicago Press, 1989); Lydia Goehr, *The Quest for Voice: Music, Politics, and the Limits of Philosophy* (New York: Oxford University Press, 1998); Berthold Hoeckner, *Programming the Absolute: Nineteenth-Century German Music and the Hermeneutics of the Moment* (Princeton, NJ: Princeton University Press, 2002); Daniel K. L. Chua, *Absolute Music And the Construction of Meaning* (Cambridge, UK: Cambridge University press, 2006); and Richard D. Leppert, *The Sight of Sound: Music, Representation, and the History of the Body* (Berkeley, CA: University of California Press, 1993).
9. Schopenhauer, *The World as Will and Representation*, Vol. II, 448.
10. For an informative discussion of Platonism in Schopenhauer's aesthetics, see Whitney Davis, *Queer Beauty: Sexuality and Aesthetics from Winckelmann to Freud and Beyond* (New York: Columbia University Press, 2010), chap. 3. My account of Schopenhauer is indebted to Davis.

11. Schopenhauer, *The World as Will and Representation*, Vol. I, 199.
12. Christopher Janaway, "Knowledge and Tranquility: Schopenhauer on the Value of Art," in *Schopenhauer, Philosophy, and the Arts*, ed. Dale Jacquette (Cambridge, UK: Cambridge University Press, 1996), 46.
13. Schopenhauer, *The World as Will and Representation*, Vol. II, 367.
14. Ibid., Vol. I, 185–186.
15. Ibid., Vol. II, 367–368.
16. Ibid., 368.
17. Wilhelm Heinrich Wackenroder, "The Strange Musical Life of the Musical Artist Joseph Berglinger," in *Confessions and Fantasies*, 149 (emphasis added).
18. Ibid.
19. Schopenhauer, *The World as Will and Representation*, Vol. II, 447.
20. Wackenroder, "The Strange Musical Life of the Musical Artist Joseph Berglinger," 149.
21. George Bernard Shaw, *The Perfect Wagnerite* (New York: Dover Publications, 1967), 128. Shaw also recommended that the English build a theater that follows the designs of Bayreuth, noting that "any English enthusiasm for Bayreuth that does not take the form of a clamor for a Festival playhouse in England may be set aside as a mere pilgrimage mania" (129).
22. See Geoffrey Skelton, "The Idea of Bayreuth," in *The Wagner Companion*, eds. Burbidge and Sutton (New York: Cambridge University Press, 1979), 390–391.
23. A similar account is related in Kierkegaard's *Either/Or*, where, at a performance of *Don Giovanni*, a concatenation of various bodily techniques for listening to music finally gives way to musical transcendence once the listener is separated by a partition. "I have sat close up, I have sat farther and farther back, I have tried a corner in the theater where I could completely lose myself in the music. The better I understood it...the farther away I was....I stand outside in the corridor; I lean up against the partition which divides me from the auditorium, and then the impression is most powerful: it is a world by itself, separated from me; I can see nothing, but I am near enough to hear, and yet so infinitely far away." Søren Kierkegaard, *Either/Or*, Vol. 1, trans. David F. Swenson and Lillian Marvin Swenson (Princeton, NJ: Princeton University Press, 1944), 119.
24. Richard Wagner, "The Festival-Playhouse at Bayreuth," in *Richard Wagner's Prose Works*, Vol. 5, trans. W. Ashton Ellis (London: Kegan Paul, Trench, Trübner & Co., 1896), 333.
25. Camille Saint-Saëns, *Harmonie et Mélodie* (Paris: Calmann Lévy, 1885), 52–53.
26. André-Ernest-Modeste Grétry, *Mémoires, ou essais sur la musique*, Vol. 3 (Paris: Pluviôse, 1797), 32.
27. Alexandre Choron and J. Adrien de La Fage, *Nouveau manuel complet de musique vocale et instrumentale, ou, Encyclopédie musicale*, Part. 2, tome 3 (Paris: Roret, 1838), 117–120.
28. Ignaz Theodor Ferdinand Cajetan Arnold, *Der angehende Musikdirektor; oder, Die Kunst ein Orchester zu bilden, in Ordnung zu erhalten, und überhaupt allen Forderungen eines guten Musikdirektors Genüge zu leisten* (Erfurt, DE: Henning, 1806), 299.
29. Quoted in Edward F. Kravitt, *The Lied: Mirror of Late Romanticism* (New Haven, CT: Yale University Press, 1996), 25–26.

30. For Marsop's account, see "*Zur Bühnen- und Konzertreform*," *Die Musik*, V. Jahr, Heft 16 (1905/06): 255.
31. Henry-Louis de la Grange, *Gustav Mahler*, Vol. 3 (New York: Oxford University Press, 1999 [1983]), 218.
32. Quoted in de la Grange, ibid.
33. Marsop, "*Zur Bühnen- und Konzertreform*," 254.
34. Carl Dahlhaus, *Nineteenth-Century Music*, trans. J. Bradford Robinson (Berkeley, CA: University of California Press, 1989), 394.
35. Ibid.
36. Ibid.
37. Paolo Morigia, *La nobilita' di Milano* (Milano, IT: appresso Gio. Battista Bidelli, 1619), 306.
38. For more on Borromeo and his discourse, see Robert Kendrick, *Celestial Sirens: Nuns and Their Music in Early Modern Milan* (Oxford, UK: Clarendon Press, 1996), 158–159.
39. Craig Monson, "Putting Bolognese Nun Musicians in their Place," 119.
40. Gregory Martin, *Roma Sancta* (1581), ed. George Bruner Parks (Roma: Edizioni di storia e letteratura, 1969), 141–142.
41. Monson, "Putting Bolognese Nun Musicians in their Place," 122. On the vocalic body, see my introduction and Connor, *Dumbstruck*.
42. Jean-Jacques Rousseau, *The Confessions*, Book 7, trans. J. M. Cohen (London: Penguin Books, 1953), 295.
43. Ibid., 295–296.
44. Robert Zimmerman, "Allgeimein Aesthetik als Formwissenschaft," *in Music in European Thought*, ed. Bojan Bujić (Cambridge, UK: Cambridge University Press, 1988), 46–49.
45. Anonymous, "Our Concerts," *Musikalische Eilpost* 4 (March 1826), in *The Critical Reception of Beethoven's Compositions by his Contemporaries*, Vol. 2, eds. Wayne Senner and William Meredith, trans. Robin Wallace (Lincoln, NE: University of Nebraska Press, 2001), 122.
46. Ibid., 123.
47. Ibid., 124.
48. Ibid., 125.
49. Ibid., 126.
50. Ibid.
51. Ibid., 127.
52. Anonymous ["a student"], "Another Evidence of the Wide-Spread Influence of Theosophical Ideals," *Century Path* (Jan. 19, 1908): 13.
53. Philippe Lacoue-Labarthe, "Diderot: Paradox and Mimesis," in *Typography*, 248–266; Aristotle, *Physics* in *The Basic Works of Aristotle*.
54. Ibid., 255.
55. Ibid.
56. Jacques Derrida, *Of Grammatology*, 215. Also see Lacoue-Labarthe, *Heidegger: Art and Politics*, trans. Chris Turner (Cambridge, UK: Basil Blackwell, 1990), 83: "The structure of original supplementarity is the very structure of the relations between *technê* and *physis*."
57. Derrida, *Of Grammatology*, 144–145. One target of Derrida's criticism might be Heidegger, whose views on technology are utterly in step with the metaphysical

expulsion of *technê*. Heidgger writes, "*Technê* can merely cooperate with *physis*, can more or less expedite the cure; but as *technê* it can never replace *physis*, and in its stead become the *arché*...." This view is prolonged in *The Question Concerning Technology*, where the perversion of modern technology is due to its attempt to displace or reverse its subordination to *physis*. See Martin Heidegger, "On the Essence and Concept of Φύσις in Aristotle's *Physics* B, I" in *Pathmarks*, ed. William McNeil (Cambridge, UK: Cambridge University Press, 1998), 197.

58. Richard Wagner, *Opera and Drama. Richard Wagner's Prose Works*, Vol. 2, trans. W. Ashton Ellis (London: Kegan Paul, Trench, Trübner & Co., 1893), 76.
59. Richard Wagner, "Beethoven," in *Richard Wagner's Prose Works*, Vol. 5, 73.
60. Ibid., 74–75.
61. Carl Dahlhaus, "The Idea of Absolute Music," 34.
62. Wagner, "The Festival-Playhouse at Bayreuth," 333.
63. Cosima Wagner, *Cosima Wagner's Diaries: Volume 2, 1878–1883*, eds. Martin Gregor-Dellin and Dietrich Mack, trans. Geoffrey Skelton (New York: Harcourt Brace Jovanovich, 1980), entry of Sept. 23, 1878.
64. Joseph Holbrooke and Herbert Trench, *Apollo and the Seaman: An Illuminated Symphony, Op. 51* (London: Novello, 1908).
65. Anonymous, "Editorials," *The New Music Review and Church Music Review*, Vol. 7, No. 76 (March 1908): 202.
66. Ibid.
67. Bryan Gilliam, "The Annexation of Anton Bruckner: Nazi Revisionism and the Politics of Appropriation," 596. For further discussion of the *Dunkelkonzerte*, see Friedrich C. Heller, "Von der Arbeiterkultur zur Theatersperre," 101ff.
68. Joseph Goebbels, "Appendix: Joseph Goebbels's Bruckner Address in Regensburg (6 June 1937)," trans. John Michael Cooper, *The Musical Quarterly*, Vol. 78, No. 3 (Autumn 1994): 607.
69. Ibid., 608. As Gilliam notes (595), the relationship between Nazism and the Church is a complicated topic, one that culminated in the Vatican's official denunciation of Nazism with the publication and reading of Pius XI's *Mit brennender Sorge* in March 1937.
70. Friedrich Bayer, "Musik in Wien," *Die Musik*, 34. Jg., No. 1 (1941), 22.
71. Philippe Lacoue-Labarthe, *Heidegger, Art and Politics*. "National-aestheticism" first appears on page 58 of the English translation, but it is a concept developed throughout his text.
72. Gilliam, "The Annexation of Anton Bruckner," 598.

Interlude

1. Schaeffer, *Traité*, 91 ff.
2. Ibid., 63–64.
3. Gayou, *Le GRM*, 110.
4. *La revue musicale*, No. 244 (Paris: Richard-Masse, 1959), 63. Also in *Pierre Schaeffer: texts et documents*, ed. François Bayle (Paris: INA-GRM, 1990), 83.
5. The translation of allure is not standardized. Allure is sometimes translated as speed (see the liner notes to Schaeffer's *Solfège*) but is left in French in John Dack's translation of Chion's *Guide*.

6. Schaeffer, *Traité*, 488. "Les *études de composition*... se proposent, à partir d'un matériel sonore donné, convenablement *limité*, de réaliser des structures authentiques, qui mettraient en valeur pour autrui les critères que le compositeur s'est efforcé, d'après un schéma personnel, de 'donner à entendre.'"
7. Jacqueline Caux, *Almost Nothing with Luc Ferrari*, trans. Jérôme Hansen (Berlin: Errant Bodies Press, 2012), 28.
8. Ibid., 33.
9. For Schaeffer's stigma on "anecdotal" sounds in *musique concrète*, see chap. 1.
10. Caux, *Almost Nothing*, 34.
11. Ibid. (translation modified).
12. Ibid.
13. Gayou, *Le GRM*, 124.
14. Caux, *Almost Nothing*, 34.
15. Eric Drott, "The Politics of *Presque rien*" in *Sound Commitments: Avant-Garde Music and the Sixties*, ed. Robert Aldington (Oxford, UK: Oxford University Press, 2009), 145.
16. Ibid.
17. Ibid., 146. Also see Michel Chion and Guy Riebel, *Les musiques electroacoustiques* (Aix-en-Provence, FR: Édisud, 1976), 67.
18. Drott, "The Politics of *Presque rien*," 145.
19. See John Levack Drever, "Soundscape Composition: The Convergence of Ethnography and Acousmatic Music," *Organised Sound*, Vol. 7, No.1 (April 2002): 21–27; and Barry Truax, "Genres and Techniques of Soundscape Composition as Developed at Simon Fraser University," *Organised Sound*, Vol. 7, No.1 (April 2002): 5–14.
20. On "found sound," see Truax, "Genres and Techniques," 7. On the "*objet trouvé*," see Brigitte Robindoré, "Luc Ferrari: Interview with an Intimate Iconoclast," *Computer Music Journal*, Vol. 22, No. 3 (Fall 1998): 11.
21. Truax, "Genres and Techniques," 7.
22. Ibid., 8.
23. Robindoré, "Luc Ferrari," 11.
24. Ibid., 13.
25. Dan Warburton, interview with Luc Ferrari, July 22, 1998 [accessed 04/20/2011: http://www.paristransatlantic.com/magazine/interviews/ferrari.html].
26. Brandon LaBelle, *Background Noise: Perspectives on Sound Art* (New York: Continuum, 2006), 31–32.
27. One could quickly get an overview of the debate concerning the terms *écouter* and *entendre* by a comparative reading of Schaeffer's *Traité*, Nancy's *Listening*, and Barthes's essay "Listening." See Jean-Luc Nancy, *Listening*, trans. Charlotte Mandell (New York: Fordham University Press, 2007); and Roland Barthes, "Listening," in *The Responsibility of Forms*, trans. Richard Howard (New York: Hill and Wang, 1985), 245–260.
28. Nancy, *Listening*, 1. For the French edition, see Nancy, *À l'écoute* (Paris: Galilée, 2002), 13. When references to Nancy's text are followed by the French, I cite the English translation first, followed by the French edition.
29. Ibid., 1, 13.
30. Ibid., 2, 14.
31. Ibid., 32.

32. Ibid., 32, 61–62.
33. Jean-Luc Nancy, "The Forgetting of Philosophy," in *The Gravity of Thought*, trans. François Raffoul and Gregory Recco (Atlantic Highlands, NJ: Humanities Press, 1997), 22.
34. Ibid., 23.
35. Ibid.
36. Nancy, *Listening*, 1–2.
37. For a thorough treatment of Nancy on *Darstellung*, see Alison Ross, *The Aesthetic Paths of Philosophy: Presentation in Kant, Heidegger, Lacoue-Labarthe, and Nancy* (Stanford, CA: Stanford University Press, 2007).
38. Nancy, "The Forgetting of Philosophy," 22.
39. Jean-Luc Nancy, *The Muses*, trans. Peggy Kamuf (Stanford, CA: Stanford University Press, 1996), 88.
40. Ibid., 89.
41. Ibid., 92.
42. Aquinas, *Summa Theologica: Complete English Edition in Five Volumes* (Notre Dame, IN: Christian Classics, 2000), Pt. 1, question 45, art. 7.
43. Nancy, *The Muses*, 94.
44. In response to François Delalande's questions about *Presque rien*, Ferrari tries to make his interlocutors resist the irresistible temptation. Ferrari sees *Presque rien* as part of his development of a "proto-minimalist aesthetic." "LF: My own minimalism consisted in bringing a minimum of musical elements into the musical world.... FD: ...as well as the minimum of intervention on the composer's part. LF: No, there is just as much composition, but it is concealed. If one hears the intervention, that means the reality has been distorted. It's like a hyperrealist painting which masks the photograph's intervention behind the act of painting. The same goes for the *Presque rien*. It is a composition, the composer intervenes all the time." Caux, *Almost Nothing*, 46.
45. For an illuminating comparison, I recommend auditioning in close succession Ferrari's *Presque rien* and R. Murray Schafer's *The Vancouver Soundscape*, LP record (Epn 186, 1972).
46. Jonathan Sterne, *The Audible Past*, chap. 2.
47. Ibid., 33.
48. Ibid., 32. For another account of the phonautograph, see Friedrich Kittler, *Gramophone, Film, Typewriter*, trans. Geoffrey Winthrop-Young and Michael Wutz (Stanford, CA: Stanford University Press, 1999), chap. 1 and the image on p. 84.
49. Sterne, *The Audible Past*, 33.
50. Kittler, *Gramophone, Film, Typewriter*, 23.
51. Other features of the work also bring the recorded character into audibility, aside from mixing: (1) the title; (2) the sudden ending, as if the tape has simply run out; (3) the length of the work (approx. 20 minutes), which is also a common length for reels of magnetic tape in the late 1960s; (4) the form of the work, which begins with a fade-in and ends suddenly, to introduce the possibility of hearing it as a soundscape and then neutralize this possibility.
52. The argument of this paragraph and the following discussion are greatly indebted to Thierry de Duve, *Kant after Duchamp* (Cambridge, MA: MIT Press, 1996), chap. 6. It should be mentioned that Nancy's aesthetic theory has been shaped by a reading

of de Duve and vice-versa. Compare de Duve, 351, n. 10, and Nancy, *The Muses*, 110, n. 44, and 188, n. 22.
53. Caux, *Almost Nothing*, 34 (translation modified).
54. de Duve, *Kant after Duchamp*, 352.
55. Ibid.
56. Nancy, *Listening*, 1–2.

Chapter 5

1. *Traité*, 269. I refer the reader to my critique of this aspect of Schaeffer's thinking in chap. 1 in the section titled "Originary experience and the problem of history."
2. Hans Jonas, "The Nobility of Sight," in *The Phenomenon of Life: Toward a Philosophical Biology* (Evanston, IL: Northwestern University Press, 2001), 135–151. Harper & Row first published Jonas's text in 1961, but all references are to the 2001 edition.
3. Ibid., 137.
4. The argument recalls Plato's rejection of artistic mimesis as producing only copies of copies, twice removed from the reality of the forms. In *The Republic*, Book X, Plato compares the cabinetmaker and the painter, both imitators, with God, the true creator of the forms. The imitator "does not make that which really is, he could not be said to make real being but something that resembles real being.... [His work] is only a dim adumbration in comparison with reality" (597, a–b).
5. Jonas, "The Nobility of Sight," 137–138.
6. Erwin Straus, "The Forms of Spatiality," in *Phenomenological Psychology*, trans. Erling Eng (New York: Basic Books, 1966), 3–37. This analysis is cited and incorporated into Straus's magnum opus, *The Primary World of the Senses*. For the original German edition, see Erwin Straus, *Vom Sinn der Sinne, ein Beitrag zur Grundlegung der Psychologie* (Berlin: J. Springer, 1935). For the revised English translation see, Erwin Straus, *The Primary World of the Senses: A Vindication of Sensory Experience*, trans. Jacob Needleman (New York: Free Press, 1963).
7. Straus, "The Forms of Spatiality," 8.
8. Ibid.
9. Ibid., 9.
10. Jonas, "The Nobility of Sight," 138.
11. Scruton, *Aesthetics of Music*, 221.
12. Ibid., 2–3.
13. Ibid., 3.
14. Husserl, *Ideas I*, §32.
15. Although the term "acousmatic" is not deployed, I would assert that Kafka is one of the great thinkers of acousmatic sound. As I argued in chapters 2 and 3, histories of the term acousmatic typically constrain the discourse on acousmatic sound to the appearance of the word in the historical record: a history of Pythagorean veils and acousmates. Rather than confine the historiography of acousmatic sound to a criterion that only accepts instances where some term (acousmatic, akousma, acousmatique, acousmate, etc.) appears, my approach has been to investigate instances where the sensory conditions of acousmatic sound are invoked, in all of their cultural and historical specificity, regardless of what descriptive terms are present.
16. Franz Kafka, "The Burrow," in *The Complete Stories*, ed. Nahum N. Glatzer (New York: Schocken Books, 1971), 325–359.

17. Ibid., 343.
18. Ibid., 344.
19. Walter Benjamin, "Franz Kafka," in *Illuminations*, ed. Hannah Arendt, trans. Harry Zohn (New York: Schocken Books, 1969), 132.
20. For recent work on Kafka from the perspective of the auditory turn, see David Copenhafer, "'Is it even singing at all?' Kafka's Musicology," in *Invisible Ink: Philosophical and Literary Fictions of Music* (Ph.D. diss., University of California–Berkeley, 2004); Gilles Deleuze and Felix Guattari, *Kafka: Toward a Minor Literature*, trans. Dana Polan (Minneapolis, MN: University of Minnesota Press, 1986); Mladen Dolar, *A Voice and Nothing More*, chap. 7, and "The Burrow of Sound," *Differences: A Journal of Feminist Cultural Studies*, Vol. 22, Nos. 2–3 (2011): 112–139; Kata Gellen, "Hearing Spaces: Architecture and Acoustic Experience in Modernist German Literature," *Modernism/Modernity*, Vol. 17, No. 4 (2011): 799–818; John Hamilton, "'*Ist das Spiel vielleicht unangenehm?*' Musical Disturbances and Acoustic Space in Kafka," *Journal of the Kafka Society of America*, Nos. 1–2 (2004), 23–27 (a special issue dedicated to Kafka and music); and Peter Szendy, *Sur écoute: esthétique de l'espionnage* (Paris: Éditions de Minuit, 2007), 71–79.
21. Heinz Politzer, *Franz Kafka: Parable and Paradox* (Ithaca, NY: Cornell University Press, 1962), 319.
22. Rosemary Arrojo, "Writing, Interpreting and the Control of Meaning," in *Translation and Power*, eds. Edwin Gentzler and Maria Tymoczko (Amherst, MA: University of Massachusetts Press, 2002), 66.
23. Ibid.
24. In line with the textualist account, literary theorist Henry Sussman writes this about "Der Bau": "the construction is already a deconstruction to the same extent that it has been constructed." The composition of the burrow becomes an allegory for the impossible construction of the text. See Sussman, *Franz Kafka: Geometrician of Metaphor* (Madison, WI: Coda Press, 1979), 149.
25. Stanley Corngold, *Franz Kafka: the Necessity of Form* (Ithaca, NY: Cornell University Press, 1988), 282.
26. Corngold's reading focuses on the role played by the figure of chiasmus (or reversal) in Kafka's texts. For Corngold, reversal is the central trope that characterizes the uncontrollable passages of "The Burrow." He offers this bit of the mole's internal soliloquy as exemplary: "And it is not only by external enemies that I am threatened. There are also enemies in the bowels of the earth. I have never seen them, but legend tells of them and I firmly believe them.... Their very victims can scarcely have seen them; they come, you hear the scratching of their claws just under you...and you are already lost. Here it is of no avail to console yourself with the thought that you are in your own house; far rather are you in theirs" (Kafka, "The Burrow," 326). Notice how the passage inverts the ownership of the burrow; the territorial speaker has been turned into the territorialized trespasser. According to Corngold, Kafka's chiasms "annihilate the opposition between the categories of victim/enemy, inside/outside, legend/knowledge, destroying the specificity of self-identical terms" (Corngold, *Franz Kafka*, 125). The text produces a series of inversions where terms on either side of the equation can never be brought back to equivalence. Like the passages, rooms, and labyrinth of the burrow, the reader follows a circuitous route that doubles back, reverses, leads nowhere, and returns back to where it started, but

never simply back to where it began. Such maddening circuits mark the experience of traversing Kafka's constructed passages.

27. On the sound: As the existential self, see Wilhelm Emrich, *Franz Kafka: A Critical Study of His Writings*, trans. Sheema Zeben Buehne (New York: Ungar, 1968), 215 ff.; as the impending other, see Politzer, *Franz Kafka: Parable and Paradox*, 328–330; as complusion, see Walter Sokel, *Kafka—Tragik und Ironie: Zur Struktur seiner Kunst* (München and Wien, DE: Albert Langen/Georg Müller, 1964), 371–387; as the fear of solitude, see Heinrich Henel, "'The Burrow,' or How to Escape from a Maze," in *Franz Kafka*, ed. Harold Bloom (New York: Chelsea House Publishers, 1986), 121; as mental illness, see Hermann J. Weigand, "Franz Kafka's 'The Burrow': An Analytical Essay," in *Franz Kafka: A Collection of Criticism*, ed. Leo Hamalian (New York: McGraw Hill, 1974), 104; as trench warfare, see Wolf Kittler, "*Grabenkrieg—Nervenkrieg—Medienkrieg. Franz Kafka und der 1. Weltkrieg*," in *Armaturen der Sinne*, eds. Jochen Hörisch and Michael Wetzel (München, DE: Wilheml Fink Verlag, 1990), 289–309; as tuberculosis, see Britta Maché, "The Noise in the Burrow: Kafka's Final Dilemma," *The German Quarterly*, Vol. 55, No. 4 (Nov. 1982): 526–540; and Mark Boulby, "Kafka's End: A Reassessment of the Burrow," *The German Quarterly*, Vol. 55, No. 2 (Mar. 1982): 184. Some evidence for the tubercular reading of the sound is offered by Kafka's friend and executor Max Brod in his epilogue to *Beschreibung eines Kampfes: Novellen, Skizzen, Aphorismen aus dem Nachlass* (Frankfurt, DE: S. Fischer, 1954), 350. However, Brod's claims are often inconsistent and disputed by Kafka scholars.
28. Deleuze and Guattari, *Kafka: Toward a Minor Literature*, 3.
29. Ibid., 4.
30. Ibid., 5.
31. Ibid., 6.
32. Ibid.
33. Deleuze and Guattari base their argument on Kafka's diaries and letters, and his well-known claim to be "unmusical." The diary entry of Dec. 13, 1911, offers evidence for Deleuze and Guattari's association of music with the problems of signification: "Brahms concert by the Singing Society. The essence of my unmusicalness consists in my inability to enjoy music connectedly, it only now and then has an effect on me, and how seldom it is a musical one. The natural effect of music on me is to circumscribe me with a wall, and its only constant influence on me is that, confined in this way, I am different from what I am when free." Later, Kafka ontologizes his unmusicality into a way of being, an "*Unmusikalisch-Sein*" in a letter to Milena of July 17, 1920. The unique feature of Deleuze and Guattari's reading of Kafka is the distinction they draw between music and sound, which turns Kafka's *Unmusikalisch-Sein* into a virtue when considering sound and sonic phenomena. Kafka's sensitivity to sound is legendary. See Franz Kafka, *The Diaries of Franz Kafka 1910–1913*, ed. Max Brod, trans. Joseph Kresh (New York: Schocken Books, 1948), 176; and *Letters to Milena*, trans. Philip Boehm (New York: Schocken Books, 1990), 92.
34. Deleuze and Guattari, *Kafka: Toward a Minor Literature*, 13.
35. Ibid.
36. And, to give away the punch line of Deleuze and Guattari's book, so does Kafka's own use of the German language.

37. In fact, it is only the small fry that might be producing the high-pitched sound (not the protagonist of "The Burrow") that are described as "becoming-molecular," an extreme version of becoming-animal. See Deleuze and Guattari, *Kafka: Toward a Minor Literature*, 37.
38. One should note that it is precisely these tunnels and passages that are the mole's productions.
39. Kafka, *Complete Stories*, 362.
40. Ibid., 281.
41. Ibid.
42. Wagner, "Beethoven," 68.
43. Ibid.
44. Pliny, *Natural History*, 10.88. The translation is taken from *The Natural History of Pliny the Elder, Vol. II*, trans. John Bostock and H. T. Riley (London: Henry G. Bohn, 1855), 547.
45. Alexander Pope, "An Essay on Man," in *Poetry and Prose of Alexander Pope*, ed. Aubrey Williams (Boston: Houghton Mifflin Company, 1969), Epistle I, 211–212.
46. Kafka, *Complete Stories*, 343.
47. Ibid., 345.
48. Ibid.
49. Ibid., 347.
50. Ibid.
51. Ibid., 348.
52. Ibid., 353.
53. Ibid.
54. Ibid., 356.
55. Ibid., 344, 353.
56. Dolar, *A Voice and Nothing More*, 166, n. 3.
57. P. F. Strawson, *Individuals: An Essay in Descriptive Metaphysics* (Garden City, NY: Anchor Books, 1963), 64.
58. Ibid., 56–57.
59. Ibid., 22.
60. Ibid.
61. Ibid., 62.
62. By employing the term *spacing*, I am invoking Derrida's use of the term *espacement*. Derrida describes spacing as "not only the interval, the space constituted between two things (which is the usual sense of spacing), but also spa*cing*, the operation, or…movement of setting aside.…It marks what is set aside from itself, what interrupts every self-identity." See Jacques Derrida, *Positions*, trans. Alan Bass (Chicago: University of Chicago Press, 1981), 106–107, n. 42. In this sense, spacing is a term that marks the stoppage of any closed system of auto-affection or identity claims based on the ground of presence. Something must be spaced from itself for it to be capable of affecting itself. For Derrida, spacing appears as one of many terms associated with *différance*, which does not simply mean difference but plays on the idea of something being, within itself, both temporally deferred or delayed and spatially demarcated or differentiated. A paradigmatic example appears in *Speech and Phenomena*, where, through a reading of Husserl on time consciousness, Derrida exposes Husserl's own thinking of the spacing at the heart of the "living present" and

thus opens up the tight circle of auto-affection that grounds Husserlian subjectivity. By invoking the term in the context of sonic source, cause, and effect, I am arguing that acousmatic sound should be characterized in terms of spacing. Acousmatic sound necessitates thinking of the relation of source, cause, and effect as both unified into a single sound and simultaneously differing as distinct parts of a sound. The spacing of source, cause, and effect forces recognition of acousmatic sound as reducible to neither the synthesis of the chain of three terms nor to the privilege of any single term. Acousmatic sound must be seen as a constant deferral or referral of the effect toward a demarcated but underdetermined source and/or cause and back again. Spacing "is the index of an irreducible exterior, and at the same time of a *movement*, a displacement that indicates an irreducible alterity" (*Positions*, 81). In its underdetermination, the effect demands a movement toward its exterior, to the source and cause. And those terms cannot account for the accomplishment of the effect. The neither-autonomous-nor-heteronomous effect is the result of a perpetual displacement, a game of hide and seek among the three terms, none of which are integral on their own. Concerning the question of reduction, Derrida writes, "Spacing also signifies, precisely, the impossibility of reducing the chain to one of its links or of absolutely privileging one—or the other" (*Positions*, ibid.). In the case of acousmatic sound, the impossibility of reducing it to one of its terms (in what I call either a materialist reduction to source/cause or an eidetic reduction to the effect) has not precluded many authors from attempting such reduction. In fact, the virulence of such reductions may indeed evince spacing's operation. In spacing's movement, "the emancipation of the sign constitutes in return the desire of presence" (*Of Grammatology*, 69). In the case of the acousmatic voice, the "emancipation" of the voice constitutes the demand for the return of presence, here, in the form of two reductions—material and eidetic. The strange absence that inhabits acousmatic sound, especially in the form of the acousmatic voice, cannot be filled by an appeal to source/cause or a hypostatization of it as "sound object." Below, I address the attempts at reduction in more detail. On spacing, see *Positions*, 27, 80–81, 91–94, and 106, n. 42; *Speech and Phenomena*, 86; "Différance," in *Margins of Philosophy*, trans. Alan Bass (Chicago: University of Chicago Press, 1982), 8, 13; and *Of Grammatology*, Pts. I.2, I.3, and II.3. For a very good account of Derrida on *espacement*, see Martin Hägglund, *Radical Atheism: Derrida and the Time of Life* (Stanford, CA: Stanford University Press, 2008), 18–19 and passim.

63. Acousmatic underdetermination plays a role not only within Kafka's narrative, but also in its reception. The glut of divergent and incompatible readings of the sound in the burrow offered by literary critics evinces the mysterious sound's underdetermination of source and cause.
64. Schaeffer, *Traité*, 93 (emphasis added).
65. Roland Barthes, "Listening," 247.
66. Chion has written widely on the *acousmêtre*, but the introduction of the term appears in his *The Voice in Cinema*, 17–29. While Chion borrows Schaeffer's acousmatic reduction to describe the *acousmêtre*, relying on the difference between the image track and the soundtrack, he does not invoke Schaeffer's eidetic reduction to sound objects or defense of reduced listening in his explanation.
67. Ibid., 24.
68. Dolar, *A Voice and Nothing More*, 67.

69. Ibid., 70.
70. Ibid.
71. Simon Emmerson, *Living Electronic Music* (London: Ashgate, 2007), 6.
72. For an informative comparison, see Emmerson, *Living Electronic Music*, 3–14, and Joanna Demers, *Listening Through the Noise* (New York: Oxford University Press, 2010), 29–37. In addition, insights from ecological listening, based on J. J. Gibson's work, have been applied to acousmatic sound in Luke Windsor, *A Perceptual Approach to the Description and Analysis of Acousmatic Music* (Ph.D. diss., City University, 1995); and Eric Clarke, *Ways of Listening*.
73. See R. Murray Schafer, *The Soundscape*; his LP, *The Vancouver Soundscape*; World Soundscape Project (Sonic Research Studio, Communication Studies Dept., Simon Fraser University, 1974); and World Soundscape Project and Barry Truax, *The World Soundscape Project's Handbook for Acoustic Ecology* (Vancouver, CA: A.R.C. Publications, 1978).
74. R. Murray Schafer, *The Soundscape*, 130–131.
75. Ibid. Schafer defines a keynote as the anchor or fundamental sound in a soundscape, one that need not be consciously recognized, yet acts as the ground upon which other sounds are noted. See *The Soundscape*, 9–10 and passim.
76. Ibid., 90.
77. Adriana Cavarero, *For More than One Voice: Toward a Philosophy of Vocal Expression*, trans. Paul A. Kottman (Stanford, CA: Stanford University Press, 2005).
78. Jacques Derrida, "Aphorism Countertime," in *Acts of Literature*, ed. Derek Attridge (New York: Routledge, 1992), 423.
79. Cavarero, *For More than One Voice*, 215.
80. Ibid., 222.
81. Ibid., 235, 238.
82. Derrida, "Aphorism Countertime," 423.
83. Italo Calvino, "A King Listens," in *Under the Jaguar Sun*, trans. William Weaver (San Diego, CA: Harcourt Brace Jovanovich, 1988).
84. Ibid., 38.
85. Ibid., 44.
86. Ibid.
87. Ibid., 49.
88. Ibid., 52.
89. Ibid., 54–55.
90. Cavarero, *For More than One Voice*, 3.
91. Ibid., 4.
92. Ibid.
93. Ibid., 7.
94. Calvino, "A King Listens," 53.
95. Cavarero, *For More than One Voice*, 8.
96. Calvino, "A King Listens," 57.
97. There is a strange dissymmetry between the two reductive routes found in music and in philosophy and filmic dis-acousmatization. After much reflection, I cannot think of a counterexample, where a filmic disembodied voice is simply detached from its source or cause and ontologically secured. Perhaps this option is foreclosed in film because, since the introduction of sound, film is predicated on the co-presence of

both image and soundtracks. Thus, the two sense modalities of the eye and the ear are forced into a perpetual negotiation in film and can be neither simply separated nor dissolved.
98. Chion, *The Voice in Cinema*, 23–24.
99. Ibid., 133.
100. Earl J. Hess and Pratibha A. Dabholkar, *Singin' in the Rain: The Making of an American Masterpiece* (Lawrence, KS: University of Kansas Press, 2009), 145.
101. Ibid. Also see *Singin' in the Rain*, DVD, 2 discs (Warner Home Video, 2002). Betty Noyes's voice replaced Debbie Reynolds's in the scene where Kathy Selden is seen dubbing Lena Lamont's voice for "The Dancing Cavalier" and sings "Would You?"
102. Kafka, *Complete Stories*, 353.
103. Ibid., 356.
104. Edgar Allan Poe, "The Tell-Tale Heart," in *Poetry and Tales* (New York: Library of America, 1984), 559.
105. Ibid.
106. Ibid.
107. Michel Chion, *Film: A Sound Art*, trans. Claudia Gorbman (New York: Columbia University Press, 2009), 453 ff.
108. John Cage, *Silence* (Middletown, CT: Wesleyan University Press, 1961), 8.
109. The addition of Poe to the analysis of acousmatic sound brings up a question concerning the relationship between acousmatic sound and the literary genre of the fantastic. Both Kafka's and Poe's tales could be thought to inhabit this genre, as noted by Tzvetan Todorov in his classic study of the topic. There are suggestive similarities and noteworthy differences concerning acousmatic listening and the fantastic. Todorov analyzes the "heart of the fantastic" as follows: "In a world which is indeed our world, the one we know, a world without devils, sylphides, or vampires, there occurs an event which cannot be explained by the laws of this same familiar world. The person who experiences the event must opt for one of two possible solutions: either he is the victim of an illusion of the senses, of a product of the imagination—and the laws of the world remain what they are; or else the event has indeed taken place, it is an integral part of reality—but then this reality is controlled by laws unknown to us.... The fantastic occupies the duration of this uncertainty. Once we choose one answer or the other, we leave the fantastic for a neighboring genre, the uncanny or the marvelous." Acousmatic sound, like the fantastic, could be described as occupying the duration of an uncertainty, an uncertainty about the source of the sound. This is in contrast to the fantastic uncertainty about the natural or supernatural status of the event. Acousmatic sounds are *not necessarily* dependent on a potential supernatural source—for instance, the sound in the burrow—although there may be specific practices involving acousmatic sound that ascribe to the source a supernatural origin—for instance, the angels heard in the voices of the hidden singing nuns. Similarly, when the source of the sound is disclosed, we leave acousmatic listening for another mode of listening (typically *entendre* or *écouter*) as shown in the analyses of sound studies, philosophy, and film discussed above. Todorov also claims that the experience of the fantastic "[does] not specify whether it is the reader or the character who hesitates." This is also a feature of the acousmatic sounds heard in Kafka and Poe, where the reader remains doubtful about the reliability of the narrator's testimony. See Tzvetan Todorov, *The Fantastic: A Structural Approach to a*

Literary Genre, trans. Richard Howard (Ithaca, NY: Cornell University Press, 1973), 25, 27.

CHAPTER 6

1. At first glance, it may seem surprising to think of Les Paul's work as relevant to questions of acousmatic sound or even acousmatic music. The term "acousmatic music," popularized by François Bayle, usually refers to a style of prerecorded electronic music that stems from the tradition of Schaeffer's *musique concrète*, where a prerecorded track is diffused live on a multi-channel speaker array. As a term that practitioners of a style use to designate their own practices, one cannot argue. However, the term is often deployed to designate a category within the large heading of electronic or electroacoustic music. For instance, Leigh Landy draws a contrast between acousmatic music and live electronic music; in the latter, technology is used to manipulate and transform the sounds of live performers, while the former "is intended for loudspeaker listening and exists only in recorded tape form (tape, compact disk, computer storage)." Given the broad historical and philosophical context in which I have tried to situate acousmatic sound, I think one can question Landy's definition. Rather than think about acousmatic music as music that reconfigures the sensory ratios of seeing and hearing, and exploits imaginative and aesthetic results that stem from the underdetermination of a sound's source by its effect, Landy's definition determines acousmatic music on the basis of its recording medium, its fixity, and lack of live performers. Thus no music composed before the age of sound recording would count as acousmatic. I think such a definition undercuts the possibility of understanding Schaeffer's practice and the practice of Bayle's *musique acousmatique* as inheritors of the tradition of 19th-century absolute music and Wagnerian operatic reforms. And vice-versa, it disallows the possibility of understanding music before the age of recording as acousmatic in the sense that I have tried to define it. In other words, I do not see acousmatic music as dependent on a particular kind of audio technology or medium. It is a technique for splitting the senses and making the underdetermination of a sonic source by its effect palpable. See Leigh Landy, *Understanding the Art of Sound Organization* (Cambridge, MA: MIT Press, 2007), 14.
2. Of course, he was not the first to use overdubbing. Before Paul, there were other instances. In 1931, baritone Lawrence Tibbett sang a duet with himself in the title song of the MGM film *The Cuban Love Song*. It was released as an RCA Victor recording with a label that read "Lawrence Tibbett—baritone, with orchestra. Mr. Tibbett also sings the tenor part." In 1935, Elisabeth Schumann accomplished a similar feat, singing the duet "Abendsegen" from Humperdinck's *Hänsel und Gretel* for the HMV label. In the jazz world, multi-instrumentalist Sidney Bechet cut a famous 1941 recording of "Sheik of Araby" in which he played harmonized lead and rhythm parts. Others followed suit, like Nelson Eddy, whose voice could be heard singing with itself in the segment "The Whale Who Wanted to Sing at The Met," from the Disney film *Make Music Mine*. According to the opening titles, Eddy "does *all* the voices for the tragic story." And, perhaps most significantly for Paul and Ford, in 1947, the singer Patti Page recorded "Confess," in which she sang with herself in call and response, and later, "With My Eyes Wide Open," in which she overdubbed four-part vocal harmonies in a style that would often be compared with Paul and Ford.

3. "Jazzorama," *Jazz Today* 3 (April 1957): 5. Quoted in Steve Waksman, *Instruments of Desire: The Electric Guitar and the Shaping of Musical Experience* (Cambridge, MA: Harvard University Press, 1999), 36.
4. Waksman, *Instruments of Desire*, 38.
5. For more on the theory of mimetic rivalry, see René Girard, *Deceit, Desire, and the Novel*, trans. Yvonne Freccero (Baltimore, MD: Johns Hopkins University Press, 1965).
6. In Waksman's narrative, Paul seeks out greater and greater mastery over his sound through two strategies: increasing technological control and a blurring of professional and domestic spaces. Paul's desire to find a "pure sound" that filters out all traces of noise is read both literally, as a search for high-fidelity guitar pickup technology, and as a metaphor for Paul's appropriation and homogenization of various kinds of racially coded music.
7. Amy Porter, "Craziest Music You Ever Heard," *Saturday Evening Post* (Jan. 17, 1953): 100.
8. Incidentally, this is the same recording technique that was later popularized in recordings by The Chipmunks, where vocal harmonies, sung over slowed-down accompaniment tracks, were transposed and spectrally shifted, producing a novel effect.
9. This cover is from the 1950 release, Capitol H-226, a 10" LP. It is worth noting that the album was re-released in 1955 on a 12" LP, with additional tracks and a new cover. That image, in bright orange, features a horizontal illustration of a guitar now superimposed by five white vertical iterations of a guitar's silhouette. The image no longer conveys a sense of multiple instruments being simultaneously played but, by 1955, Paul's strategies for maintaining an air of mystery around the new sound had changed.
10. Jon Sievert, "Les Paul," *Guitar Player* (Dec. 1977): 43–60 [accessed 06/19/2013: http://www.gould68.freeserve.co.uk/LPGPInter.htm].
11. Mary Alice Shaughnessy, *Les Paul: An American Original* (New York: William Morrow and Co., 1993), 197.
12. Ibid.
13. Sievert, "Les Paul."
14. Porter, "Craziest Music You Ever Heard," 98.
15. *The Les Paul Show*, radio program, NBC. This program aired July 11, 1950. The original dates of broadcasts for other episodes are noted in the text.
16. Ibid.
17. R. Buskin, "Classic Tracks: Les Paul and Mary Ford, 'How High the Moon,'" *Sound on Sound*, Jan. 2007 [accessed 01/05/2009: http://www.soundonsound.com/sos/jan07/articles/classictracks_0107.htm].
18. "The Case of the Missing Les Paulverizer," on Les Paul and Mary Ford, *The Legend and the Legacy*, audio CD (Capitol Records, 1991). This is an episode from NBC's *The Les Paul Show*, although I have been unable to discover its original airdate.
19. *The Les Paul Show*, Aug. 11, 1950.
20. "Les Paul and Mary Ford," *Omnibus*, Season 2, Ep. 4 (Oct. 23, 1953), CBS Television. On this episode, Paul and Ford were in the third and final slot, following a performance of "The Gold Dress," based on a play by Stephen Vincent Benet, and an interview with Frank Lloyd Wright.

Notes

21. Les Paul and Mary Ford, "Don'cha Hear Them Bells/The Kangaroo," Capitol single #2614. "The Kangaroo" was a new release (Oct. 12, 1953) at the time *Omnibus* aired and was the B-side of "Don'cha Hear Them Bells," the song that Mary sang into the fictional Paulverizer. Paul penned both songs. Perhaps the television spot helped record sales: "The Kangaroo" charted on Oct. 31, 1953, a week after *Omnibus* aired, eventually reaching #25; "Don'cha Hear Them Bells" charted the week after (Nov. 7, 1953), reaching #13. See Les Paul and Mary Ford, *The Legend and the Legacy*, liner notes to Disc 2.
22. Although the duo did indeed describe and demonstrate their process, if one closely watches the footage, it is clear that various technical tricks are being used to simplify and demonstrate the process without actually making a "multiple" recording. For instance, the first passage that Paul plays, the opening lick from their famous recording of "How High the Moon," is not the same as what is played back in the headphones immediately afterward; this is acknowledged by Cooke as being "practically the same," followed by a laconic "almost" from Paul. When Mary starts to record her vocals, we hear the vocal tracks without the guitar accompaniment, even though she is ostensibly adding her vocals to the guitar tracks already recorded. Unlike later multitrack technology, it would have been impossible to isolate Ford's vocals using the Ampex machines shown in the studio. Rather, it appears that the various sounds required to demonstrate the duo's recording process in all of its various stages were prepared and cued up in the television studio, a recreation of their process rather than documentation of the process itself.
23. Shaughnessy, *Les Paul*, 231.
24. Buskin, "Classic Tracks."
25. Sievert, "Les Paul."
26. Buskin, "Classic Tracks."
27. Ibid.
28. If George Barnes was the mimetic rival who set the "new sound" in motion, Patti Page was perhaps Mary Ford's sonic-mimetic rival. Often in a competition to score hits on the charts, Page and the duo recorded and released many of the same songs, sometimes within days of each other. Page's version of "Tennessee Waltz," which featured the singer overdubbing vocal harmonies, was covered by Paul and Ford; similarly, Paul and Ford's version of "Mockin' Bird Hill" was covered by Page.
29. Roland Barthes, "Listening," 255.

Chapter 7

1. This line comes from a fictional Thomas Edison, in Auguste Villiers de L'Isle-Adam, *Tomorrow's Eve*, trans. Robert Martin Adams (Urbana, IL: University of Illinois Press, 2001), 10.
2. This recording is available online at http://archive.org/details/iamed1906 [accessed 09/28/2012]. According to Jason Camlot, the voice on the recording belongs to Len Spencer, recorded in West Orange in 1906. "This two-minute wax cylinder was distributed to dealers for the purpose of demonstrating the Edison phonograph but was not sold commercially." See Camlot, "Early Talking Books: Spoken Recording and Recitation Anthologies, 1880–1920," *Book History*, Vol. 6 (2003): 170, n. 43.
3. This advertisement is in the collection of the Bodleian Library. See Henri-Louis Jacquet-Droz, *A Description of Several Pieces of Mechanism, Invented by the Sieur*

Jacquet Droz, of... Switzerland. And Which Are Now To Be Seen at the Great Room, No. 6, in King-Street, Covent-Garden (London: n.p., 1780?).
4. For a detailed technical explanation of the workings of the Scribe, see Alfred Chapius and Edmond Droz, *Automata: A Historical and Technological Study*, trans. Alex Reid (Neuchâtel, CH: Editions de Giffon, 1958), 292–297.
5. For other takes on the Scribe's Cartesian proposition, see Gaby Wood, *Edison's Eve* (New York: Alfred A. Knopf, 2002), xiv; Paul Hoffman, "Enchantments of a Swiss Town," *New York Times* (April 18, 1993), XX8; and Kara Reilly, "Automata: A Spectacular History of the Mimetic Faculty" (Ph.D. diss., University of Washington, 2006), 113. I am unsure if the Scribe wrote the statement in Latin or French, but given the limitation to 40 characters, Descartes's "*Je pense donc je suis*" or "*Cogito ergo sum*" would both potentially work. A much larger automaton would be required to write out Kant's categorical imperative!
6. Jacques Derrida, "My Chances/Mes Chances: A Rendezvous with Some Epicurean Stereophonies," in *Taking Chances: Derrida, Psychoanalysis, and* Literature, eds. Joseph H. Smith and William Kerrigan (Baltimore, MD: Johns Hopkins University Press, 1984), 16.
7. Otto Jespersen, *Language: Its Nature, Development and Origin* (New York: Holt, 1921), 123. Although Jespersen introduced the term, the most influential account of the shifter is found in Benveniste, who writes: "Language is possible only because each speaker sets himself up as a *subject* by referring to himself as *I* in his discourse. Because of this, *I* posits another person, the one who, being, as he is, completely exterior to 'me,' becomes my echo to whom I say *you* and who says *you* to me. This polarity of persons is the fundamental condition in language, of which the process of communication, in which we share, is only a mere pragmatic consequence." Emile Benveniste, *Problems in General Linguistics*, trans. Mary Elizabeth Meek (Coral Gables, FL: University of Miami Press, 1971), 225. Also see Roman Jakobson, "Shifters, Verbal Categories, and the Russian Verb" in Jakobsen, *Selected Writings*, Vol. 2 (The Hague, NL: Mouton, 1971), 130–147.
8. Ibid. Given the important role of the "I" in Fichte's philosophy, this anecdote might itself be another "philosophical joke."
9. Ibid.
10. The quotation is taken from Horatio Nelson Powers's poem "The Phonograph's Salutation," Powers, "The First Phonogramic Poem," Thomas A. Edison Papers, D8850 124:705. It is available for viewing online at http://edison.rutgers.edu [accessed 09/23/2012]. This version of the poem is also reproduced in Francis Arthur Jones, *Thomas Alva Edison: Sixty Years of an Inventor's Life* (New York: Thomas Y. Crowell & Co., 1908), 155. Both are renderings of Powers's poem as played at the Crystal Palace in 1888. A slightly different version of the poem, which features the line "I am the youngest born of Edison" appears (with other small alterations) in Powers's own published version; see Horatio Nelson Powers, *Lyrics of the Hudson* (Boston: D. Lothrop, 1891), 69.
11. Thomas A. Edison, "The Perfected Phonograph," *North American Review*, Vol. 146, No. 379 (June 1888): 649.
12. For a fine discussion of Gouraud's activities, see John M. Picker, *Victorian Soundscapes* (New York: Oxford University Press, 2003), 117.
13. Ibid., 116.

14. See note 10 above for publication information on the "The Phonograph's Salutation." The recording can be heard today at the British Library's sound archive, C98/1 C7.
15. Dyson, *Sounding New Media*, 46.
16. Picker, *Victorian Soundscapes*, 117.
17. For a reading of the poem along these lines, see Camlot, "Early Talking Books," 157.
18. William Kennedy Laurie Dickson and Antonia Dickson, "The Life and Inventions of Edison," *Cassier's Magazine*, Vol. 3, No. 18 (April 1893): 457.
19. Edmund Husserl, *Logical Investigations*, Vol. 1, §2, 270.
20. Ibid., §5, 275.
21. Ibid., §7, 277.
22. Ibid.
23. Ibid., §1, 269.
24. To think otherwise would be nonsense, according to Husserl. He writes, "Shall we say that, even in solitary mental life, one still uses expressions to intimate [i.e., indicate] something, though not to a second person? Shall one say that in soliloquy one speaks to oneself, and employs words as signs, i.e., as indications, of one's own inner experiences? I cannot think such a view acceptable" (§8, 279). If speaking to oneself is not an act of communication, then Husserl is in a position to argue that "an expression's meaning, and whatever else pertains to it essentially, cannot coincide with its feats of intimation [indication]."
25. Ibid., §9, 280–281.
26. We can illuminate Husserl's ideas on the three parts of any expression (the physical phenomenon of the sign, the meaning-intention, and the fulfillment) by comparing it with Schaeffer's work on the sound object. In chapter 1, I argued that Schaeffer's brand of phenomenology was highly influenced by Husserl. When describing how Schaeffer delimited the acousmatic field into modes of listening and the sound object, I put this in terms of Husserl's descriptions of noesis and noema. (Even though Husserl's theory of sense-giving intentions and sense-fulfilling intuitions was written long before he formulated the noetic/noematic distinction, commentators on Husserl's *Logical Investigations* such as Derrida have often used the later terminology to describe the former.) Schaeffer's various modes of listening (*ouïr, comprendre, écouter, entendre*) correspond to different kinds of sense-giving intentions; the appropriate sound objects, which are either absent or present as the case may be, can intuitively fulfill those intentions. (1) On the correlation between modes of listening in Schaeffer and sense-giving acts in Husserl, it should be noted that I can listen to the same sound object in different modes, or I can intend the same object in different ways. For example, I can listen to a recording of Pierre Schaeffer's voice and I can listen for the sense of his discourse (in the mode *comprendre*), that is, I can listen with the meaning-intention of understanding the discourse he utters; or I can listen to the same recording and only pay attention to the intrinsic qualities of his voice, that is, I can listen with the meaning-intention directed toward the timbre of his French (in the mode *entendre*). (2) Schaeffer differentiates the physical phenomenon of a sign from my meaning-giving and fulfilling acts, just as Husserl separated the physical phenomenon of a sign from the meaning-giving intention that animates it, and the intuition that fulfills that intention. In a famous passage from *Traité*, Schaeffer is clear that the sound object is irreducible to the physical medium that holds it, viz., it is irreducible to the sounds on the tape. The sound object, the

intentional object that can fulfill any act of meaning-intention (no matter what the mode of listening may be) is not something subjective. Because it is an ideal object, it is shareable and intersubjective. Various acts of listening by various listeners can all intend the same object, just as various speech acts by various speakers can all intend the same object. Schaeffer writes, "To avoid confusing [the sound object] with its physical cause of a 'stimulus,' we seemed to have grounded the sound object on our subjectivity. But...the sound object is not modified...with the variations in listening from one individual to another, nor with the incessant variations in our attention and our sensibility. Far from being subjective...[sound objects] can be clearly described and analyzed" (*Traité*, 97). Moreover, "we must therefore stress emphatically that [a sound] object is something real [i.e., objective], in other words that something in it endures through these changes and enables different listeners (or the same listener several times) to bring out as many aspects of it as there have been ways of focusing the ear, at the various levels of attention or intention of listening [... *les multiple attentions ou intentions d'entendre*]" (*Traité*, 59). The English translation loses the close association between attention, intention, and reduced listening.

27. In *Speech and Phenomena*, Derrida notes that Husserl's investigations of expression and indication are consistent with his later formulation of the phenomenological reduction. Derrida claims that the "extrinsic character of the indicative sign is inseparable from the possibility of all the forthcoming reductions, be they eidetic or transcendental" (30).
28. Husserl, *Ideas*, §49, 110.
29. Derrida, *Speech and Phenomena*, 16. A parallel argument could be made concerning Schaeffer's work on the sound object, as presented in note 26 above. Since the sound object, as an ideal objectivity, does not rely on *any actual* worldly manifestation, it too would pass Husserl's test of the "annihilation of the world." Thus one might follow Derrida and argue that Schaeffer's real interest is in a phenomenological form of listening, animated by an intention to hear, which transforms the sounds of the world into ideal sound objects; this is a listening that continues to hear even in the absence of a world, since the imagination is just as valid a source of fulfilling intuitions as actual perception.
30. Husserl, *Logical Investigations*, Vol. 1, §10, 283.
31. Husserl's concept of "originary experience," which appears in late works like *The Crisis of the European Sciences* and related material like the essay "The Origin of Geometry," prolongs his views concerning meaning and meaning-intention first presented in the *Logical Investigations*. The meaning-intention, which may or may not be recovered in any indicative act of communication, is like the originary experience that founds an ontological region. In both cases, Husserl's emphasis is on the recovery of an intentional act or noesis that risks distortion or loss as it passes to the exterior. See chapter 1, passim, for more on "The Origin of Geometry" essay's influence on Schaeffer.
32. Husserl, *Logical Investigations*, Vol. 1, §5, 275.
33. Derrida, *Speech and Phenomena*, 43
34. Ibid., 75.
35. The closed circuitry of speaking-to-oneself is easier to demonstrate in French, where the verb "*entendre*" means both to hear and to understand and has etymological affiliations to the word "intention." I intend [*j'ai l'intention*], hear [*j'entends*], and understand [*j'entends*] all at once.

36. That is not to say that playback does not affect the inscription, for we know that wax cylinders could only be played a handful of times before they would wear out. Playback indeed altered the inscription. However, the degradation of the wax inscription was due to stress on the grooves caused by the friction of the needle on the soft medium, not to the phonograph recording at the same time that it reproduced. Thus it is not comparable to the feedback loop of the subject that speaks-to-oneself.
37. For the source of this argument, see Derrida, *Speech and Phenomena*, chap. 5.
38. Husserl, *Logical Investigations*, Vol. 1, §26, 316.
39. Ibid.
40. Ibid.
41. Ibid.
42. Ibid.
43. Ibid., 318.
44. Recall, Husserl claims that "...*expressions* function meaningfully even in *isolated mental life, where they no longer serve to indicate anything*" (Ibid., Vol. 1, §1, 269).
45. Martin Heidegger, *Being and Time*, trans. Joan Stambaugh (Albany, NY: State University of New York Press, 1996), §34, 163. Note: All page numbers to *Being and Time* refer to Heidegger's original pagination, which is reproduced not only in Stambaugh's translation, but also in Macquarrie and Robinson's. See Heidegger, *Being and Time*, trans. John Macquarrie and Edward Robinson (New York: Harper and Row, 1962). I prefer Stambaugh's translation to Macquarrie and Robinson's, although I have consulted both throughout and occasionally altered the translations for clarity.
46. Ibid., 164.
47. Ibid.
48. Edmund Husserl, "Philosophy as a Rigorous Science," in *Phenomenology and the Crisis of Philosophy*, ed. and trans. Quentin Lauer (New York: Harper and Row, 1965), 116.
49. "The 'essence' of Dasein lies in its existence. The characteristics to be found in this being are thus not objectively present 'attributes' of an objectively present being which has such and such an 'outward appearance,' but rather possible ways for it to be, and only this." Heidegger, *Being and Time*, §9, 42.
50. Hubert Dreyfus, *Being-in-the-World: A Commentary on Heidegger's Being and Time, Division 1* (Cambridge, MA: MIT Press, 1991), 13.
51. Heidegger, *Being and Time*, §4, 12. This quotation is taken from the Macquarrie and Robinson translation.
52. This way of describing Husserl's project as starting with my world and moving outward is indebted to Dreyfus, *Being-in-the-World*, 142.
53. Martin Heidegger, *History of the Concept of Time: Prolegomena*, trans. Theodore Kisiel (Bloomington, IN: Indiana University Press, 1985), 188. Quoted in Dreyfus, *Being-in-the-World*, 142.
54. I owe this distinction to Dreyfus, *Being-in-the-World*, 154. For a fuller account, see his *Being-in-the-World*, chap. 8.
55. Heidegger, *Being and Time*, §35, 167.
56. Ibid., §35, 168–169.
57. Ibid., 169.
58. Ibid.

59. Ibid., §38, 175.
60. Ibid., 177.
61. Ibid., §54, 268.
62. Ibid., 267.
63. Ibid., §56, 273.
64. Ibid.
65. Ibid.
66. Ibid., §55, 271. Stambaugh's edition includes Heidegger's marginalia.
67. Ibid.
68. Ibid., §56, 274.
69. Ibid., §57, 274.
70. Ibid., 275.
71. Ibid.
72. Ibid., 277.
73. Ibid., 275.
74. Ibid.
75. Ibid.
76. Ibid., 276.
77. Ibid., 277.
78. Ibid.
79. Ibid.
80. For Husserl, the voice that speaks to itself in solitary mental life is a voice that guarantees the meaningfulness of one's language; in Heidegger's case, the voice that speaks to itself is a voice that brings one back to the possibility of an authentic life. In both, exteriority must be expelled for the sake of the security of auto-affection. Hetero-affection must be eliminated. For Husserl, exteriority, in the form of the indicative sign that refers without a guarantee that a fulfilling intuition will be recovered, is rejected in the name of the interior co-presentation of intention and intuition at the heart of the subject's solitary mental life. For Heidegger, exteriority, in the form of idle talk and the "they," is rejected in the name of an interior call and response that brings *Dasein* back to its constitutive uncanniness, its *ownmost* potentiality for being—a potentiality that cannot be dependent on any other being.
81. Heidegger reiterates this point in his sixth lecture in *The Principle of Reason*. "Of course," he writes, "we hear a Bach fugue with our ears, but if we leave what is heard only at this, with what strikes the tympanum as sound waves, then we can never hear a Bach fugue. *We* hear, not the ear. We certainly hear through the ear, but not with the ear if 'with' here means the ear is a sense organ that conveys to us what is heard... rather, what the ear perceives and how it perceives will already be attuned and determined by what *we* hear...." Heidegger, *The Principle of Reason*, trans. Reginald Lilly (Bloomington, IN: Indiana University Press, 1991), 47 (translation modified).
82. Heidegger, *Being and Time*, §34, 163. For commentary on the "voice of the friend" in *Being and Time*, see Jacques Derrida, "Heidegger's Ear: Philopolemology (*Geschlecht* IV)," trans. John P. Leavey, Jr., in *Reading Heidegger: Commemorations*, ed. John Sallis (Bloomington, IN: Indiana University Press, 1993), 163–218; and Stephen Mulhall, *Heidegger and Being and Time* (London: Routledge, 1996), 130–136.

83. If Husserl and Heidegger ultimately discovered the source of the acousmatic voice within the subject, Levinas could be described as forcing a break with this position. In his critique of totality, Levinas argues that the subject is constituted by ethical engagement with an unpresentable Other. The ethical voice, the voice of conscience, is no longer *Dasein* speaking to itself authentically, but the Other who speaks to the subject through the subject, breaking the closed circuit of subjectivity. For Levinas, the mark of the subject is not inwardness; rather, the subject is constituted by being concerned with and circumscribed by exteriority. In *Otherwise than Being*, Levinas writes, "Inwardness is not a secret place somewhere in me; it is that reverting in which the eminently exterior, precisely in virtue of this eminent exteriority, this impossibility of being contained and consequently entering into a theme, forms, as infinity, an exception to essence, concerns me and circumscribes me and orders me by my own voice. The command is stated by the mouth of him it commands. The infinitely exterior becomes an 'inward' voice..." (147). At the same time, Levinas's penetration of the self by the other keeps both terms separate, privileging the other as the starting point for the formation of the subject. Extending and ultimately reversing Heidegger, Levinas's authentic subject could be understood as one that is integrated into a circuit of exteriority, a subject that is properly expropriated. See Emmanuel Levinas, *Totality and Infinity: An Essay on Exteriority*, trans. Alphonso Lingis (Pittsburgh, PA: Duquesne University Press, 1969); and *Otherwise than Being; Or, Beyond Essence*, trans. Alphonso Lingis (Pittsburgh, PA: Duquesne University Press, 1998).

84. Slavoj Žižek, *Looking Awry*, 125. Although Lacan spent less time theorizing the voice than the gaze, he added both to an extended list of "partial objects" that included the breast, excrement, and the phallus. Lacan's suggestions concerning the voice are scattered throughout his writings and seminars, but were consolidated and theorized by his son-in-law and literary executor, Jacques-Alain Miller, as well as Žižek and Dolar. For a good overview, see Miller, "*Jacques Lacan et la voix*," in *La voix: actes du colloque d'Ivry* (Paris: La lysimaque, 1989), 175–184. A particularly suggestive and often cited essay dealing with the voice is Jacques Lacan, "The Subversion of the Subject and the Dialectic of Desire in the Freudian Unconscious," in *Écrits*, trans. Bruce Fink (New York: W. W. Norton, 2006).

85. See Žižek, *Looking Awry*, 126 ff.; and Mladen Dolar, *A Voice and Nothing More*, passim.

86. Žižek, *Looking Awry*, 126. Žižek notes that "the voice as object was developed by Michel Chion apropos of the notion of the *la voix acousmatique*, the voice without a bearer, which cannot be attributed to any subject and thus hovers in some indefinite interspace." On the very first page of Chion's *The Voice in Cinema*, he writes, "A serious elaboration of the voice as an object did become possible with Lacan, when he placed the voice—along with the gaze, the penis, the feces, and nothingness—in the ranks of "*objet (a)*," these *part[ial] objects*..." (Chion, *The Voice in Cinema*, 1). Chion also cites Denis Vasse, *L'ombilic et la voix deux enfants en analyse* (Paris: Éd. du Seuil, 1974).

87. Slavoj Žižek, *Interrogating the Real: Selected Writings*, eds. Rex Bulter and Scott Stevens (London: Continuum, 2005), 194.

88. Bruce Fink, *The Lacanian Subject: Between Language and Jouissance* (Princeton, NJ: Princeton University Press, 1995), 7. For a thorough overview of Lacan's

changing views on "alienation in language," see Lorenzo Chiesa, *Subjectivity and Otherness: A Philosophical Reading of Lacan* (Cambridge, MA: MIT Press, 2003), 36 ff.

89. Žižek's quick reference to the voice of conscience connects his view of the voice with the Heidegerrian ontological voice. Instead of the voice of conscience functioning as *Dasein* speaking to itself, this voice is heard as the voice of another—perhaps the voice of someone less amiable than the "friend" that Heidegger mentions. By way of a detour through Lacan, Žižek transforms the Heideggerian ontological voice into a permanently acousmatic voice.

90. Although I focus less on Žižek than Dolar, it should be noted that both are in close agreement about the nature of the object voice, as evidenced by the reliance of these thinkers on each other's work. A large portion of *A Voice and Nothing More* reproduces and develops an earlier article entitled "The Object Voice," published in *Gaze and Voice as Love Objects*, eds. Renata Salecl and Slavoj Žižek (Durham, NC: Duke University Press, 1996).

91. Quoted in Dolar, *A Voice and Nothing More*, 18.

92. Slavoj Žižek, *The Indivisible Remainder: An Essay on Schelling and Other Related Matters* (London: Verso, 1996), 147; Žižek, "The Eclipse of Meaning: On Lacan and Deconstruction," in *Interrogating the Real*, 197.

93. Ibid.

94. Dolar, *A Voice and Nothing More*, 61–62.

95. Ibid., 61.

96. These are two of the powers of the *acousmêtre* named by Chion in *The Voice in Cinema*, 24.

97. Dolar, *A Voice and Nothing More*, 62.

98. Ibid.

99. Ibid., 67.

100. Slavoj Žižek, *Less Than Nothing: Hegel and the Shadow of Dialectical Materialism* (London: Verso, 2012), 91.

101. Dolar, *A Voice and Nothing More*, 68.

102. Ibid., 66.

103. Ibid., 68.

104. See Slavoj Žižek, *The Sublime Object of Ideology* (London: Verso, 1989), 18.

105. Dolar, *A Voice and Nothing More*, 67.

106. Jacques Lacan, *Four Fundamental Concepts of Psychoanalysis*, ed. Jacques-Alain Miller, trans. Alan Sheridan (New York: W. W. Norton, 1978), 103.

107. Jacques-Alain Miller, "The Prisons of *Jouissance*," quoted in Žižek, *Less than Nothing*, 694.

108. Hegel, *Phenomenology of Spirit*, trans. A. V. Miller (Oxford, UK: Clarendon Press, 1977), 103. Pliny's fable and the Hegel passage are juxtaposed in Žižek, *The Sublime Object of Ideology*, 196. Dolar's discussion of the passages appears in *A Voice and Nothing More*, 77.

109. Žižek, *The Sublime Object of Ideology*, 196.

110. Dolar, *A Voice and Nothing More*, 66.

111. Ibid., 70.

112. Ibid.

113. For a clear exposition of Lacan's *objet a*, see Fink, *The Lacanian Subject*, chap. 7. Fink also includes a list of places in Lacan's œuvre where the *objet a* is elaborated (Fink, 188, n. 2).
114. Dolar, *A Voice and Nothing More*, 70. Žižek also says as much: "An unbridgeable gap separates forever a human body from 'its' voice. The voice always displays a spectral autonomy, it never quite belongs to the body we see...it is as if the speaker's own voice hollows out and in a sense speaks 'by itself,' through him." Žižek, *On Belief* (London: Routledge, 2001), 58.
115. Ibid., 75. According to Dolar, "there is a strong etymological link between the two [listening and obeying] in many languages: to obey, obedience, stems from French *obéir*, which in turn stems from Latin *ob-audire*, derivative of *audire*, to hear; in German *gehorchen*, *Gehorsam* stem from *hören*; in many Slav languages *slušati* can mean both to listen and to obey; the same goes apparently for Arabic, and so on" (Dolar, 75–76).
116. Ibid., 77.
117. Ibid., 72.
118. Ibid.
119. Ibid., 73.
120. Ibid., 74.
121. Ibid., 22–23 (emphasis added).
122. Bruce Fink, *The Lacanian Subject*, 41.
123. Dolar, *A Voice and Nothing More*, 98.
124. Ibid., 98–99.
125. Ibid., 99.
126. Ibid., 122.
127. Ibid.
128. Ibid., 122–123.
129. The supposed origin of this Freudian practice is related by Freud's famous patient Serge Pankejeff, the "Wolf Man," in his memoir. "Freud told me that he had originally sat at the opposite end of the couch, so that analyst and analysand could look at each other. One female patient, exploiting this situation, made all possible—or rather impossible—attempts to seduce him. To rule out anything similar, once and for all, Freud moved from his earlier position to the opposite end of the couch." See *The Wolf-Man by the Wolf-Man*, ed. Muriel Gardiner. (New York: Basic Books, 1971), 142. Edmund Engelman's fascinating photographs of Freud's office in Vienna document the seating arrangement. See Engelman, *Berggasse 19*, plates 12–13.
130. Dolar, *A Voice and Nothing More*, 161.
131. Ibid., 153.
132. Ibid., 157–158.
133. Žižek, *Enjoy your Symptom!: Jacques Lacan in Hollywood and Out* (New York: Routledge, 1992), 53.
134. On the act, Sarah Kay writes, "In Žižek's terms, we accomplish the political equivalent of 'traversing the fantasy'—a phrase referring to the outcome of Lacanian therapy, in which we glimpse that what we had taken for reality was all along an illusion masking the space of the real, and so have an opportunity to build 'reality' afresh." Kay, *Žižek: A Critical Introduction* (Cambridge, UK: Polity, 2003), 5. Žižek says that

the act is "the negative gesture that clears a space for creative sublimation." Žižek, *The Ticklish Subject* (New York: Verso, 1999), 159.
135. Dolar, *A Voice and Nothing More*, 158.
136. Ibid., 158.
137. Ibid., 160.
138. Ibid., 153.
139. Ibid., 160.
140. Ibid.
141. Ibid., 162.
142. Ibid., 160.
143. Both words share the same Latin root, *fides*, which is the source of the canine name Fido ("I am faithful"). The most famous Fido of the 20th century was an Italian dog who became famous during WWII. Rescued from the streets, on workdays, Fido would accompany his master to the San Lorenzo bus stop and greet him there in the evening upon his return. After his master was killed during Allied bombardment, the dog would go to the same bus stop every day, awaiting his master, for the next 14 years. See "Fido," *Time*, Vol. 69, Issue 13 (April 1, 1957), 30.
144. Dolar, *A Voice and Nothing More*, 77 (emphasis added).
145. Ibid., 122–123.
146. Ibid.
147. Both voices gain their power from the atopical dislocation of the voice from its source, in the one case, the anonymity of the big Other, in the other, the sitelessness of the subject of enunciation.
148. Ibid., 70.
149. Ibid., 79.
150. I say "typically" since I claimed earlier that Luc Ferrari had to go to extraordinary lengths to break the grip that phantasmagoria had within *musique concrète*.

CONCLUSION
1. Bertrand Russell, *The Philosophy of Logical Atomism* (Chicago: Open Court, 1985), 53.
2. Stanley Cavell, *Must We Mean What We Say? A Book of Essays* (Cambridge, UK: Cambridge University Press, 1969), 52.
3. Consequently, all acousmatic sounds must be identified de facto.
4. I do not mean that the source and cause are simply identified with the loudspeaker itself and the electrical current, respectively, but rather the kind of instrument or voice recorded and its manner of sonic production.
5. See chap. 3, n. 7.
6. This is why I have used the terms "acousmatic sound" and "acousmatic listening" almost interchangeably.
7. Deleuze often discusses the materiality of an artwork, whether a musical composition, painting, or film, in terms of its "direct action on the nervous system." Although this thesis is implied in his *Kafka* as a claim about sound's ability to evade systems of signification, Deleuze presents it most directly in his *Francis Bacon: The Logic of Sensation*, trans. Daniel W. Smith (Minneapolis, MN: University of Minnesota Press, 2005), 31 ff. By "direct action on the nervous system," Deleuze means a material effect that immediately affects a subject but bypasses conceptual mediation

through "the head" or "the brain." Deleuze has influenced much recent work in sound studies; in particular, I refer to Christoph Cox, "Beyond Representation and Signification: Toward a Sonic Materialism," *Journal of Visual Culture* Vol. 10, No. 2 (2011): 145–161; Steve Goodman, *Sonic Warfare: Sound, Affect, and the Ecology of Fear* (Cambridge, MA: MIT Press, 2010); Greg Hainge, *Noise Matters: Towards an Ontology of Noise* (New York: Bloomsbury Academic, 2013); and Paul Hegarty, *Noise/Music* (New York: Continuum, 2007).

8. Moreover, some instances of acousmatic sound, like the silent psychoanalytic voice, do not vibrate at all.
9. For a thorough treatment of the philosophical issues, see Casey O'Callaghan, *Sounds: A Philosophical Theory* (New York: Oxford University Press, 2007).

Bibliography

Abbate, Carolyn. "Debussy's Phantom Sounds." *Cambridge Opera Journal* 10/1 (1988): 67–96.
———. *In Search of Opera*. Princeton, NJ: Princeton University Press, 2001.
Académie française. *Dictionnaire de l'Académie françoise, nouvelle édition. Vol. 1, A–K*. Paris: Brunet, 1762.
Adorno, Theodor. *Against Epistemology: A Metacritique*, trans. Willis Domingo. Cambridge, MA: MIT Press, 1982.
———. *In Search of Wagner*, trans. Rodney Livingstone. London: Verso, 1985.
Adorno, Theodor, and Ernst Krenek. *Briefwechsel*. Frankfurt, DE: Suhrkamp Verlag, 1974.
Anonymous ["a student"]. "Another Evidence of the Wide-Spread Influence of Theosophical Ideals." *Century Path* (Jan. 19, 1908): 13.
Anonymous. "Account of the Performances of Different Ventriloquists, with Observations on the Art of Ventriloquism." *Edinburgh Journal of Science* 9 (1928): 252–259.
Anonymous. *Connecticut Gazette* (Aug. 20, 1790): 3.
Anonymous. "Editorials." *The New Music Review and Church Music Review* 7/76 (March 1908): 201–204.
Anonymous. "Extrait d'une Lettre de M. l'Abbé le...au sujet des Pierres de foudre tombées en Arsois." *Mercure de France* (Sept. 1738): 1986–1987.
Anonymous. "Extrait d'une Lettre écrite de Bourgogne à M. D. L. R. le 4. Fevrier 1731, contenant quelques Reflexions sur l'Akousmate d'Ansacq, dont il est parlé dans le second Volume du Mercure de Decembre dernier." *Mercure de France* (Feb. 1731): 333–338.
Anonymous. "Fido." *Time* 69/13 (April 1, 1957): 30.
Anonymous. "Lettre sur les bruits Aeriens entendus près le Village d'Ansacq, écrite de Paris ce 15. Fevrier 1731." *Mercure de France* (March 1731): 447–457.
Anonymous. "Lettre écrite à M. de le R. par M. P**** Commissaire des Poudres à C**** le 25. Mars 1731 sur le bruit d'Ansacq." *Mercure de France* (May 1731): 1028–1034.
Anonymous [Francis Dhomont?]. "New Media Dictionary." *Leonardo* 43/3 (2001): 261–264.
Anonymous. "Our Concerts." *Musikalische Eilpost* 4 (March 1826), in *The Critical Reception of Beethoven's Compositions by His Contemporaries, Vol. 2*, eds. Wayne Senner and William Meredith, trans. Robin Wallace, 121–127. Lincoln, NE: University of Nebraska Press, 2001.
Apollinaire, Guillaume. *Œuvres en prose complètes*. Paris: Gallimard, 1977.
———. *Œuvres Poétiques*. Paris: Gallimard, 1965.
———. *The Poet Assassinated*, trans. Ron Padgett. San Francisco: North Point Press, 1984.
Aquinas, St. Thomas. *St. Thomas Aquinas' Summa Theologica: Complete English Edition in Five Volumes*. Notre Dame, IN: Christian Classics, 2000.
Aristotle. *The Basic Works of Aristotle*, ed. Richard McKeon. New York: Random House, 1941.

Aristoxenus. *Die Schule des Aristoteles: texte und kommentar. Heft II. Aristoxenus*, ed. Fritz Wehrli. Basel, CH: Schwabe, 1967.

Arnold, Ignaz Theodor Ferdinand Cajetan. *Der angehende Musikdirektor; oder, Die Kunst ein Orchester zu bilden, in Ordnung zu erhalten, und überhaupt allen Forderungen eines guten Musikdirektors Genüge zu leisten*. Erfurt, DE: Henning, 1806.

Arnott, W. Geoffrey. *Alexis: The Fragments. A Commentary*. Cambridge, UK: Cambridge University Press, 1996.

Arrojo, Rosemary. "Writing, Interpreting and the Control of Meaning," in *Translation and Power*, eds. Edwin Gentzler and Maria Tymoczko, 63–79. Amherst, MA: University of Massachusetts Press, 2002.

Athenaeus of Naucratis. *The Deipnosophists*, trans. C. B. Gulick. Cambridge, MA: Harvard University Press, 1927.

Barber, John Warner. *Connecticut Historical Collections Containing a General Collection of Interesting Facts, Traditions, Biographical Sketches, Anecdotes, &c., Relating to the History and Antiquities of Every Town in Connecticut, with Geographical Descriptions*. New Haven, CT: J. W. Barber, 1836.

Barthes, Roland, and Roland Havas. "Listening," in *The Responsibility of Forms*, trans. Richard Howard, 245–260. New York: Hill and Wang, 1985.

——. *Mythologies*, trans. Annette Lavers. New York: Hill and Wang, 1972.

Battier, Marc. "What the GRM Brought to Music: From *Musique Concrète* to Acousmatic Music." *Organised Sound* 12/3 (2007): 189–202.

Bayer, Friedrich. "Musik in Wien." *Die Musik*, 34/1 (1941): 21–22.

Bayle, François. "Image-of-Sound, or i-Sound: Metaphor/Metaform." *Contemporary Music Review* 4 (1989): 165–170.

——. *Musique Acousmatique: Propositions—Positions*. Paris: INA, 1993.

Béclard, A., et al. *Nouveau dictionnaire de médecine, chirurgie, pharmacie, physique, chimie, histoire naturelle, etc., Vol. 1*. Paris: L.-T. Cellot, 1821.

Benjamin, Walter. "Franz Kafka," in *Illuminations*, ed. Hannah Arendt, trans. Harry Zohn, 111–140. New York: Schocken Books, 1969.

Benveniste, Emile. *Problems in General Linguistics*, trans. Mary Elizabeth Meek. Coral Gables, FL: University of Miami Press, 1971.

Berthelin, Pierre-Charles. *Abrégé du dictionnaire universel françois et latin, vulgairement appelé dictionnaire de Trevoux: contenant la signification, la définition & l'explication de tous les termes de sciences et arts*. Paris: Libraires Associes, 1762.

Borel, Petrus. "Les Pressentiments," in *Album de la mode: chroniques du monde fashionable, ou, Choix de morceaux de littérature contemporaine*, 313–342. Paris: Louis Janet, 1833.

Boulby, Mark. "Kafka's End: A Reassessment of the Burrow." *The German Quarterly* 55/2 (Mar. 1982): 175–185.

Bregman, Albert S. *Auditory Scene Analysis: The Perceptual Organization of Sound*. Cambridge, MA: MIT Press, 1990.

Brod, Max, ed. *Franz Kafka: Beschreibung eines Kampfes: Novellen, Skizzen, Aphorismen aus dem Nachlass*. Frankfurt, DE: S. Fischer, 1954.

Brown, Peter. *The Making of Late Antiquity*. Cambridge, MA: Harvard University Press, 1978.

Bujić, Bojan, ed. *Music in European Thought*. Cambridge, UK: Cambridge University Press, 1988.

Burkert, Walter. *Lore and Science in Ancient Pythagoreanism*, trans. Edwin L. Minar, Jr. Cambridge, MA: Harvard University Press, 1972.

Cage, John. *Silence*. Middletown, CT: Wesleyan University Press, 1961.

Calvino, Italo. "A King Listens," in *Under the Jaguar Sun*, trans. William Weaver, 31–64. San Diego, CA: Harcourt Brace Jovanovich, 1988.
Camlot, Jason. "Early Talking Books: Spoken Recording and Recitation Anthologies, 1880–1920." *Book History* 6 (2003): 147–173.
Carnoy, E. Henri. "Les Acousmates et les chasses fantastiques." *Revue de l'histoire des religions* 9 (1884): 370–378.
Caux, Jacqueline. *Almost Nothing with Luc Ferrari*, trans. Jérôme Hansen. Berlin: Errant Bodies Press, 2012.
Cavarero, Adriana. *For More than One Voice: Toward a Philosophy of Vocal Expression*, trans. Paul A. Kottman. Stanford, CA: Stanford University Press, 2005.
Cavell, Stanley. *Must We Mean What We Say? A Book of Essays*. Cambridge, UK: Cambridge University Press, 1969.
Chapius, Alfred, and Edmond Droz. *Automata: A Historical and Technological Study*, trans. Alex Reid. Neuchâtel, CH: Editions de Giffon, 1958.
Chiesa, Lorenzo. *Subjectivity and Otherness: A Philosophical Reading of Lacan*. Cambridge, MA: MIT Press, 2003.
Chion, Michel. *Audio-Vision*, trans. Claudia Gorbman. New York: Columbia University Press, 1994.
———. *Film: A Sound Art*, trans. Claudia Gorbman. New York: Columbia University Press, 2009.
———. *Guide des objets sonores*. Paris: Buchet/Chastel, 1983. Translated as *Guide to the Sound Object*, trans. John Dack [accessed 1/9/2013: http://www.ears.dmu.ac.uk/spip.php?page=articleEars&id_article=3597].
———. *The Voice in Cinema*, trans. Claudia Gorbman. New York: Columbia University Press, 1999.
Chion, Michel, and Guy Riebel. *Les musiques electroacoustiques*. Aix-en-Provence, FR: Édisud, 1976.
Choron, Alexandre, and J. Adrien de La Fage. *Nouveau manuel complet de musique vocale et instrumentale, ou, Encyclopédie musicale*. Part 2, Tome 3. Paris: Roret, 1838.
Chua, Daniel K. L. *Absolute Music and the Construction of Meaning*. Cambridge, UK: Cambridge University Press, 2006.
Clarke, Eric F. *Ways of Listening: An Ecological Approach to the Perception of Musical Meaning*. Oxford, UK: Oxford University Press, 2005.
Clement of Alexandria. *Stromateis: Books 1–3*, trans. John Ferguson. Washington, DC: Catholic University of America Press, 1991.
———. *Les Stromates*, ed. Alain Le Boulluec. Paris: Éditions de Cerf, 1981.
Comotti, Giovanni. "Pitgora, Ippaso, Laso e il metodo sperimentale," in *Harmonica mundi. Musical e philosophic nell'antichità*, eds. R. W. Wallace and B. MacLauchlan, 20–29. Rome: Biblioteca di Quarderni Urbinati di Cultura Classica 5, 1991.
Connor, Steven. *Dumbstruck: A Cultural History of Ventriloquism*. Oxford, UK: Oxford University Press, 2000.
Copenhafer, David. "Invisible Ink: Philosophical and Literary Fictions of Music," Ph.D. diss., University of California–Berkeley, 2004.
Corngold, Stanley. *Franz Kafka: The Necessity of Form*. Ithaca, NY: Cornell University Press, 1988.
Courchene, Kim S. "A Conversation with Beatriz Ferreyra." *Computer Music Journal* 25/3 (Fall 2001): 14–21.
Cox, Christoph. "Beyond Representation and Signification: Toward a Sonic Materialism," *Journal of Visual Culture* 10/2 (2011): 145–161.

Crary, Jonathan. *Techniques of the Observer*. Cambridge, MA: MIT Press, 1990.
Dack, John. "Acoulogie: An Answer to Lévi-Strauss." *Proceedings of EMS07* (2007) [accessed 12/28/2010: http://www.ems-network.org/IMG/pdf_DackEMS07.pdf].
——. "Ear-Training Using the Computer and PROGREMU" [accessed 12/1/2010: http://www.cea.mdx.ac.uk/local/media/downloads/Dack/Ear_training.pdf].
Dahlhaus, Carl. *The Idea of Absolute Music*, trans. Roger Lustig. Chicago: University of Chicago Press, 1989.
——. *Nineteenth-Century Music*, trans. J. Bradford Robinson. Berkeley, CA: University of California Press, 1989.
Davies, Margaret. *Apollinaire*. New York: St. Martin's Press, 1964.
Davis, Whitney. *A General Theory of Visual Culture*. Princeton, NJ: Princeton University Press, 2011.
——. *Queer Beauty: Sexuality and Aesthetics from Winckelmann to Freud and Beyond*. New York: Columbia University Press, 2010.
Dawson, David. *Allegorical Readers and Cultural Revision in Ancient Alexandria*. Berkeley, CA: University of California Press, 1992.
de Boer, Jelle Zeilinga. *Stories in Stone*. Middletown, CT: Wesleyan University Press, 2009.
de Duve, Thierry. *Kant after Duchamp*. Cambridge, MA: MIT Press, 1996.
de Fontenelle, Bernard Le Bouyer. *Œuvres complètes*, ed. G.-B. Depping. Geneva, CH: Slatkine Reprints, 1968.
Degenaar, Marjolein. *Molyneux's Problem*. Boston: Kluwer, 1996.
de la Grange, Henry-Louis. *Gustav Mahler, Vol. 3*. New York: Oxford University Press, 1999.
Deleuze, Gilles, and Felix Guattari. *Kafka: Toward a Minor Literature*, trans. Dana Polan. Minneapolis, MN: University of Minnesota Press, 1986.
della Mirandola, Giovanni Pico. *De Hominis Dignitate, Heptaplus, De Ente Et Uno, E Scritti Vari*, ed. Eugenio Garin. Florence, IT: Vallecchi, 1942.
——. *On the Dignity of Man*, trans. Charles Glenn Wallis, Paul J. W. Miller, Douglas Carmichael. Indianapolis, IN: Bobbs-Merrill, 1965.
de Mairan, J.-J. Dortous. *Traité physique et historique de l'aurore boréale*. Paris: n.p., 1754.
Demers, Joanna. *Listening Through the Noise*. New York: Oxford University Press, 2010.
de Ptoncour, Treüillot. "Lettre de M. Treüillot de Ptoncour, Curé d'Ansacq, à Madame la Princesse de Conty, troisiéme Doüairiere, & Relation d'un Phénomene très extraordinaire, &c." *Mercure de France* (Dec. 1730, Vol. 2): 2804–2833.
Derrida, Jacques. "Aphorism Countertime," in *Acts of Literature*, ed. Derek Attridge, 414–433. New York: Routledge, 1992.
——. "Différance," in *Margins of Philosophy*, trans. Alan Bass, 1–28. Chicago: University of Chicago Press, 1982.
——. *Edumund Husserl's Origin of Geometry: An Introduction*, trans. John. P. Leavey, Jr. Lincoln, NE: University of Nebraska Press, 1982.
——. "Heidegger's Ear: Philopolemology (*Geschlecht* IV)," trans. John P. Leavey, Jr., in *Reading Heidegger: Commemorations*, ed. John Sallis, 163–218. Bloomington, IN: Indiana University Press, 1993.
——. "My Chances/Mes Chances: A Rendezvous with Some Epicurean Stereophonies," in *Taking Chances: Derrida, Psychoanalysis, and Literature*, eds. Joseph H. Smith and William Kerrigan, 1–32. Baltimore, MD: Johns Hopkins University Press, 1984.
——. *Of Grammatology*, trans. Gayatri Chakravorty Spivak. Baltimore, MD: Johns Hopkins University Press, 1976.

——. "Plato's Pharmacy," in *Dissemination*, trans. Barbara Johnson, 61–171. Chicago: University of Chicago Press, 1981.

——. *Positions*, trans. Alan Bass. Chicago: University of Chicago Press, 1981.

——. *Speech and Phenomena*, trans. David B. Allison. Evanston, IL: Northwestern University Press, 1973.

Descartes, René. "Treatise on Light," in *The World and Other Writings*, ed. and trans. Stephen Gaukroger, 3–75. Cambridge, UK: Cambridge University Press, 1998.

de Soulaines, Laloüat. "Lettre de M. Laloüat de Soulaines, écrite à M. L. B. C. D. au sujet de l'Akousmate d'Ansacq, et d'un autre pareil dont il a été le témoin." *Mercure de France* (June 1731, Vol. 2): 1516–1531.

Dhomont, Francis. "Acousmatic, what is it?" Liner notes to *Cycle D'errance*, sound recording. Montréal: Diffusion i Média, 1996.

——. "Is there a Québec sound?" *Organised Sound* 1/1 (1996): 24–28.

Dickson, William Kennedy Laurie, and Antonia Dickson. "The Life and Inventions of Edison." *Cassier's Magazine* 3/18 (April 1893): 445–459.

Diderot, Denis. "Acousmatiques." *Encyclopédie ou dictionnaire raisonné des sciences, des arts et des métiers*, 1:111. Paris: Chez Briasson, 1751.

——. "Letter on the Blind," in *Diderot's Early Philosophical Works*, trans. and ed. Margaret Jourdain, 142–157. Chicago: Open Court Press, 1916.

Diels, Herman, and Walther Kranz. *Die Fragmente der Vorsokratiker*. Berlin: Weidmann, 1972.

Dillon, John. "Iamblichus of Chalcis." *Aufstieg und Niedergang der Römischen Welt, Band II*, 36.2 (1987): 862–909.

Diogenes Laertius. *Lives of Eminent Philosophers*, trans. R. Drew Hicks. Cambridge, MA: Harvard University Press, 1950.

Dolar, Mladen. "The Burrow of Sound." *Differences: A Journal of Feminist Cultural Studies* 22/2–3 (2011): 112–139.

——. "The Object Voice," in *Gaze and Voice as Love Objects*, eds. Renata Salecl and Slavoj Žižek, 7–31. Durham, NC: Duke University Press, 1996.

——. *A Voice and Nothing More*. Cambridge, MA: MIT Press, 2006.

Dorland, W. A. Newman. *The American Illustrated Medical Dictionary*. Philadelphia: Saunders, 1907.

Drake, Samuel Adams. *A Book of New England Legends and Folk Lore*. Rutland, VT: C. E. Tuttle, 1971.

Drever, John Levack. "Soundscape Composition: The Convergence of Ethnography and Acousmatic Music." *Organised Sound* 7/1 (April 2002): 21–27.

Dreyfus, Hubert. *Being-in-the-World: A Commentary on Heidegger's Being and Time, Division 1*. Cambridge, MA: MIT Press, 1991.

Drott, Eric. "The Politics of *Presque rien*," in *Sound Commitments: Avant-garde Music and the Sixties*, ed. Robert Aldington, 145–166. New York: Oxford University Press, 2009.

DuBos, Abbé. *Critical Reflections on Poetry, Painting, and Music*, trans. Thomas Nugent. London: Printed for J. Nourse, 1748.

Dufourt, Hugues. "Pierre Schaeffer: le son comme phénomène de civilization," in *Ouïr, entendre, écouter, comprendre après Schaeffer*, 69–82. Paris: Buchet/Chastel, 1999.

Dyson, Frances. *Sounding New Media: Immersion and Embodiment in the Arts and Culture*. Berkeley, CA: University of California Press, 2009.

Eidsheim, Nina Sun. "Voice as a Technology of Selfhood: Towards an Analysis of Racialized Timbre and Vocal Performance," Ph.D. diss., University of California– San Diego, 2008.

Emmerson, Simon. *Living Electronic Music*. London: Ashgate, 2007.

Emrich, Wilhelm. *Franz Kafka: A Critical Study of His Writings*, trans. Sheema Zeben Buehne. New York: Ungar, 1968.

Engelman, Edmund. *Berggasse 19: Sigmund Freud's Home and Offices, Vienna, 1938.* New York: Basic Books, 1976.

Eskilson, Stephen. *Graphic Design: A New History*. New Haven, CT: Yale University Press, 2007.

Fara, Patricia. "Lord Derwentwater's Lights: Prediction and the Aurora Polaris Astronomers." *Journal for the History of Astronomy* 27 (1996): 239–258.

Federal Writer's Project for the State of Connecticut. *Connecticut; A Guide to Its Roads, Lore, and People*. Boston: Houghton Mifflin Co., 1938.

Ferreyra, Beatriz. "CD Program Notes." *Computer Music Journal* 25/4 (Winter 2001): 118–123.

Feydel, Gabriel. *Remarques Morales, Philosophiques et Grammaticales, sur le Dictionnaire de L'académie Françoise*. Paris: Antoine-Augustin Renouard, 1807.

Fink, Bruce. *The Lacanian Subject: Between Language and Jouissance*. Princeton, NJ: Princeton University Press, 1995.

Foucault, Michel. *The Order of Things: An Archaeology of the Human Sciences*. New York: Pantheon Books, 1971.

Freud, Sigmund. *The Ego and the Id*, trans. James Strachey. New York: Norton, 1989.

———. "Recommendations for Physicians on the Psycho-Analytic Method of Treatment," in *Collected Papers, Vol. 2*, trans. under the supervision of Joan Riviere, 323–333. New York: Basic Books. 1959.

Gardiner, Muriel, ed. *The Wolf-Man by the Wolf-Man*. New York: Basic Books, 1971.

Gayou, Évelyne. *Le GRM: le groupe de recherches musicales*. Paris: Éditions Fayard, 2007.

Gellen, Kata. "Hearing Spaces: Architecture and Acoustic Experience in Modernist German Literature." *Modernism/Modernity* 17/4 (2011): 799–818.

Gilliam, Bryan. "The Annexation of Anton Bruckner: Nazi Revisionism and the Politics of Appropriation." *The Musical Quarterly* 78/3 (Autumn 1994): 584–604.

Girard, René. *Deceit, Desire, and the Novel*, trans. Yvonne Freccero. Baltimore, MD: Johns Hopkins University Press, 1965.

Goble, Mark. *Beautiful Circuits: Modernism and the Mediated Life*. New York: Columbia University Press, 2010.

Goebbels, Joseph. "Appendix: Joseph Goebbels's Bruckner Address in Regensburg (6 June 1937)," trans. John Michael Cooper. *The Musical Quarterly* 78/3 (Autumn 1994): 605–609.

Goehr, Lydia. *The Quest for Voice: Music, Politics, and the Limits of Philosophy*. New York: Oxford University Press, 1998.

Goodman, Steve. *Sonic Warfare: Sound, Affect, and the Ecology of Fear*. Cambridge, MA: MIT Press, 2010.

Grétry, André-Ernest-Modeste. *Mémoires, ou essais sur la musique, Vol. 3*. Paris: Pluviôse, 1797.

Hägg, Henny Fiska. *Clement of Alexandria and the Beginning of Christian Apophaticism*. New York: Oxford University Press, 2006.

Hägglund, Martin. *Radical Atheism: Derrida and the Time of Life*. Stanford, CA: Stanford University Press, 2008.

Hainge, Greg. *Noise Matters: Towards an Ontology of Noise*. New York: Bloomsbury Academic, 2013.

Hamilton, John. "'*Ist das Spiel vielleicht unangenehm?*' Musical Disturbances and Acoustic Space in Kafka." *Journal of the Kafka Society of America*, Nos. 1–2 (2004): 23–27.

Hartford, Robert, ed. *Bayreuth, the Early Years*. Cambridge, UK: Cambridge University Press, 1980.

Hasher, Lynn, David Goldstein, and Thomas Toppino. "Frequency and the Conference of Referential Validity." *Journal of Verbal Learning and Verbal Behavior* 16/1 (1977): 107–112.

Hegarty, Paul. *Noise/Music*. New York: Continuum, 2007.

Hegel, Georg Wilhelm Friedrich. *Phenomenology of Spirit*, trans. A. V. Miller. Oxford, UK: Clarendon Press, 1977.

Heidegger, Martin. *The Basic Problems of Phenomenology*, trans. Albert Hofstadter. Bloomington and Indianapolis, IN: Indiana University Press, 1982.

——. *Being and Time*, trans. John Macquarrie and Edward Robinson. New York: Harper and Row, 1962.

——. *Being and Time*, trans. Joan Stambaugh. Albany, NY: State University of New York Press, 1996.

——. *History of the Concept of Time: Prolegomena*, trans. Theodore Kisiel. Bloomington, IN: Indiana University Press, 1985.

——. "On the Essence and Concept of Φύσις in Aristotle's *Physics* B, I," in *Pathmarks*, ed. William McNeil, 183–230. Cambridge, UK: Cambridge University Press, 1998.

——. *Principle of Reason*, trans. Reginald Lilly. Bloomington, IN: Indiana University Press, 1991.

——. *The Question Concerning Technology*, trans. William Lovitt. New York: Harper Colophon Books, 1977.

Heller, Feriedric C. "Von der Arbeiterkultur zur Theatersperre," in *Das Wiener Konzerthaus*, ed. Friedrich C. Heller and Peter Revers, 87–109. Vienna: Wiener Konzerthausgesellscaft, 1983.

Henel, Heinrich. "'The Burrow,' or How to Escape from a Maze," in *Franz Kafka*, ed. Harold Bloom, 119–132. New York: Chelsea House Publishers, 1986.

Henry, Pierre. *Variations pour une porte et un soupir* (1963), *Mix 03.1*, sound recording, Philips Music Group, 2001.

Hess, Earl J., and Pratibha A. Dabholkar. *Singin' in the Rain: The Making of an American Masterpiece*. Lawrence, KS: University of Kansas Press, 2009.

Hesselberg, Erik. "'Moodus Noises' Strike Again." *Hartford Courant*, March 24, 2011 [accessed 12/27/2012: http://www.courant.com/news/connecticut/hc-moodus-quake-0325-20110324,0,6968586.story].

Hipparchus. *Hipparchou Tōn Aratou kai Eudoxou Phainomenōn exēgēseōs vivlia tria = Hipparchi in Arati et Eudoxi Phaenomena commentariorum libri tres*, ed. and trans. Karl Mantiius. Lipsiae, DE: In aedibus B. G. Teubneri, 1894.

Hoeckner, Berthold. *Programming the Absolute: Nineteenth-Century German Music and the Hermeneutics of the Moment*. Princeton, NJ: Princeton University Press, 2002.

Hoffman, Paul. "Enchantments of a Swiss Town." *New York Times* (April 18, 1993): XX8 [accessed 06/20/13: http://www.nytimes.com/1993/04/18/travel/enchantments-of-a-swiss-town.html?pagewanted=all&src=pm].

Holbrooke, Joseph, and Herbert Trench. *Apollo and the Seaman: An Illuminated Symphony, Op. 51*. London: Novello, 1908.

Horky, Philip Sidney. "Aristotle's Description of Mathematical Pythagoreanism in the 4th Century B.C.E." [accessed 12/27/2010: http://www.princeton.edu/~pswpc/pdfs/horky/051002.pdf].

Hosmore, Stephen. "Letter to Thomas Prince, August 13, 1729." *Collections of the Connecticut Historical Society* 3 (1890): 280–281.

Huffman, Carl. *Philolaus of Croton: Pythagorean and Presocratic: A Commentary on the Fragments and Testimonia with Interpretive Essays*. Cambridge, UK: Cambridge University Press, 1993.

———. "Pythagoreanism," in *Stanford Encyclopedia of Philosophy* [accessed 12/28/2010: http://plato.stanford.edu/entries/pythagoreanism].

Husserl, Edmund. *Cartesian Meditations*, trans. Dorion Cairns. The Hague, NL: Martinus Nijhoff, 1970.

———. *The Idea of Phenomenology*, trans. William P. Alston and George Nakhnikian. The Hague, NL: Martinus Nijhoff, 1964.

———. *Ideas*, trans. W. R. Boyce Gibson. New York: Collier-Macmillan, 1962.

———. *Ideas Pertaining to a Pure Phenomenology and to a Phenomenological Philosophy, First Book*, trans. F. Kersten. The Hague, NL: Martinus Nijhoff, 1982.

———. *Idées directrices pour une phénoménologie*, trans. Paul Ricoeur. Paris: Gallimard, 1950.

———. *Logical Investigations*, trans. J. N. Findlay. New York: Humanities Press, 1970.

———. *Logique formelle et logique transcendantale: essai d'une critique de la raison logique*, trans. Suzanne Bachelard. Paris: Presses Universitaires de France, 1957.

———. *Méditations cartésiennes: introduction à la phénoménologie*, trans. Gabrielle Peiffer and Emmanuel Levinas. Paris: Librairie Philosophique J. Vrin, 2001.

———. *On the Phenomenology of the Consciousness of Internal Time (1893–1917)*, trans. John B. Brough. Dordrecht, NL: Kluwer Academic Publishers, 1991.

———. "The Origin of Geometry," in *The Crisis of the European Sciences and Transcendental Phenomenology*, trans. David Carr, 353–378. Evanston, IL: Northwestern University Press, 1970.

———. *L'origine de la geometrie*, trans. and ed. Jacques Derrida. Paris: Presses Universitaires de France, 1962.

———. "Philosophy as a Rigorous Science," in *Phenomenology and the Crisis of Philosophy*, ed. and trans. Quentin Lauer, 71–147. New York: Harper and Row, 1965.

Husserl, Edmund, and Jean Wahl. *L'ouvrage posthume de husserl: la "Krisis": la crise des sciences européennes et la phénoménologie transcendantale*. Paris: Centre de Documentation Universitaire, 1957.

Husserl, Edmund, and Ludwig Landgrebe. *Experience and Judgment*, trans. James S. Churchill and Karl Ameriks. Evanston, IL: Northwestern University Press, 1973.

Iamblichus. *De communi mathematica scientia*, ed. Nicolas Festa. Stutgardiae, DE: Teubner, 1975.

———. *De Mysteriis*, trans. Emma C. Clarke, John M. Dillon, and Jackson P. Hershbell. Leiden, NL: Brill, 2004.

———. "The Life of Pythagoras," in *The Pythagorean Sourcebook and Library: An Anthology of Ancient Writings Which Relate to Pythagoras and Pythagorean Philosophy*, trans. Kenneth Sylvan Guthrie, 57–122. Grand Rapids, MI: Phanes Press, 1987.

———. *De vita pythagorica. On the Pythagorean Way of Life*, trans. John Dillon and Jackson Hershbell. Altanta, GA: Scholars Press, 1991.

Itter, Andrew C. *Esoteric Teaching in the Stromateis of Clement of Alexandria*. Leiden, NL: Brill, 2009.

Jacoby, Felix. *Die Fragmente der Griechischen Historiker, Dritter Teil, B*. Leiden, NL: Brill, 1950. Online resource with English translations: http://www.brill.nl/brillsnewjacoby [accessed 12/29/2010].

Jacquet-Droz, Henri-Louis. *A Description of Several Pieces of Mechanism, Invented by the Sieur Jacquet Droz, of... Switzerland. And which Are Now To Be Seen at the Great Room, No. 6, in King-Street, Covent-Garden.* London: Bodleian Library, 1780(?).

Jakobson, Roman. "Shifters, Verbal Categories, and the Russian Verb," in *Selected Writings, Vol. 2*, 130–147. The Hague, NL: Mouton, 1971.

Janaway, Christopher. "Knowledge and Tranquility: Schopenhauer on the Value of Art," in *Schopenhauer, Philosophy, and the Arts*, ed. Dale Jacquette, 39–61. Cambridge, UK: Cambridge University Press, 1996.

Jespersen, Otto. *Language: Its Nature, Development and Origin.* New York: Holt, 1921.

Jonas, Hans. "The Nobility of Sight," in *The Phenomenon of Life: Toward a Philosophical Biology*, 135–151. Evanston, IL: Northwestern University Press, 2001.

Jones, Francis Arthur. *Thomas Alva Edison: Sixty Years of an Inventor's Life.* New York: Thomas Y. Crowell & Co., 1908.

Joost-Gaugier, Christiane L. *Pythagoras and Renaissance Europe: Finding Heaven.* Cambridge, UK: Cambridge University Press, 2009.

Kafka, Franz. "The Burrow," in *The Complete Stories*, ed. Nahum N. Glatzer, 325–359. New York: Schocken Books, 1971.

———. *The Diaries of Franz Kafka, 1910–1913*, ed. Max Brod, trans. Joseph Kresh. New York: Schocken Books, 1948.

———. *Letters to Milena*, trans. Philip Boehm. New York: Schocken Books, 1990.

Kahn, Charles H. *Pythagoras and the Pythagoreans: A Brief History.* Indianapolis, IN: Hackett, 2001.

Kahn, Douglas. "Introduction: Histories of Sound Once Removed," in *Wireless Imagination*, eds. Douglas Kahn and Gregory Whitehead, 1–29. Cambridge, MA: MIT Press, 1994.

Kay, Sarah. *Žižek: A Critical Introduction.* Cambridge, UK: Polity, 2003.

Kendrick, Robert. *Celestial Sirens: Nuns and Their Music in Early Modern Milan.* Oxford, UK: Clarendon Press, 1996.

Kermode, Frank. *The Art of Telling: Essays on Fiction.* Cambridge, MA: Harvard University Press, 1983.

———. *The Genesis of Secrecy: On the Interpretation of Narrative.* Cambridge, MA: Harvard University Press, 1979.

Kierkegaard, Søren. *Either/Or*, trans. David F. Swenson and Lillian Marvin Swenson. Princeton, NJ: Princeton University Press, 1944.

Kittler, Friedrich. *Gramophone, Film, Typewriter*, trans. Geoffrey Winthrop-Young and Michael Wutz. Stanford, CA: Stanford University Press, 1999.

Kittler, Wolf. "Grabenkrieg—Nervenkrieg—Medienkrieg. Franz Kafka und der 1. Weltkrieg," in *Armaturen der Sinne*, eds. Jochen Hörisch and Michael Wetzel, 289–309. München, DE: Wilhelm Fink Verlag, 1990.

Kravitt, Edward F. *The Lied: Mirror of Late Romanticism.* New Haven, CT: Yale University Press, 1996.

LaBelle, Brandon. *Background Noise: perspectives on sound art.* New York: Continuum, 2006.

Lacan, Jacques. *Écrits*, trans. Bruce Fink. New York: W. W. Norton, 2006.

———. *Four Fundamental Concepts of Psychoanalysis*, ed. Jacques-Alain Miller, trans. Alan Sheridan. New York: W. W. Norton, 1978.

Lacoue-Labarthe, Philippe. *Heidegger: Art and Politics*, trans. Chris Turner. Cambridge: Basil Blackwell, 1990.

———. *Typography*, trans. Christopher Fynsk. Stanford: Stanford University Press, 1989.
Landy, Leigh. *Understanding the Art of Sound Organization*. Cambridge: The MIT Press, 2007.
Langlois, Eustache-Hyacinthe. *Essai historique, philosophique et pittoresque sur les danses des morts: Accompagné de cinquante-quatre planches et de nombreuses vignettes*. Rouen: A. Lebrument, 1852.
Langwith, B. "An Account of the Aurora Borealis That Appear'd Oct. 8. 1726. In a Letter to the Publisher from the Reverend Dr. Langwith, Rector of Petworth in Sussex." *Philosophical Transactions of the Royal Society* 34 (1726-7): 132-6.
Larousse, Pierre. *Grand dictionnaire universel du XIXe siècle: Français, historique géographique, biographique, mythologique, bibliographique, etc.*, in 15 volumes. Paris: Administration du Grand dictionnaire universel, 1900-.
Lawler, James R. "Apollinaire inédit: Le Séjour à Stavelot," *Mercure de France* 1098 (Feb. 1, 1955): 296-309.
Le Gendre, Charles. *Traité historique et critique de l'opinion, vol. 6*. Paris: Briason, 1758.
Le Roy, P. Charles. *Traité de l'orthographie Françoise en forme de dictionnaire*. Poitiers: chez J. Félix Faulcon, 1764.
Leibowitz, René. *Schoenberg and his School*, trans. Dika Newlin. New York: Da Capo Press, 1979.
Leppert, Richard D. *The Sight of Sound: Music, Representation, and the History of the Body*. Berkeley: University of California Press, 1993.
Lessing, Gotthold Ephraim. *Laocoön*, trans. Robert J. Phillimore. London: Routledge, 1905.
Levinas, Emmanuel. *Otherwise than Being; or, Beyond Essence*, trans. Alphonso Lingis. Pittsburgh, PA: Duquesne University Press, 1998.
———. *Totality and Infinity: An Essay on Exteriority*, trans. Alphonso Lingis. Pittsgurgh, PA: Duquesne University Press, 1969.
Levi-Strauss, Claude. *The Raw and The Cooked: Introduction to a Science of Mythology I*, trans. John and Doreen Weightman. New York: Harper & Row, 1969.
Liddell, Henry George, and Robert Scott, eds. *A Greek-English Lexicon*. Oxford, UK: Clarendon Press, 1979.
Locke, John. "An Essay on Human Understanding," in *The Works of John Locke, in Nine Volumes*. London: C. and J. Rivington, 1823.
Maché, Britta. "The Noise in the Burrow: Kafka's Final Dilemma." *The German Quarterly* 55/4 (Nov. 1982): 526-540.
Marsop, Paul. "Der Musiksaal der Zukunft." *Die Musik* 2/1 (1902): 3-21.
———. "Zur Bühnen- und Konzertreform." *Die Musik* 5/16 (1905/06): 247-258.
Martin, Gregory. *Roma Sancta (1581)*, ed. George Bruner Parks. Roma: Edizioni di storia e letteratura, 1969.
Marx, Karl. *Capital*, trans. Ben Fowkes. New York: Vintage Books, 1976.
McKirahan, Richard D. *Philosophy before Socrates: An Introduction with Texts and Commentary*. Indianapolis, IN: Hackett, 1994.
Merleau-Ponty, Maurice. *Phenomenology of Perception*, trans. Colin Smith. London: Routledge & Kegan Paul, 1962.
Miller, Jacques-Alain. "Jacques Lacan et la voix," in *La voix: actes du colloque d'Ivry*, 175-184. Paris: La lysimaque, 1989.
Monson, Craig. "Putting Bolognese Nun Musicians in their Place," in *Women's Voices across Musical Worlds*, ed. Jane Bernstein, 118-141. Boston: Northeastern University Press, 2004.

Morellet, André. *Observations sur un ouvrage anonyme intitulé: Remarques morales, philosophique et grammaticales, sur le Dictionnaire de l'Académie françoise.* Paris: n.p., 1807.
Morgan, Michael J. *Molyneux's Question: Vision, Touch, and the Philosophy of Perception.* Cambridge, UK: Cambridge University Press, 1977.
Morigia, Paolo. *La nobilita' di Milano.* Milano, IT: appresso Gio. Battista Bidelli, 1619.
Mulhall, Stephen. *Heidegger and Being and Time.* London: Routledge, 1996.
Nancy, Jean-Luc. *À l'écoute.* Paris: Galilée, 2002.
——. "The Forgetting of Philosophy," in *The Gravity of Thought*, trans. François Raffoul and Gregory Recco, 7–71. Atlantic Highlands, NJ: Humanities Press, 1997.
——. *The Inoperative Community*, trans. Peter Collins, Lisa Garbus, Michael Holland, and Simon Sawhney. Minneapolis, MN: University of Minnesota Press, 1991.
——. *Listening*, trans. Charlotte Mandell. New York: Fordham University Press, 2007.
——. *The Muses*, trans. Peggy Kamuf. Stanford, CA: Stanford University Press, 1996.
Nietzsche, Friedrich. "On Music and Words," in Carl Dahlhaus, *Between Romanticism and Modernism,* trans. Mary Whittall, 103–119. Berkeley, CA: University of California Press, 1980.
Niles, Hosford B. *The Old Chimney Stacks of East Haddam.* New York: Lowe & Co., 1887.
Obadike, Mendi. "Low Fidelity: Stereotyped Blackness in the Field of Sound," Ph.D diss., Duke University, 2005.
O'Callaghan, Casey. *Sounds: A Philosophical Theory.* New York: Oxford University Press, 2007.
O'Meara, Dominic. *Pythagoras Revived: Mathematics and Philosophy in Late Antiquity.* Oxford, UK: Clarendon Press, 1989.
Palombini, Carlos. "Pierre Schaeffer and Pierre Henry," in *Music of the Twentieth-Century Avant-Garde: A Biocritical Sourcebook,* ed. Larry Sitsky, 432–445. Westport, CT: Greenwood Press, 2002.
——. "Technology and Pierre Schaeffer: Walter Benjamin's *technische Reproduzierbarkeit,* Martin Heidegger's *Ge-Stell* and Pierre Schaeffer's *Arts-Relais*." *Organised Sound* 3/1 (1998): 35–43.
Paul, Les. "The Case of the Missing Les Paulverizer." *Les Paul and Mary Ford, The Legend and the Legacy,* sound recording, Capitol Records, 1991.
——. *The Les Paul Show.* NBC Radio, 1950.
"Les Paul and Mary Ford." *Omnibus,* Season 2, Episode Ep. 4 (Oct. 23, 1953), CBS Television.
Peignot, Jérôme. "De la musique concrete à l'acousmatique." *Esprit* 280 (Jan. 1960): 111–120.
Philips, David E. *Legendary Connecticut.* Hartford, CT: Spoonwood Press, 1984.
Piaget, Jean. *The Psychology of Intelligence,* trans. Malcolm Piercy and D. E. Berlyne. London: Routledge, 2003.
Picker, John M. *Victorian Soundscapes.* New York: Oxford University Press, 2003.
Pierret, Marc. *Entretiens avec Pierre Schaeffer.* Paris: P. Belfond, 1969.
Plato. *The Collected Dialogues of Plato, Including the Letters,* eds. Edith Hamilton and Huntington Cairns. Princeton, NJ: Princeton University Press, 1961.
Pliny the Elder. *The Natural History of Pliny the Elder, Vol. II,* trans. John Bostock and H. T. Riley. London: Henry G. Bohn, 1855.
Poe, Edgar Allan. "The Tell-Tale Heart," in *Poetry and Tales,* 555–559. New York: Library of America, 1984.
Politzer, Heinz. *Franz Kafka: Parable and Paradox.* Ithaca, NY: Cornell University Press, 1962.

Pope, Alexander. *Poetry and Prose of Alexander Pope*, ed. Aubrey Williams. Boston: Houghton Mifflin Company, 1969.

Porphyry. "Life of Pythagoras," in *Heroes and Gods: Spiritual Biographies in Antiquity*, eds. Moses Hadas and Morton Smith, 105–128. New York: Harper and Row, 1965.

Porter, Amy. "Craziest Music You Ever Heard." *Saturday Evening Post* (Jan. 17, 1953): 25, 98, 100.

Power, Henry, and Leonard W. Sedgwick. *The New Sydenham Society's Lexicon of Medicine and the Allied Sciences*. London: New Sydenham Society, 1881.

Powers, Horatio Nelson. "The First Phonogramic Poem." Thomas A. Edison Papers, D8850 124:705.

———. *Lyrics of the Hudson*. Boston: D. Lothrop, 1891.

Prévost, Abbé. *Manuel lexique ou dictionnaire portatif des mots françois dont la signification n'est pas familière a tout le monde*. Paris: Chez Didot, 1755.

Price, Carl F. *Yankee Township*. East Hampton, NY: Citizens' Welfare Club, 1941.

Rath, Richard Cullen. *How Early America Sounded*. Ithaca, NY: Cornell University Press, 2003.

Rée, Jonathan. *I See a Voice: Deafness, Language, and the Senses*. New York: H. Holt and Co., 1999.

Reilly, Kara. "Automata: A Spectacular History of the Mimetic Faculty," Ph.D. diss., University of Washington, 2006.

Reuchlin, Johann. *On the Art of the Kabbalah = De Arte Cabalistica*, trans. Martin and Sarah Goodman. New York: Abaris Books, 1983.

Ricoeur, Paul. "Husserl and the Sense of History," in *Husserl: An Analysis of His Phenomenology*, trans. Edward G. Ballard and Lester E. Embree, 143–174. Evanston, IL: Northwestern University Press, 1967.

———. *A Key to Edmund Husserl's Ideas I*, trans. Bond Harris and Jacqueline Bouchard Spurlock. Milwaukee, WI: Marquette University Press, 1996.

Riedweg, Christoph. *Pythagoras: His Life, Teaching, and Influence*, trans. Steven Rendall. Ithaca, NY: Cornell University Press, 2005.

Roberts, Alexander, and James Donaldson, eds. *Ante-Nicene Fathers: Translations of the Writings of the Fathers Down to A.D. 325. American Reprint of the Edinburg Edition, revised and arranged by A. Cleveland Coxe, Vol. II*. Grand Rapids, MI: W. B. Eerdmanns, 1978–1981.

Robindoré, Brigitte. "Luc Ferrari: Interview with an Intimate Iconoclast." *Computer Music Journal* 22/3 (Fall 1998): 8–16.

Ross, Alison. *The Aesthetic Paths of Philosophy: Presentation in Kant, Heidegger, Lacoue-Labarthe, and Nancy*. Stanford, CA: Stanford University Press, 2007.

Rousseau, Jean-Jacques. *The Confessions*, trans. J. M. Cohen. London: Penguin Books, 1953.

Russell, Bertrand. *The Philosophy of Logical Atomism*. Chicago: Open Court, 1985.

Saint-Saëns, Camille. *Harmonie et Mélodie*. Paris: Calmann Lévy, 1885.

Schaeffer, Pierre. *A la recherche d'une musique concrète*. Paris: Éditions du Seuil, 1952 (*In Search of a Concrete Music*, trans. Christine North and John Dack. Berkeley, CA: University of California Press, 2012).

———. "Lettre à Albert Richard." *La revue musicale* 236 (1957): iii–xvi.

———. *Machines à communiquer*. Paris: Éditions du Seuil, 1970.

———. *La musique concrète*. Paris: Presses Universitaires de France, 1967.

———. *L'œuvre musicale*, ed. François Bayle. Paris: INA-GRM, 1990.

———. *Pierre Schaeffer: texts et documents*, ed. François Bayle. Paris: INA-GRM, 1990.

——. *Solfège de l'objet sonore*, sound recording. Paris: INA-GRM, 1998.
——. *Traité des objets musicaux*. Paris: Éditions du Seuil, 1966.
Schafer, R. Murray. *The Soundscape: Our Sonic Environment and the Tuning of the World*. Rochester, NY: Destiny Books, 1994.
——. *The Vancouver Soundscape*, sound recording, Epn 186, 1972.
Schopenhauer, Arthur. *The World as Will and Representation*, two vols., trans. E. F. J. Payne. New York: Dover Publications, 1969.
Schwab, Heinrich W. *Konzert: öffentliche Musikdarbietung vom 17. bis 19. Jahrhundert*. Leipzig, DE: Deutscher Verlag für Musik, 1971.
Scruton, Roger. *The Aesthetics of Music*. New York: Oxford University Press, 1997.
Sexton, Jamie. *Music, Sound and Multimedia: From the Live to the Virtual*. Edinburgh, UK: Edinburgh University Press, 2007.
Shaughnessy, Mary Alice. *Les Paul: An American Original*. New York: William Morrow and Co., 1993.
Shaw, George Bernard. *The Perfect Wagnerite*. New York: Dover Publications, 1967.
Shaw, Gregory. *Theurgy and the Soul: The Neoplatonism of Iamblichus*. University Park, PA: Pennsylvania State University Press, 1995.
Shepard, Odell. *Connecticut Past and Present*. New York: Alfred A. Knopf, 1939.
Sievert, Jon. "Les Paul." *Guitar Player* (Dec. 1977): 43–60 [accessed 06/19/2013: http://www.gould68.freeserve.co.uk/LPGPInter.htm].
Silverman, Kaja. *The Acoustic Mirror: The Female Voice in Psychoanalysis and Cinema*. Bloomington, IN: Indiana University Press, 1988.
Singin' in the Rain. Dir. Stanley Donen and Gene Kelly. DVD, two discs. Warner Home Video, 2002.
Skelton, Geoffrey. "The Idea of Bayreuth," in *The Wagner Companion*, eds. Peter Burbidge and Richard K. Sutton, 389–411. New York: Cambridge University Press, 1979.
Smalley, Denis. "The listening imagination: listening in the electroacoustic era." *Contemporary Music Review*, 13/2 (1996): 77–107.
Smith, A. D. *Husserl and the Cartesian Meditations*. New York: Routledge, 2003.
Smith, F. Joseph. *The Experiencing of Musical Sound: Prelude to a Phenomenology of Music*. New York: Gordon and Breach, 1979.
Société académique d'archéologie, sciences et arts du département de l'Oise. *Compte-rendu des séances*. Beauvais, FR: Moniteur de L'Oise, 1901.
Sokel, Walter. *Kafka—Tragik und Ironie: Zur Struktur seiner Kunst*. München and Wien, DE: Albert Langen/Georg Müller, 1964.
Solomos, Makis. "Schaeffer phénoménologue," in *Ouïr, entendre, écouter, comprendre après Schaeffer*, 53–67. Paris: Buchet/Chastel, 1999.
Stanley, Thomas. *The History of Philosophy: Containing the Lives, Opinions, Actions and Discourses of the Philosophers of Every Sect*. London: printed for A. Millar et al., 1743.
Sterne, Jonathan. *The Audible Past: Cultural Origins of Sound Reproduction*. Durham, NC: Duke University Press, 2003.
Straus, Erwin. "The Forms of Spatiality," in *Phenomenological Psychology*, trans. Erling Eng, 3–37. New York: Basic Books, 1966.
——. *Vom Sinn der Sinne, ein Beitrag zur Grundlegung der Psychologie*. Berlin: J. Springer, 1935 (revised English translation: Erwin Straus, *The Primary World of Senses: A Vindication of Sensory Experience*, trans. Jacob Needleman. New York: Free Press, 1963).
Strawson, P. F. *Individuals: An Essay in Descriptive Metaphysics*. Garden City, NY: Anchor Books, 1963.
Sussman, Henry. *Franz Kafka: Geometrician of Metaphor*. Madison, WI: Coda Press, 1979.

Szendy, Peter. "Abstraite musique concrète ou « tout l'art d'entendre," in *Le comentaire et l'art abstrait*, eds. Murielle Gagnebin and Christine Savinel, 73–96. Paris: Presses de la Sorbonne Nouvelle, 1999.

——. *Sur écoute: esthétique de l'espionnage*. Paris: Éditions de Minuit, 2007.

Szlezák, Thomas A. *Reading Plato*, trans. Graham Zanker. London: Routledge, 1999.

Taylor, Timothy. *Strange Sounds*. New York: Routledge, 2001.

Todd, Charles Burr. *In Olde Connecticut; Being a Record of Quaint, Curious and Romantic Happenings There in Colonial Times and Later*. New York: Grafton Press, 1906.

Todorov, Tzvetan. *The Fantastic: A Structural Approach to a Literary Genre*, trans. Richard Howard. Ithaca, NY: Cornell University Press, 1973.

Truax, Barry. "Genres and Techniques of Soundscape Composition as Developed at Simon Fraser University." *Organised Sound* 7/1 (April 2002): 5–14.

Vartanian, Aram. *Science and Humanism in the French Enlightenment*. Charlottesville, VA: Rookwood Press, 1999.

Vasse, Denis. *L'ombilic et la voix: deux enfants en analyse*. Paris: Éditions du Seuil, 1974.

Villiers de L'Isle-Adam, Auguste. *Tomorrow's Eve*, trans. Robert Martin Adams. Urbana, IL: University of Illinois Press, 2001.

von Fritz, Kurt. *Pythagorean Politics in Southern Italy*. New York: Columbia University Press, 1940.

von Kleist, Heinrich. "Holy Cecilia or the Power of Music," in *Music in German Romantic Literature*, ed. and trans. Linda Siegel, 205–217. Novato, CA: Elra Publications, 1983.

Wackenroder, Wilhelm Heinrich. *Confessions and Fantasies*, ed. and trans. Mary Hurst Schubert. University Park, PA: Pennsylvania State University Press, 1971.

Wagner, Cosima. *Cosima Wagner's Diaries: Volume 2, 1878–1883*, eds. Martin Gregor-Dellin and Dietrich Mack, trans. Geoffrey Skelton. New York: Harcourt Brace Jovanovich, 1980.

Wagner, Richard. *Opera and Drama. Richard Wagner's Prose Works, Vol. 2*, trans. W. Ashton Ellis. London: Kegan Paul, Trench, Trübner & Co., 1893.

——. *Richard Wagner's Prose Works, Vol. 5*, trans. W. Ashton Ellis. London: Kegan Paul, Trench, Trübner & Co., 1896.

Waksman, Steve. *Instruments of Desire: The Electric Guitar and the Shaping of Musical Experience*. Cambridge, MA: Harvard University Press, 1999.

Warburton, Dan. Interview with Luc Ferrari, July 22, 1998 [accessed 04/20/2011: http://www.paristransatlantic.com/magazine/interviews/ferrari.html].

Webster, Clarence M. *Town Meeting Country*. New York: Duell, Sloan & Pearce, 1945.

Weigand, Hermann J. "Franz Kafka's 'The Burrow': An Analytical Essay," in *Franz Kafka: A Collection of Criticism*, ed. Leo Hamalian, 85–108. New York: McGraw Hill, 1974.

White, Glenn. *Folk Tales of Connecticut*. Meriden, CT: Journal Press, 1977.

Windsor, Luke. "A Perceptual Approach to the Description and Analysis of Acousmatic Music," Ph.D. diss., City University of New York, 1995.

Wong, Mandy-Suzanne. "Sound Objects: Speculative Perspectives," Ph.D. diss., University of California–Los Angeles, 2012.

Wood, Gaby. *Edison's Eve*. New York: Alfred A. Knopf, 2002.

World Soundscape Project. *The Vancouver Soundscape*. Sonic Research Studio, Communication Studies Dept., Simon Fraser University, 1974.

World Soundscape Project and Barry Truax. *The World Soundscape Project's Handbook for Acoustic Ecology*. Vancouver: A.R.C. Publications, 1978.

Zahavi, Dan. *Husserl's Phenomenology*. Stanford, CA: Stanford University Press, 2003.

Zhmud, Leonid. *Wissenschaft, Philosophie and Religion im frühen Pythagoreismus*. Berlin: Akademie Verlag 1997.

Žižek, Slavoj. *Enjoy Your Symptom!: Jacques Lacan In Hollywood and Out*. New York: Routledge, 1992.

———. *The Indivisible Remainder: An Essay on Schelling and Other Related Matters*. London: Verso, 1996.

———. *Interrogating the Real: Selected Writings*, eds. Rex Bulter and Scott Stevens. London: Continuum, 2005.

———. *Less than Nothing: Hegel and the Shadow of Dialectical Materialism*. London: Verso, 2012.

———. *Looking Awry: An Introduction to Jacques Lacan through Popular Culture*. Cambridge, MA: MIT Press, 1991.

———. *On Belief*. London: Routledge, 2001.

———. *The Sublime Object of Ideology*. London: Verso, 1989.

———. *The Ticklish Subject*. New York: Verso, 1999.

Index

4'33" (Cage), 125, 126

A la recherche d'une musique concrète (Schaeffer). *See In Search of a Concrete Music*
Abbate, Carolyn, 6, 48, 74
absolute music, 99, 105, 110
absorption, listening and, 10, 100, 105, 112, 115
abstract music. *See musique concrète*
acousmate, 73–94
   acousmatique and *acousmate*, comparison of terms, 75–77
   d'Ansacq, 82–91
   baptism of the word, 84–85
   Battier's argument, 77–79
   debate between Feydel and Morellet, 78–81
   definitions, 54, 75–76, 78–81, 92
   enchantment, state of, 78, 79
   meaning in scientific, psychological, and literary contexts, 91–93
   origin of term, 10, 46, 47, 73–94
   rarity of term, 10, 46, 50, 75, 81, 84, 91
   use of term in Schaefferian tradition, 46–47, 49, 93
acousmate d'Ansacq, 81–91
   comparison with aurorae, 83, 85
   depositions concerning, 81–83
   explained by natural science and philosophy, 84–88, 90–91
   as supernatural event, 82, 88
   ventriloquism, 88–90
"Acousmate" (Apollinaire), 46–50, 74–78, 93
"acousmatic"
   defined, 3, 24, 78
   mythic origin of term in Schaefferian tradition, 41, 46–51, 73, 74–75

   Pythagorean origin of the term, 54–57
   rarity of term, 4, 8, 75, 186, 206
   relationship between "acousmatic" and "acousmate," 47, 54, 73–77
acousmatic experience, 4–6, 17, 23–25, 26, 30, 36–7, 39, 40, 52, 102, 137, 152, 234n108
   See also acousmatic reduction
acousmatic listening
   acousmaticity, 148
   "The Burrow," 138, 142
   explained, 7–9
   "form of life," acousmatic sound within, 223, 224, 226
   identified with acousmatic sound, 225, 229n27
   originary experience of, 52, 60–61, 79, 119,
   practice, acousmatic listening as, 7–9, 17, 93–94, 223–6
   underdetermination and uncertainty, 148, 150
   See also acousmate; acousmatic experience; acousmatic reduction; acousmatic sound; acousmaticity; bodily techniques; modes of listening; phantasmagoria; *physis*; source, cause, and effect; *technê*; transcendence; underdetermination
acousmatic phantasmagoria. *See* phantasmagoria
acousmatic reduction, 15–41, 137–138, 143
   bracketing. *See* bracketing
   conflation with reduced listening, 5, 137–8
   counter-reduction, 152, 156
   disclosure of modes of listening, 26–30
   followed by eidetic reduction, 31. *See* eidetic reduction

acousmatic reduction (*Cont.*)
  objections to Schaeffer's theory, 36–41
  relationship to *epoché*, 22–26
  relationship to sound object and reduced listening, 17, 22–3, 30, 36, 37
  See also bracketing; eidetic reduction; *epoché*; phenomenology; reduced listening; sound object
"acousmatic silence," 216–218, 222
  See also drive; psychoanalytic voice; silence
acousmatic sound
  definitions, 3, 4, 224
  history of acousmatic sound, critique of standard history, 43–94
  identified with acousmatic listening, 225
  and Les Paul, 165–179
  limitations of aesthetic approach to, 6–7, 8, 150, 263n1
  live performance of. *See* live performances
  nontology, 148–149
  ontology of, 134–161
  relationship with history of phantasmagoria, 97–133
  Schaeffer's discovery of, 1–41
  and voice, 180–222
  See also acousmate; acousmatic experience; acousmatic listening; acousmatic reduction; acousmaticity; bodily techniques; modes of listening; phantasmagoria; *physis*; source, cause, and effect; *technê*; transcendence; underdetermination
acousmatic voice, 149, 165, 185, 186, 193–195, 206–213, 216, 217, 219–222
  vs. acousmatic silence, 216–218, 222
  "who speaks," 165, 166, 195, 202–206, 217
acousmaticity, 11, 148–50, 165, 171, 174, 176, 179, 204–205, 207, 220, 224–225, 229n27
  explained, 148–9
l'acousmatique (acoustics), 22, 24, 30, 74, 75–77

acousmate as synonym or homonym, 93
  vs. acoustics (*l'acoustique*), 24, 37
  etymology, 76
  fixation on word, 93
  musique acousmatique, 46, 47, 51, 52, 120, 224
acousmêtre, 6, 39, 149, 157, 177, 207, 224
  powers of (omnipotence, omnipresence, panopticism, ubiquity), 149, 224
  in Žižek and Dolar, 186, 207, 271n86
  See also acousmatic silence; acousmatic voice
acousmonium, 51, 52, 105, 112, 137
acoustics (*l'acoustique*); *see l'acousmatique*
act *vs.* activity, 217
The Acts of the Martyrs, 70
Adorno, Theodor, 38, 98
adumbrations (*Abschattungen*). *See* phenomenology
advertising, 170, 180, 184, 213
aesthetics
  aesthetic contemplation, 99–101, 123, 124
  of music, 10, 28, 29, 52, 79, 99, 110, 137, 143, 209
  musique concrète, 28, 29, 51, 52, 119–123, 126, 132, 150, 255n44
  "national aestheticism," 118
  presentation (*Darstellung*), image and vestige, 129–33
  Schopenhauer, 99–101
  surface, 130–132
  trace, 130–132
  Wagner, 102–103, 114–115
  See also phantasmagoria; transcendence
Aesthetics of Music (Scruton), 5, 137
akousmata, 54–65, 68–72, 84, 239n63, 241n92, 248n53
  appropriation by Plato, Aristotle, *et al.*, 63
  encryption/coding of messages, 63–65, 69, 85
  vs. mathemata, 56–59, 69–70
  "method of concealment," 68–71
  in Pythagorean school, 54–61
  oracular utterances, 62, 67, 68
  veiled utterances, 63–65
  See also allegory

Index 295

akousmatikoi and *mathematikoi*, 4, 24, 41, 54–61, 72, 212, 213, 220
  aptitudes, 63
  asceticism, 56
  audience as *akousmatikoi*, 52
  "auditors" and "students," 55
  distinguishing *akousmatikoi* from *mathematikoi*, 56, 59, 63, 64, 210
  genuine Pythagoreans, 55–61
  juxtaposition of, 65
  mathematical and scientific teachings *vs.* shamanistic and religious teachings, 56, 58–62
  silence/secrecy of, 46, 48–50, 55, 57, 62, 63, 66, 74, 76, 210
  understandings of *akousmata*, 63
  veil between. *See* Pythagorean curtain/veil
  See also Pythagoras and Pythagoreanism
Alexis, 56
alienation in language. *See* Lacan, Jacques and Lacanian Theory
allegory, 62–68, 72, 85, 98, 139, 154
  figural language, 65–67
  origin of the Pythagorean veil, 64–66
Allgemein Aesthetik als Formwissenschaft (Zimmerman), 110
Allgemeiner Deutscher Musikverein, 104
allure. See sound object, morphological features of
"almost nothing," 131–133
Andrews Sisters, 165
Androcydes, 63
androids. *See* automatons
anecdotal listening/anecdotal sounds, 28–30, 122, 123, 125
anechoic chamber, 161, 202
Der angehende Musikdirektor (Arnold), 103
angelic choir, 108–113, 185
annihilation of world. *See* epoché
"Another Evidence of the Wide-Spread Influence of Theosophical Ideals," 112
Ansacq, 81–91
anxiety/anxious listening, 83, 147–150, 154, 155, 157, 159
"Aphorism Countertime" (Derrida), 152

Apollinaire, Guillaume, 46, 47, 49, 50, 53, 54, 74–78, 93
Apollo and the Seaman (Holbrooke), 116
Aquinas, Thomas, 130
The Arcades Project (Benjamin), 98
architecture, 6, 102–116
Archives of the Voice, 77
Aristophon, 56
Aristotle, 23, 56–64, 113, 114
Aristoxenus, 57
Arndt, Felix, 168
Arnold, Ignaz Ferdinand, 103
Arrojo, Rosemary, 139
Arthuys, Philippe, 73, 74
Arts et métiers graphiques, 75
astronomy and astronomical events, 56, 59, 60, 83, 84
audio technology. *See* technology
"audiovisual complex," 4, 5
auditory culture, 226
aurora borealis, 83–85, 92
authenticity vs. inauthenticity
  discourse, 199–203, 215–217
  ethical life, 200, 204–206
  they-self, 201, 204, 205
  See also call of conscience
auto-affection, 160, 186, 187, 205, 206, 218, 259n62, 270n80
automatons, 180–185, 192
autonomy of sound, 8, 30, 39, 74, 99, 134, 136, 148–150, 179, 185, 195, 220, 260n62, 273n114
  See also sound objects
averted glance. *See* bodily techniques
awe, state of, 4, 61, 92

Babbitt, Milton, 137
Background Noise: Perspectives on Sound Art (LaBelle), 125
badger. *See* "The Burrow"
Balla, Giacomo, 170
Barber, John Warner, 2, 3
Barnes, George, 166, 167, 177
Barraud, Francis, 213
Barthes, Roland, 40, 148, 177
The Basic Problems of Phenomenology (Heidegger), 34

Batignolles, 15
Battier, Marc, 49, 74–79, 81, 93, 203
"Der Bau" (Kafka). *See* "The Burrow"
Baudelaire, Charles, 98
Bayle, François, 41, 46, 47, 51–53, 74–76, 112
Bayreuth, *Festspeilhaus*, 6, 102, 103, 105, 110, 112
beating or pulsing
　allure. See sound object, morphological features of
　heartbeat, 159, 160
Beethoven, Ludwig van, 92, 102, 103, 110, 115, 116
Being and Time (Heidegger), 185, 195, 197, 198, 201, 205, 216
"being-in-the-world," 22, 27, 34, 198–201, 204
　See also Dasein
being-with (*Mitsein*), 198–199. *See also* Dasein
Benjamin, Walter, 98, 139
Berglinger, Joseph, 108, 110, 221
Bessarion, Cardinal, 71
Bidule en ut (Henry and Schaeffer), 233n82
big Other. *See* Lacan, Jacques and Lacanian theory
black box (Les Paulverizer), 172–179, 226
Blavatsky, Madame, 113
blindness, 86, 112, 143
bodily techniques, 99–105, 108, 111, 112
　averted glance, 101, 102
　as preparatory for aesthetic contemplation, 101–102, 111
　relation to architecture, 102–105, 108, 112
　See also phantasmagoria; *technê*
Borel, Petrus, 92
Borromeo, Federigo, 108, 109, 116, 185
bracketing, 8, 23–25, 32, 36, 37, 39, 120, 143, 150, 159, 189, 190, 195, 197, 220, 221, 226
　explained, 23
　See also acousmatic reduction; *epoché*
"Brazil" (Les Paul), 166, 167
broadcast media, 4, 5, 26, 46, 49, 52, 73, 74, 121, 137, 150, 151

acousmatic reduction, 24, 25, 35, 39, 45
　Les Paul's radio performances, 165–176
　Marconi's wireless telegraphy, 112
　Radiodifusion Française, 15, 35
Brown, Peter, 70
Bruckner, Anton, 117, 118
Burkert, Walter, 56, 58, 60–63
"The Burrow" (Kafka), 134–161
　animal in, 143–148
　chiasmic sonic ascription in, 159–161
　ear, burrow as, 145
　interpretation of title ("Der Bau"), 139, 145
　interpretations of sound in, 138–141
　intensity of sound, 140–141
　as labyrinth, 139, 141
　mysterious sound in, 139, 142–148, 159, 160
　paranoia, 140, 154
　See also nontology; *Pfeifen*; source, cause, and effect; underdetermination; Zischen

C-Minor Symphony (Beethoven), 110
Cage, John, 124, 125, 161, 202, 203
Cagean non-intentionality, 125, 126
call of conscience, 200–206, 215
　as address, 202, 203
　alien power entering *Dasein*, 203
　naturalistic explanation, 203
　"nothing" heard in, 205
　silence, 204
　source of, 203–206
　spacing, 203
　See also authenticity *vs*. inauthenticity; commands/orders; *Dasein*; Heidegger, Martin; nothing; ontological difference; ontological voice; silence; voice
Calligrammes (Apollinaire), 75
Calvino, Italo, 152, 154–156
Capitol Records, 166–168, 170
Capture éphémère (Parmegiani), 122
Cartesian Meditations (Husserl), 18, 22, 31
The Castle (Kafka), 140
Catholicism, 6, 108, 117, 118. *See also clausura*

cause of sound. *See* source, cause, and effect
Cavarero, Adrianna, 152–156
Cavell, Stanley, 223
Cecilia, St., 78–81, 93
Century Path, 112
certainty, 24, 147–159, 172, 195, 224
  concerning source, cause, and effect of sounds, 147–159
  and uncertainty, 154, 159, 172, 224
  See also authenticity *vs.* inauthenticity; epistemology; intentional object; intentionality; source, cause, and effect; underdetermination
chatter, 199–201, 204, 215–217
  See also authenticity *vs.* inauthenticity; *Dasein*; discourse
chiasm, 159–161, 202
  See also "The Burrow"
Chion, Michel, 4–6, 27, 39, 46, 74, 76, 124, 149, 150, 156, 157, 160, 161, 186, 207, 229n31, 236n20, 241n93, 271n86
choral music, 108–110
Choron, Alexandre, 102, 103
Christian traditions and teachings, 6
  Catholicism, 6, 108, 117, 118
  "friends of God," 70
  Jesus, parables and miracles of, 68–71
cinema. *See* film
Cinq études de bruits (Schaeffer), 73, 120
clairvoyance, 115, 116, 221
clausura, 6, 108, 109, 185
Clement of Alexandria, 46, 64–72, 84, 92
Clermont en Beauvoisis, 81, 91
cloche coupée. See musique concrète
clocks. *See* horology
Cocteau, Jean, 75
coding/encryption of messages, 63–65, 69, 85. *See also* Pythagoras and Pythagoreanism
color, 31, 32, 51, 83, 86, 103, 115, 136
commands/orders, 194, 204, 207, 214, 215, 217, 219.
  See also enunciation *vs.* statement; Lacan, Jacques and Lacanian theory

commodity fetishism, 97
De communi mathematica scientia (Iamblichus), 58
communicative signs/speech, 28, 29, 37, 38, 188, 192, 193. *See also* index/indication
comprendre. See modes of listening
concealment, method of, 66–71. *See also akousmata;* allegory; Clement of Alexandria
concentrated listening, 41, 45, 46, 50, 52, 61, 120
concert performances, 5, 101–108, 137, 142
concert reform movement, 103–105, 112
  aesthetic shaping of the listening public, 103
  flowers used to obscure musicians, 103, 104
Concret PH (Xenakis), 120
concrete music. *See musique concrète*
"Confess" (Patti Page), 177
confessional, 6, 108, 109
Connecticut Gazette, 2
Connor, Steven, 8, 88, 89
conscience, call of. *See* call of conscience
Continuo (Schaeffer and Ferrari), 120, 122, 124
convents, 6, 108–110, 185
Cooke, Alistair, 173, 174
Copenhagen. *See* Konzertpalais
Corngold, Stanley, 139, 140
Council of Trent, 108
counter-reduction. *See* acousmatic reduction
Crary, Jonathan, 97
Crosby, Bing, 165
crying, 81, 83, 85
Crystal Palace, 183
cubism, 75
curiosity, state of, 4, 25, 110, 150, 170
curtain. *See* veil
cylinders, wax. *See* phonograms

Dabholkar, Pratibha, 157
Dack, John, 29
Dahlhaus, Carl, 105, 115
d'Alembert, Jean, 46

Dalmatia, 123–126, 129, 131–133. *See also* *Presque rien*
dance of the dead, 93
darkness
  occultation of production, 98, 99, 119, 220, 221
  Pythagoras teaching in darkness, 41, 45, 46, 51
  as technique for transcendence, 110–112, 115, 118, 139, 153
  See also bodily techniques; phantasmagoria; *technê*
Darstellung. *See* aesthetics
Dasein, 197–203, 215, 216
  authenticity, 199–203, 206
  being-in-the-world, 198–201, 204
  being-with (*Mitsein*), 198–199
  call of conscience. *See* call of conscience
  conformism, 199
  equipment, 196–198
  existence and essence, relationship of, 197
  explained, 197
  friend, voice of, 205, 206, 216, 221
  projects and goals, 196–199
  ready-to-hand *vs*. present-at-hand, 196–198, 200, 203–205
  as sender and receiver of call, 200, 205
  silence, 205
  solicitude, 198
  "they" and they-self, 198, 200, 201, 204, 205
  underdetermination of source of voice, 202, 203, 206
  See also authenticity *vs*. inauthenticity; call of conscience; chatter; Heidegger, Martin; idle talk; ontological difference; ontological voice; subject; voice
Dawson, David, 68
"Debussy's Phantom Sounds" (Abbate), 48, 49
decorations in concert halls, 92, 103–105, 107, 116
Decsey, Ernst, 105
de Duve, Thierry, 133
delay. *See* Paul, Les

Deleuze, Gilles, 140, 141, 144, 225
demonic events, 88, 89
Derrida, Jacques, 113, 127, 152–154, 181, 186, 189–193, 206, 207
Descartes, René, 87, 197, 198
desire, 19, 21
  vs. drive, 213
  fable of Parrhasios and Zeuxis, 211, 212
  See also drive; Lacan, Jacques and Lacanian theory; object voice; *objet (a)*; veil; voice
Dessau. *See Hoftheater-Orchester*
destruction of world. *See epoché*
Dhomont, Francis, 47, 51, 74
Diamorphoses (Xenakis), 120
Dictionnaire (Académie Française), 49, 50, 75, 76, 78–81, 91, 92
Diderot, Denis, 46–48, 50, 53, 55, 65, 93
"Diderot: Paradox and Mimesis" (Lacoue-Labarthe), 113
Diogenes Laërtius, 48, 66, 209
direct listening, 4, 5, 39
dis-acousmatization, 148, 150, 157–159, 210–212
  Dolar's denial of, 150, 212, 220–222
  and Les Paulverizer, 176
disciples of Pythagoras. *See akousmatikoi* and *mathematikoi*
discourse, 4, 5, 23, 40, 41, 45, 52, 53, 55, 65–68, 98, 108, 114, 120, 123, 138, 160, 184, 187, 188, 192, 199–201, 205, 209, 210, 214, 216, 217, 218, 221
  See also allegory; authenticity *vs*. inauthenticity; chatter; communicative signs/speech; expression *vs*. indication; idle talk; Lacan, Jacques and Lacanian theory; linguistics; logos; shifters; voice; writing
distance. *See* spacing (*espacement*)
Divertissements typographiques, 75
dodecaphonic music. *See* serialism
dogs, 135
  canine fidelity, 213, 214, 219, 274n143
  "The Investigations of a Dog" (Kafka) 139, 142, 143

Index

Dolar, Mladen, 6, 48, 74, 145, 149, 150, 156, 186, 206–212, 214–222, 225
"Don'cha Hear Them Bells" (Les Paul and Mary Ford), 174
Dräseke, Felix, 102
dread. *See* fearful sounds
dreaming, 98, 111, 115, 126, 143
drive
  vs. desire, 213
  silence of, 217, 222
  voice as object of, 213, 214, 217, 218, 222
  See also acousmatic silence; desire; Lacan, Jacques and Lacanian theory; object voice; psychoanalytic voice; voice
Drott, Eric, 123–126
dubbed recordings. *See* recordings, overdubbed
Duchamp, Marcel, 124
Dumbstruck (Connor), 88
Dunkelkonzerte, 6, 117, 118
duration. *See* sound object
Dyson, Frances, 37, 183

eardrum, 131, 132
echo, 91, 114, 167, 168, 177. *See also* Paul, Les
ecological listening, 127, 229n29, 261n72
écoute réduite. See reduced listening
écouter. See modes of listening
Edinburgh Journal of Science, 88
Edison, Thomas, 165, 179, 182–184, 206
effect, sound as. *See* source, cause, and effect
ego, 113, 128, 182, 216
  superego, 6, 203, 206, 207, 209
  See also auto-affection; "I"; self; subject; voice
eidetic reduction, 18, 30–33, 36–39, 119, 129, 138, 147, 150, 156
  explained, 30, 154n104, 260n62
  fact *vs.* essence, 31, 35–6, 40
  fiction, role of, 32
  imaginative variation, 31–3, 37, 121, 156, 196
  invariant properties of objects, 31–33
  See also phenomenology

Eisenhower, Dwight, 175, 176
Eisenhower, Mamie, 176
electroacoustic music, 24, 47–49, 51, 53, 74, 126, 263n1
  See also musique concrète
elektronische Musik, 17
Ellison, Ralph, 7
enchantment. *See* acousmate
encryption/coding of messages. *See akousmata*; allegory; concealment, method of; Pythagoras and Pythagoreanism
L'encyclopédie, 46–48, 50, 55
entendre. See modes of listening; reduced listening
enunciation *vs.* statement. *See* Lacan, Jacques and Lacanian theory
environmental situation of sound, 7, 8, 123, 124, 151, 152, 225
Epicharides, 56
Epicureans, 64
epistemology, 86, 87, 128, 148, 159, 206, 224, 225
  Cartesian, 87–88, 197
  Lockean, 86–87
  qualities, primary *vs.* secondary, 86, 87
  of the senses, 45, 86, 87, 148
  See also certainty; phenomenology; sound object; source, cause, and effect; underdetermination
epoché, 17–19, 22–26, 30, 32, 34, 36, 38, 41, 47, 50, 52, 121, 128, 136–138, 147, 186, 191–196, 205
  affinity between *epoché* and role played by Pythagorean veil, 25
  annihilation of world, 189, 195, 197
  natural attitude/natural standpoint, 23, 24, 27, 34, 136, 195–197
  use of term, 23
  See also acousmatic reduction; bracketing; phenomenology; suspending
equipment. *See Dasein*
erotic sound, 104, 109, 110
espacement. See spacing
Esprit, 73
Essay on Man (Pope), 143

An Essay Concerning Human Understanding (Locke), 86, 87
Étirement (Henry), 29
Étude aux accidents (Ferrari), 120, 122
Étude aux allures (Schaeffer), 29, 120–123
Étude aux chemins de fer (Schaeffer), 26, 28, 29, 120–123
Étude aux objets (Schaeffer), 29, 120, 121
Étude aux sons animés (Schaeffer), 29, 120
Étude aux sons tendus (Ferrari), 120
Étude floue (Ferrari), 120, 122
everydayness/everyday life, 56, 61, 197
  acousmatic reduction, 23, 27, 36
  phantasmagoria, 111, 112, 114, 125
  voice, 191, 196–204, 217, 218
expression *vs.* indication, 8, 187–194
  characterizations and distinctions, 188, 190, 191
  expressive indications/indicative expressions, 194
  facial expressions and gestures, 178, 187–189
  meaning–intention and meaning-fulfillment. *See* intention/intentionality, meaning-intentions
  reduction of indication, 188
  solitary mental life, 192
  writing as inscription of expression, 190–192
  See also auto-affection; communicative signs/speech; index/indication; inscription; intention/intentionality; internal soliloquy; modes of listening; phenomenological voice; phenomenology; signs and signification; speaking-to-oneself; writing
exteriority, 190, 191, 193, 206, 207, 260n62, 268n31, 270n80, 271n83
eyes
  averted glance, 101, 102, 221
  blindness, 86, 112, 143, 221
  closing, 10, 20, 221, 222

faith in exterior world. *See* natural attitude/natural standpoint
Fara, Patricia, 84

fearful sounds, 1, 4, 83, 89, 92, 155
  animal in "The Burrow," 143–148
feedback loops, 192, 193
Ferrari, Luc, 119, 120, 122–126, 131–133, 150
  Continuo, 120, 122, 124
  Étude aux accidents, 120, 122
  Étude aux sons tendus, 120
  Étude floue, 120, 122
  Hétérozygote, 122, 123
  Presque rien, 11, 119, 120, 123–126, 129, 131–133
Ferreyra, Beatriz, 48
"fervor of listening," 26, 39
Festspielhaus. See Bayreuth
fetishism, 97, 210–212, 220–222
Feuerbach, Ludwig, 115
fevers, 80–1
Feydel, Gabriel, 79–81, 92, 93
Fichte, Johann, 182
Ficino, Marsilio, 71
fiction. *See* eidetic reduction
fidelity
  to acousmatic voice of master, 219–220
  to acousmatic silence of psychoanalytic voice, 219–220
  canine fidelity, 213, 214, 219, 274n143
  phantasmagoria, 218–222
  of phonographic voice, 191,192
  recovery of prior meaning-intentions, 190, 191, 199
  to split subject, 215–216, 219–220
  See also "I know very well…but nevertheless…"; Lacan, Jacques, and Lacanian theory; phantasmagoria
figural language. *See* allegory
film
  acousmêtre, 6, 39, 149, 157, 177, 207, 224
  fundamental noise in cinema, 160
  Psycho, 207
  Singin' in the Rain, 157, 158, 171, 223
  The Testament of Dr. Mabuse, 157, 207
  The Voice in Cinema, 39, 46, 156, 207
  The Wizard of Oz, 210
Film: A Sound Art (Chion), 160
Fink, Bruce, 215

flatness, 131
Fontenelle, Bernard Le Bovier de, 87, 88
For More than One Voice (Cavarero), 152
Ford, Carol, 171, 172
Ford, Mary, 165, 168–178
"The Forgetting of Philosophy" (Nancy), 128
form of life. *See* acousmatic listening
Formal and Transcendental Logic (Husserl), 18
formalism, 136
"The Forms of Spatiality" (Jonas), 136
Forte, Allen, 137
The Four Fundamental Concepts of Psychoanalysis (Lacan), 211
Foye, Wilbur Garland, 3
Frankfurter Zeitung, 103
French Enlightenment, 46, 47, 54, 55
French Pavilion, Brussels World's Fair, 120, 122
Freud, Sigmund, 6, 208, 210, 216
friends and friendship, 66, 190
    voice of friend, 205, 206, 216, 221
fugue, 29, 118, 223n82, 270n81,
"fulfilled" meaning. *See* intention/intentionality; intuition
fundamental sounds, 160, 161

Gayou, Évelyne, 123
German Enlightenment, 129
Gesamtkunstwerk, 103
gestures. *See* expression *vs.* indication
Ghil, René, 8
Gide, André, 5
Gilliam, Bryan, 117, 118
Girette, Jean, 103
Glass, Louis, 105
glee club, 173, 176
God/gods, 1, 2, 6, 61, 62, 67–70, 88, 114, 118, 203
    heavenly music, 70, 108–113, 221
    prophecies and oracles spoken in enigmas, 62, 67, 68
    tabernacles and consecrated spaces, 68, 69
    See also religion
Goebbels, Joseph, 117

Gouraud, George, 183, 184, 186, 206
grain. *See* sound object, morphological features of
Of Grammatology (Derrida), 114
gramophone, 131, 213, 219, 221
graphic design, 75
Gregory, Martin, 108
Grétry, André, 102, 103
griphos, 62, 65, 68
Groupe de recherche de musique concrète (GRMC), 46, 73, 74
Groupe de recherche musicale (GRM), 49, 73, 74, 77, 78, 119, 123
Guattari, Felix, 140, 141, 144
Guide des objets sonores (Chion), 4
guitar, 166–179, 226
    See also Paul, Les

Haas, Georg Friedrich, 6
Hagen, Jean, 157
Hägg, Henny Fiska, 67
half-speed recordings. *See* recordings, overdubbed
hallucination, 35, 75, 92, 140, 203
Handel Festival, 183
Hanslick, Eduard, 117, 137
harmony, divine, 70
hearing *vs.* listening, 127–129, 132
heartbeat, 159, 160
heavenly source of music. *See* angelic choir; harmony
Hebraic tradition, 6, 69, 71
Hegel, Georg, 128–130, 211, 212
Heidegger, Martin, 22, 34, 40, 185, 186, 195–206, 215, 216, 221
Heidelberg. *See* Stadthalle
Helmholtz, Hermann, 131
Henry, Pierre, 29, 30, 73, 119, 122
hermeneutics, 64, 65, 68, 142, 243n118
Hermes, 68
Hermes Trismegistus, 71
hermeticism, 70, 75, 93
Hesiod, 68
Hess, Earl, 157
Hétérozygote (Ferrari), 122, 123
Hicks, R. D., 66

hidden musicians, 103–105, 108
  screen illustrations, 104, 107
  See also phantasmagoria; technê
Hildegard of Bingen, 209
Hippasus of Metapontum, 57, 58
"His Master's Voice," 214–215, 217, 219.
  See also Dolar, Mladen; Lacan, Jacques
    and Lacanian Theory
His Master's Voice (painting), 213
Histoire des oracles (Fontenelle), 87
History of the Concept of Time
  (Heidegger), 198
Hitchcock, Alfred, 207
Hitler, Adolf, 117, 118
HMV (His Master's Voice) label, 214
Hobbamock, 1, 3, 8
Hoeberg, Georg, 105
Hoftheater-Orchester, Dessau, 107
Holbrooke, Joseph, 116, 117
Homer, 68
Horky, Philip Sidney, 59
horology, 180-2
  See also automatons;
    Jacquet-Droz, Pierre
Hosmore, Stephen, 1, 6
"How High the Moon" (Les Paul and Mary
    Ford), 171, 174
Husserl, Edmund, 18–26, 30–32, 35–38,
    185–199, 206, 208
  Cartesian Meditations, 18, 22, 31
  Formal and Transcendental Logic, 18
  Idea of Phenomenology, 32
  Ideas Pertaining to a Pure Phenomenology
    and to a Phenomenological
    Philosophy, 18, 20, 189
  Logical Investigations, 186–195
  Origin of Geometry, 35, 190, 234n103
  See also Husserlian phenomenology;
    phenomenology
Husserlian phenomenology, 17–19, 22, 23,
    26, 36, 38, 41, 50, 52, 121, 128, 137,
    147, 186, 191–196, 205
  See also phenomenology
hypnotic clairvoyance, 115, 116, 221

"I", 181–185, 192–195, 215
  heteronomic origin, disavowal, 219

identification with soul or subject, 185
intuition and fulfillment of
  I-presentation, 193
"non-I," 182
self-referentiality and iterability,
  181, 182
underdetermination of, 185
See also auto–affection; ego; self;
  subject; voice
"I know very well… but nevertheless…,"
  210, 211, 219, 221, 222
"I think therefore I am," 181, 194
Iamblichus, 55–59, 61, 63–66, 69–71, 84
Ich bin der Welt abhanden gekommen
  (Mahler), 105
Idea of Phenomenology (Husserl), 32
ideality, 129, 150
  ideal objects/ideal objectivity, 21,
    32–34, 36, 37
  relation to inscription, 189–191
Ideas Pertaining to a Pure Phenomenology
  and to a Phenomenological
  Philosophy (Husserl), 18, 20, 189
identifiability of sound, 27, 28, 29, 31, 33,
  122, 139, 141, 142, 146, 147, 148,
  149, 151, 224
identification, 81, 109, 125, 130, 159, 185,
  203, 206, 207, 212, 214, 215
  mythic, 41, 45
  mimetic, 51–53, 166
identity, numerical vs. qualitative, 122, 146,
  147, 154
identity theft, 166–168
ideology, 41, 210, 211
idle talk, 199–201, 204, 215–217
  See also authenticity vs.
    Inauthenticity; Dasein
illusions, 210
  aural, 88–90, 102, 105, 131
  comparison of visual and auditory
    illusions, 89, 90
  Les Paul, 165–179
  in phantasmagoric performances, 97, 98,
    102, 105
  visual, 23, 85
illustrations, doctored, 169, 170
image. See aesthetics

imagination, 8, 16, 19, 28, 21, 32, 169
  imaginary Les Paulverizer, 172–179, 226
  imaginary sounds, 47, 92, 134, 140, 154, 207
  intuition of intentional objects, 188
  paranoia, 140, 154, 207
  purely auditory world, 145–147
  See also eidetic reduction, imaginative variation; intention/intentionality; supplement; vocalic body
imitation
  art as, 99
  of Les Paul's style, 166–168
In Search of a Concrete Music (Schaeffer), 15, 19
inauthenticity. *See* authenticity *vs.* inauthenticity
incantation, 79, 80, 92, 93
  disambiguation from acousmate, 248n53
  as mental disorder, 79
incense, 104
index/indication, 25, 26, 38, 120, 127, 131, 135, 152, 179
  écouter as indexical listening for sources, 121
  entendre as stripping away indicative significations, 196
  vs. expression, 187–194
  indexical listening, 25, 27, 121, 127
  See also modes of listening; signs and signification
indirect listening, 5
  See also reduced listening
Individuals (Strawson), 145–147
inference/speculation as to causal sources, 4, 24, 202
  paranoia, 140, 154, 156, 160, 207
inner/outer ambiguity. *See* chiasm
The Inoperative Community (Nancy), 52
inscription
  permanence, 190
  recorded sound, 77, 79, 132, 183, 191, 192, 269n36
  relationship of inscription and intention, 190
  of silence, 218
  technology for, 4

  the un-inscribable, 192
  voice, 183, 189–192, 218
  See also ideality; writing
intensity. *See* "The Burrow"
intention/intentionality, 21, 24–28, 129, 137, 139, 144, 186, 187, 189
  Cagean nonintentionality, 125, 126
  intentional objects, 19, 32, 35, 128, 136, 137, 188–191, 194, 221
  meaning-intentions, 188–194, 199
  modes of presentation, 19
  names, 188–189
  phantasy, 35
  reactivation/recovery of prior meaning-intentions, 190
  relationship between meaning-intention and intuitive fulfillment, 188, 189
  unfulfilled intentions, 188, 189
  See also intuition; phenomenology
internal voice, 188–194
  interlocution *vs.* soliloquization, 188
  See also phenomenological voice; voice
interruption of myth, 54, 65, 99
intersubjectivity, 7, 20–22
intuition, 128, 188–191, 199
  meaning-intention and fulfillment, 188–189, 267n26, 268n29, 270n80
  of self, 193–195
  See also intention/intentionality; phenomenology
"The Investigations of a Dog" (Kafka), 139, 142, 143
"Is there a Québec Sound?" (Dhomont), 47
iterability, 181, 182, 184–186.
  See also "I"; ideality; inscription; writing

Jacquet-Droz, Pierre, 179–182, 184, 185, 192
Jakobsen, Roman, 208
Janaway, Christopher, 99
Jespersen, Otto, 182
Jesus, parables and miracles of, 68–71
Jewish traditions and teachings, 6, 71
Jonas, Hans, 135, 136, 138, 151, 161
"Josephine the Mouse Singer" (Kafka), 139, 142, 143

Kabbalah, 71
Kafka, Franz, 39, 134–161, 202, 203, 223, 226
  "The Burrow," 134–161
  The Castle, 140
  "The Investigations of a Dog," 139, 142, 143
  "Josephine the Mouse Singer," 139, 142, 143
Kafka: Toward a Minor Literature (Deleuze and Guattari) 140
Kamin, Wally, 171, 172
"The Kangaroo," (Les Paul) 174
Kant, Immanuel, 99, 100, 128, 132
Kelly, Gene, 157
Kermode, Frank, 67, 68
"key myth" of Pythagorean veil, 46, 49–51, 74, 76
Kindertotenlieder (Mahler), 104
"A King Listens" (Calvino), 152–156
Kittler, Friedrich, 132
Kleist, Heinrich von, 79
Konzertpalais, Copenhagen, 105, 107, 109

LaBelle, Brandon, 125, 126
labyrinth. *See* "The Burrow"
Lacan, Jacques and Lacanian theory, 206–208, 210–216, 219, 221, 222
  "acousmatic silence," 216–218, 222
  act *vs.* activity, 217
  alienation in language, 207, 208, 214, 218
  big Other, 210, 214–219, 222
  desire *vs.* drive, 213
  enunciation *vs.* statement, 214–217
  fable of Parrhasios and Zeuxis, 211, 212
  fetishism, 210–212, 220–222
  "His Master's Voice," 214–215, 217, 219.
  "I know very well... but nevertheless...," 210, 211, 219, 221, 222
  objet (a), 212-3, 271n86
  object voice, 6, 206–214, 217, 219, 272n90
  orders/commands, 194, 204, 207, 214, 215, 217, 219.
  split subject, 218, 219
  See also acousmatic silence; *acousmêtre;* authenticity *vs.* inauthenticity,
  ethical life; desire; Dolar, Mladen; drive; fidelity; *objet (a);* psychoanalysis; psychoanalytic voice; voice; Žižek, Slavoj
Lacoue-Labarthe, Philippe, 113, 118
Lang, Fritz, 207
language
  alienation in language. *See* Lacan, Jacques and Lacanian theory
  figural language, 65–67
  linguistic tricks, 180–182
  "method of concealment," 68–71
  See also akousmata; allegory; authenticity *vs.* inauthenticity; chatter; communicative signs/speech; discourse; expression *vs.* indication; idle talk; Lacan, Jacques and Lacanian theory; linguistics; logos; shifters; voice; writing
Language: Its Nature, Development and Origin (Jespersen), 182
Larousse (dictionary), 24, 47–50, 75, 76, 142
laughter, sounds of, 81, 82, 85, 89, 91, 180
layered/dubbed recordings. *See* recordings, overdubbed
Le Corbusier, 75
Le Gendre, Gilbert Charles, 92
Le Gorse, Louis, 92
Lebenswelt (lifeworld), 22
Leo X, 71
Leonardo, 47
Les Paulverizer, 175–179, 226
Levi-Strauss, Claude, 46, 52
Levinas, Emmanuel, 206
life *vs.* art, 124
lifeworld *(Lebenswelt),* 22
linguistics, 182, 214, 221
  philosophical jokes and linguistic tricks, 180–182
  shifters, 182–184, 192, 193, 195
  structural linguistics, 208
  See also logos; voice
Linus, 68
lip-synching, 157, 177
"Listening" (Barthes), 148
Listening (Nancy), 127, 128

listening *vs.* hearing, 127–129, 132
Little Menlo, 183
live performances
　concert performances, 5, 101–108, 110, 137, 142
　of Les Paul, 170–172, 176
Lives of Eminent Philosophers (Diogenes), 66
Livy, 92
Locke, John, 86, 87
Logical Investigations (Husserl), 186–195
logos, 67–71
　classical logocentrism, 67
　phoné vs. logos, 152
　psychoanalytic voice, 208–210, 213, 217, 218
　See also phoné; logos; voice
López, Francisco, 150
Lore and Science in Ancient Pythagoreanism, 58
"Lover" (Les Paul), 166, 167, 225

Macbeth (Shakespeare), 110
Machemoodus, 1
magic, 62, 71, 79, 80, 82, 83, 117
magic lanterns, 97, 111, 114
　See also phantasmagoria
Mahler, Gustav, 104, 105
das Man. See Dasein, "they" and they-self
Mandell, Charlotte 127
Mansfield, Craig, 3
Manuel de musique (Choron), 102
Marconi, Guglielmo, 112
Mark, Saint, 68
Marsop, Paul, 104, 105, 116, 118
Marx, Karl, 41, 97, 98
materialism, 113, 192, 260n62
mathemata. See akousmata
mathematikoi. See akousmatikoi and *mathematikoi*
Mauke, Wilhelm, 103
McLuhan, Marshall, 224
meaning-intentions. *See* intention/intentionality
meaning-fulfilling intuitions. *See* intention/intentionality; intuition
Le médecin malgre lui (Molière), 80

media. *See* broadcast media
Mémoires (Grétry), 102
memory, 19–21, 71, 92, 167
mental disorders, 92, 93, 140
　acousmate and incantation as, 79
　paranoia, 140, 154, 207
Mercure de France, 81–83, 88–91
Merleau-Ponty, Maurice, 17, 18, 22, 23, 27
Messiaen, Olivier, 29
Metaphysics (Aristotle), 59
metempsychosis, 56, 60, 70
microphone, 15, 77, 123–125, 132, 157, 171–174, 193
　stereo microphone, 124, 125
Miller, Jacques-Alain, 211
mimicry, 154, 166–168, 177
miracles, 70, 71, 82, 87, 92
　Pythagoras as shaman/miracle-worker, 56, 58–60, 70
misdirection. *See* Paul, Les
Mitchell, Millard, 157
modes of listening, 26–30, 36, 133, 138
　comprendre, 27, 28, 36, 127, 128, 138
　écouter, 27–30, 36, 121, 125, 126, 128, 132, 133, 196
　entendre, 27–30, 36, 121, 123, 124, 126–133, 138, 189, 196
　ouïr, 27
　See also acousmatic reduction; reduced listening
mole. *See* "The Burrow"
Molière, 80
Molyneux, William, 86, 87
monotone, 140–142
montage, 29, 75, 120, 121
Moodus noises, 1–4, 6, 8, 223, 226
Morellet, André, 80, 81, 92
Morigia, Paolo, 108
morphology. *See* sound objects, morphological features of
Moses, 71
mouth, 7, 87, 131, 134, 135, 198, 212
Mozart, Wolfgang Amadeus, 118
Mt. Tom, 1
multitrack recordings. *See* recordings, overdubbed
Musaeus, 68

musical work, 17, 38, 136, 137, 147
Die Musik, 103–105, 107, 196
Musikalische Eilpost, 110, 114–116, 221
musique acousmatique, 46, 47, 51–52, 120, 224, 263n1
Musique Acousmatique (Bayle), 46, 74
Musique animée (Arthuys and Peignot), 46
musique concrète, 15–41, 48, 73, 119–133
  vs. abstract music/serialism, 17, 51
  "acousmatic" as term for describing, 50, 51
  cessation of use of term, 51, 73, 120
  cloche coupée, 16, 18, 35, 119, 125
  mythic identification with Pythagoras, 45
  as phantasmagoric, 39, 40, 119–133
  sillon fermé, 16–18, 26, 119
  See also acousmatic listening; acousmatic sound; acousmatic reduction; broadcast media; Ferrari, Luc; Henry, Pierre; modes of listening; *musique acousmatique*; musique *expérimentale*; phantasmagoria; Schaeffer, Pierre and Schaefferian tradition; sound object; sound recording; tape recording; reduced listening
musique expérimentale, 73, 120
De Mysteriis (Iamblichus), 69, 70
mystery, 64, 67, 68, 71, 72, 85, 142
  Ansacq, 81–91
  Les Paul's performances, 171–176
  Moodus noises, 1–4, 6, 8
  mysterious sound in "The Burrow," 143–148, 159, 160
  prophecies and oracles, 62, 67, 68
mystery cults, 62
mystic gulf, 143
mystical listening, 79, 93
myths surrounding word "acousmatic," 40, 41, 47–54
  autopoiesis (self-foundation) as mythic identification, 53, 54
  interruption of mythical narratives, 43–94

Nachträglichkeit, 53
Nagra recorder. *See* tape recording

names, theory of, 188–189
  See also intention/intentionality
Nancy, Jean-Luc, 52–54, 127–131, 133
"national aestheticism," 118
Native Americans, Moodus noises, 1–4, 6, 8
natural attitude/natural standpoint. *See epoché*; phenomenology
Natural History (Pliny), 143
natural science, 2, 3, 23–24, 34, 38, 60, 69, 83–85, 87–88, 91, 135–6
Nazi Germany, 117, 118
NBC radio, 168, 173
Neoplatonism, 57, 62, 69, 70
"New Media Dictionary," 47
The New Sound (Les Paul), 170, 178
New York Symphony Orchestra, 112
Nicholson, William, 88
Nietzsche, Friedrich, 79, 139
nighttime. *See* darkness
Nipper, 219
Nixon, Pat, 176
Nixon, Richard, 175, 176
"The Nobility of Sight" (Jonas), 135
noema and *noesis*. *See* intention/intentionality; phenomenology
noise, 47–49, 75–77, 80, 91, 136, 151, 225
  as acoustic event, 132
  Ansacq, 81–91
  cinema, fundamental noise in, 160
  concert audiences, 110
  distinguished from music, 125
  Hétérozygote (Ferrari), 123
  Moodus noises, 1–4, 6, 8
  mysterious sound in "The Burrow," 143–148, 159, 160
  pure noise, 196
  "symphony of noises," 15
  See also Pfeifen; Zischen
"Nola" (Les Paul), 68
nonsense, 63, 65, 72, 189, 217
nontology (non-ontology), 149
  See also acousmatic sound; ontology; source, cause, and effect; spacing; underdetermination
notes, music bound to, 17, 38, 99, 110, 147
nothing

in call of conscience, 201, 202, 203
in Lacanian theory of voice, 212, 213
See also silence
Nouveau dictionnaire de médecine, 92
Noyes, Betty, 157
numerical identity of sound, 122, 146, 147, 154
nuns, singing, 6, 108–110, 185

obedience
    etymological link between listening and obeying, 273n115
    to habit, 34
    to psychoanalytic voice, 213–220
object. *See* sound object
object voice, 6, 206–214, 217, 219, 272n90
"The Object Voice" (Dolar), 209
objet (a), 212-3, 271n86
objet sonore. See sound object
Objets exposés (Schaeffer), 29
Objets rassemblés (Schaeffer), 29
occultation of production. *See* phantasmagoria
O'Connor, Donald, 157
Of Grammatology (Derrida), 114
"Old Cape Cod" (Patti Page), 177
Omnibus, 173–175
omnipotence and omnipresence of voice, 149, 207, 209, 211–213, 215, 221, 224.
    See also acousmêtre, powers of
on-stage performances. *See* live performances
On the Pythagorean Way of Life (Iamblichus), 62, 71
On the Pythagoreans (Aristotle), 61
Ondes Martenot, 47
ontological difference, 197–202
    explained, 198
    See also authenticity *vs.* inauthenticity; call of conscience; *Dasein;* Heidegger, Martin
ontological hearing, 205, 272n81
    See also Heidegger, Martin
ontological separation of sound object from source, 134, 147, 148
ontological voice, 185, 195–206

affected by acousmatic voice, 203–206
    See also acousmatic voice; call of conscience; phenomenological voice; phonographic voice; psychoanalytic voice
ontology
    eidetic reduction. *See* eidetic reduction
    event-based ontology, 38, 39
    nontology (non-ontology), 149
    objections to Schaeffer's ontology, 37–39
    in Schaeffer's theory of sound object, 15–17, 30–33
    See also nontology; ontological voice; source, cause, and effect; sound object; spacing; underdetermination
opera, 6, 102–3, 109, 114, 115
Opera and Drama (Wagner), 114, 115
oracular utterances. *See akousmata*
orchestras, 173
    Holbrooke's illuminated symphony, 116, 117
    invisible, 101–108
orders/commands, 194, 204, 207, 214, 215, 217, 219.
    See also Lacan, Jacques and Lacanian theory, enunciation *vs.* statement
Origin of Geometry (Husserl), 35, 190, 234n103
originary experience. *See* phenomenology
Origine des fables (Fontenelle), 87
Orpheus, 68, 71, 72
Orphic mysteries, 71, 72
the Other. *See* Lacan, Jacques and Lacanian theory
otherworldliness, 112, 220
ouïr. See modes of listening
"Our Concerts" (anonymous), 110, 113
outer/inner ambiguity. *See* chiasm
Outpourings of an Art-loving Friar (Wackenroder), 101
overdubbed recordings. *See* recordings, overdubbed

paganism, 69–71
Page, Patti, 177
Palombini, Carlos, 40

panopticism. *See acousmêtre*, powers of
Paramount Theater, 171
paranoia, 140, 154, 207
Parmegiani, Bernard, 122
Parrhasios. *See* desire
pathology of listening, 92
Paul, Les, 165–179, 223, 225, 226
  "Brazil," 166, 167
  delay effects, 167, 177
  "Don'cha Hear Them Bells," 174
  "double sound" trick, 172
  echo effect, 167, 168, 175, 177
  "How High the Moon," 171, 174
  imitation by other guitarists, 166–168
  "The Kangaroo," 174
  Les Paulverizer. *See* Les Paulverizer
  live performances of, 170–172, 176
  "Lover," 166, 167, 225
  misdirection of audience, 168–170, 176
  The New Sound, 170, 178
  "Nola," 68
  photograph, 169
  proprietary sound, 166, 168, 171, 174
  slapback effects, 167, 177
  stolen identity, 165–168
  "Tennessee Waltz," 177
  "The World Is Waiting for the Sunrise," 169
  See also broadcast media; guitar; Les Paulverizer; phantasmagoria; recordings; rivalry, mimetic; tape recording; technology; trickery; underdetermination
Paul, Saint, 88
Peignot, Charles, 74
Peignot, Jérôme, 4, 46–54, 61, 73–76, 93, 112, 120
perception, 19–27, 30–32, 34, 49, 77, 100, 134, 136–7, 182, 194
"The Perfected Phonograph" (Edison), 183
Pfeifen (piping), 139, 142, 160, 161, 202. *See also* "The Burrow"; *Zischen*
phantasmagoria, 39, 40, 97–133, 136, 218–222
  Adorno's use of word, 98
  angelic choir and, 108–110
  architectural techniques in production of, 102–105
  bodily techniques in production of, 99–102
  conditions of phantasmagoria, 116, 119
  Dunkelkonzerte, 117–1188
  explained, 99
  fidelity, 218–222
  invisible orchestras and, 110–112
  musique concrète, 119–133
  occultation of production, 98, 99, 119, 209, 220, 221
  origin of term, 97
  phantasmagoria as production of transcendence, 108
  and psychoanalytic voice, 218–222
  in theory of sound object, 39–40
  theosophy and, 112–113
  visual technology, 97–98
  See also clausura; *physis*; supplement; *technê*; transcendence
phantasy. *See* intention/intentionality, intentional objects
phénomènes, 84, 85
phenomenological reduction. *See epoché*
phenomenological voice, 186–195
  affected by acousmatic voice, 193–5
  phonographic voice, compared, 191
  See also acousmatic voice; ontological voice; phonographic voice; psychoanalytic voice
phenomenology, 18, 26, 129, 134–137, 151
  adumbrations (*Abschattungen*), 20–22, 132
  doing phenomenology *vs.* knowing phenomenology, 18, 19
  influence on Pierre Schaeffer, 17–22
  natural attitude/natural standpoint, 23, 24, 27, 34, 136, 195–197
  noema, 20–22, 26, 39, 138
  noesis, 20–22, 26, 27, 33, 35
  originary experience, 33–36, 53, 79, 212
  reactivation of, 19, 35, 39, 40, 119
  "regressive inquiry" (*Rückfrage*), 35
  See also acousmatic reduction; auto–affection; bracketing; *epoché*; eidetic reduction; expression *vs.* indication; Heidegger, Martin; Husserl, Edmund; Husserlian phenomenology; Merleau-Ponty,

Index

Maurice; phenomenological voice; Schaeffer, Pierre and Schaefferian tradition; sound object; speaking-to-oneself; suspending; voice
Phenomenology of Perception (Merleau-Ponty), 18, 211
Philips Pavilion, 120
philosophical jokes and linguistic tricks, 180–182
phoné, 152, 153, 218
   aphonic *phoné*, 213–214, 218
   disenfranchisement of, 152
   vs. logos, 152, 208–209
   silent voice, 212, 214
   vs. topos, 209–213
   See also logos; topos; voice
phonocentrism, 152, 153, 193, 207
phonogéne, 40
phonograms (wax cylinders for recording), 180, 182, 192, 269n36
phonographic voice, 180, 182–186
   affected by acousmatic voice, 185-6
   explained, 184
   indicative nature of, 192
   phenomenological voice, compared, 191
   See also acousmatic voice; ontological voice; phenomenological voice; psychoanalytic voice
phonographs/phonography, 16–18, 26, 77, 78, 180, 182, 183
   gramophone, 131, 213, 219, 221
   presence and absence in phonography, 184
   "The Phonograph's Salutation" (Powers), 184–186, 191–193
   sillon fermé, 16–18, 26, 119
phonology *vs.* phonetics, 208
photographs
   Stadthalle, Heidelberg, 105, 106
   use of doctored altered or illustrations, 158, 169, 170, 177–179
Physics (Aristotle), 113
physis, 119, 165, 175, 226
   relationship between *technê* and *physis,* 113–118, 165, 222
   See also phantasmagoria; supplement; *technê*

Picker, John M., 183
Pico della Mirandola, Giovanni, 72
Pimander (Ficino), 71
piping. *See Pfeifen*
plants strategically placed to obscure musicians, 103, 104, 116
   sketch, 104
Plato and Platonic tradition, 56–58, 60, 63–65, 67, 152
   Pythagoras as precursor to, 69
   Renaissance Platonists, 71
   Schopenhauer's Platonism, 99
Pletho, 71
Pliny the Elder, 92, 143, 211
Plotinus, 70
Poe, Edgar Allan, 39, 159–161, 202, 203
Poème électronique (Varèse), 120
Poèmes retrouvés (Apollinaire), 76
poetry
   "Acousmate" (Apollinaire), 46–50, 74–78, 93
   Holbrooke's illuminated symphony, 116, 117
   "The Phonograph's Salutation" (Powers), 184–186, 191–193
   as veil, 68
Politzer, Heinz, 139
Polycrates, 59
polyphony
   "original meaning" of, 234n103
   ventriloquial, 89–90
Pope, Alexander, 143
Pope Leo X, 71
Porphyry, 62, 63, 70
Portier, Nicolas, 83
Powers, Horatio Nelson, 183, 185, 206
practice, acousmatic listening as, 7–9, 17, 93–94, 223-6
prenatal sounds in womb. *See* sonorous envelope
presence
   and absence in phonography, 184
   metaphysics of, 152, 186, 207
   voice as guarantee of, 185
present-at-hand *vs.* ready-to-hand. See *Dasein*
presentation. *See* aesthetics
Presley, Elvis, 209

Presque rien n° 1 ou Le lever du jour au bord de la mer (Ferrari), 11, 119, 120, 123–126, 129, 131–133
Les Pressentiments (Borel), 92
Prévost, Abbé, 91
Price, C. F., 3
primary *vs.* secondary qualities. *See* epistemology
Programme de la Recherche Musicale (PROGREMU), 29
pronouns as shifters. *See* shifters, linguistic
See also "I"
prophecies. *See akousmata*
Psycho (Hitchcock), 207
psychoanalysis, 6, 206–208, 216–222
    Freudian, 6, 208, 210, 213, 216, 273n19
    Lacanian, 206–222
    session as acousmatic situation, 6, 207–208, 221
psychoanalytic voice, 186, 206, 208, 216–222
    vs. acousmatic voice, 217, 218–220
    comparison between psychoanalytic session and Pythagorean legend, 216
    explained, 208
    vs. psychoanalyst's voice, 217, 222
    See also acousmatic voice; ontological voice; phenomenological voice; phonographic voice
pulsing or beating
    allure. *See allure*
    heartbeat, 159, 160
"pure listening," 46, 52
pure perception. *See* aesthetics, Schopenhauer
pure will-less knowledge. *See* aesthetics, Schopenhauer
purely auditory world, 145–147
Pythagoras and Pythagoreanism, 4, 24, 25, 35, 41, 74, 76, 93, 171, 210, 212, 216, 220
    Academic tradition, 56, 57, 69
    assimilation into Christian tradition, 71
    coded messages. *See akousmata*
    cultic status, 62, 209
    darkness, teaching in, 41, 45, 46, 51, 66

    exoteric *vs.* esoteric disciples, 45, 48, 50, 56, 62, 63
    genuine Pythagoreans. *See akousmatikoi* and *mathematikoi*
    listening techniques, Pythagoras as teacher of, 60
    mathematical and scientific teachings *vs.* shamanistic and religious teachings, 56, 58–60
    mimetic identifications with Pythagoreanism, 51–53
    miracle worker, 70
    "philosopher," Pythagoras as, 48
    precursor to Plato, Pythagoras as, 69
    psychoanalytic session and Pythagorean legend, comparisons, 216
    Pythagorean *bios*, 56
    Renaissance Pythagoreanism, 71, 72
    shamanism, 56, 58–60
    silence by disciples of, 46, 48–50, 55, 57
    See also akousmata; *akousmatikoi vs. mathematikoi*; Pythagorean curtail/veil
Pythagorean *akousmata*. *See akousmata*
Pythagorean curtain/veil, 4–6, 33, 37, 40, 45–72, 157
    affinity between *epoché* and, 25
    historical origin of, 63–66
    "key myth" of, 46–50, 74, 76
    in Schaefferian tradition, 50–53
    as separating spirit from body, 209
Pythagorean disciples. *See akousmatikoi* and *mathematikoi*

qualitative identity of sound, 122, 146, 147, 154

radar, 171, 172, 176
radio. *See* broadcast media
Rath, Richard Cullen, 2
rattling, 3, 15, 122
re-identification of sound, 31, 33, 146, 147
    See also eidetic reduction; identity, numerical *vs.* qualitative
reactivation of originary experience, 9, 35, 39, 40, 119

Index

reactivation of meaning-intentions. *See* intention/intentionality
ready-to-hand *vs.* present-at-hand. *See* Dasein
reconstitution/reconstruction of sound, 25, 34, 38, 77, 124, 131, 135
recordings
　half-speed recordings, 167, 168, 177
　overdubbed, 157, 166–168, 175–179
　See also Paul, Les
reduced listening, 4–6, 23–25, 120–123, 138, 150, 156
　as aesthetic value, 123
　"anti-natural" effort, 5, 34
　conflation of acousmatic reduction and reduced listening, 137
　as disciplined listening, 119
　eidetic reduction, 30–33
　explained, 5, 28
　relationship between acousmatic experience, sound object, and reduced listening, 5, 17
　See also modes of listening
reduction. *See* acousmatic reduction; eidetic reduction; *epoché*; phenomenology
Rée, Jonathan, 87
Reger, Max, 29
Regina coeli (Mozart), 118
register. *See* sound object, morphological features of
"regressive inquiry" (*Rückfrage*). *See* phenomenology
reification, 98, 37, 136, 197
　See also phantasmagoria
Reinhardt, Django, 165
religion
　authorities, curtailment of music, 209
　Jewish traditions and teachings, 6, 71
　mystical voices, 203
　religious idols, 97
　vs. science, 87
　tabernacles and consecrated spaces, 68, 69
　theurgy, 70
　and transcendence, 108, 226
　See also Christian traditions and teachings
Remarques morales, philosophiques, et grammaticales, sur le dictionnaire de l'académie françoise (Feydel), 79
Renaissance Pythagoreanism, 71, 72
Renouard, Antoine-Augustin, 79
repetition, 16, 17, 25, 26, 33, 222
　of mythemes, 46
　of plausible statements, 45
reproduction of sound, 4, 24, 25, 35, 45, 47–49, 51, 53, 74, 126, 131, 132, 150, 183–185, 225
"residual signification," 25, 38
Reuchlin, Johannes, 71
revelation/unveiling. *See* veil
reversibility. *See* chiasm
La revue musicale, 73
Reynolds, Debbie, 157
Rice, William North, 3
Richard, Albert, 73
riddles, 62, 65, 68
Riebel, Guy, 124
Riedweg, Christoph, 62
rivalry, mimetic, 166, 171
Robindoré, Beatrice, 124
Le Roi-Lune (Apollinaire), 77
Romanticism, 10, 99, 110
Romeo and Juliet (Shakespeare), 152, 223
Rothstein, Arthur, 178
Rousseau, Jean-Jacques, 109
Rückfrage ("regressive inquiry"). *See* phenomenology
Russell, Bertrand, 223

Le sabbat des esprits (Le Gorse), 92
the Said *vs.* the Saying, 152
Saint-Saëns, Camille, 102
Salmon, André, 7
"sameness" of sound. *See* identity, numerical *vs.* qualitative
Santi Domenico e Sisto, 109
Saturday Evening Post, 167, 172, 173
Saussure, Ferdinand, 208
Schaeffer, Pierre and Schaefferian tradition, 3–6, 8–41, 46–54, 74–79, 93, 98, 99, 108, 112, 116, 119–129, 132–134,

Schaeffer (*Cont.*)
    136–138, 142, 147, 148, 150, 151, 154, 156, 161, 165, 186, 189, 196, 202, 207, 221, 223
  Bidule en ut, 233n82
  Cinq études de bruits, 73, 120
    Étude aux chemins de fer, 26, 28, 29, 120–123
  Continuo, 120, 122, 124
  Étude aux allures, 29, 120–123
  Étude aux objets, 29, 120, 121
    Objets exposés, 29
    Objets rassemblés, 29
  Étude aux sons animés, 29, 120
  Groupe de recherche de musique concrète (GRMC), 46, 73, 74
  Groupe de recherche musicale (GRM), 49, 73, 74, 77, 78, 119, 123
  Husserl's influence on, 17–22
  In Search of a Concrete Music, 15, 19
  "Lettre à Albert Richard," 73
  objections to Schaefferian theory of acousmatic sound, 95–161
  Programme de la Recherche Musicale (PROGREMU), 29
  Solfège de l'objet sonore, 15, 32
  sound object and acousmatic reduction, 15–41
  Symphonie pour un homme seul, 123
  Traité des objets musicaux, 4, 6, 9, 15–19, 26, 27, 36, 47–50, 60, 74, 119, 121, 133, 147, 148, 186
  See also acousmate; acousmatic reduction; *l'acousmatique*; Bayle, François; broadcast media; Chion, Michel; eidetic reduction; Ferrari, Luc; Henry, Pierre; "key myth" of Pythagorean veil; modes of listening; *musique concrète*; *musique expérimentale*; phenomenology; phonographs/phonography; Pythagorean curtain/veil; reduced listening; *solfège*; tape recording; technology
Schafer, R. Murray, 150–152, 156, 225
Schenker, Heinrich, 137
schizophonia, 77, 151, 225, 245n7
  vs. acousmatic sound, 225
Schopenhauer, Arthur, 79, 99–101, 105, 111, 114–116
screened music, 103–105, 108, 109, 114, 116, 117
  illustrations, 104, 107
the Scribe (Jacquet-Droz), 181, 182
  See also automatons
Scruton, Roger, 137, 138, 142, 143
Second Epistle (Plato), 67
secondary *vs.* primary qualities. *See* epistemology
secrecy, 155, 211
  Les Paul's techniques, 166–168, 170–172
  Pythagorean disciples, 45–50, 53, 55, 57, 62–68, 71, 72
  sacred teachings, 62, 67, 68
self
  self-consciousness, 182
  self-foundation (*autopoiesis*), 53, 54
  self-identity, 206
  self-presence, 218
  self-reference, 181, 182, 184, 193, 194
  speaking-to-oneself, 193, 205
  they-self distinguished from self, 201
  See also ego; "I"
Semper, Gottfried, 102
sensation, 23, 40, 86–87, 93, 101, 122, 127, 129, 196
sense-giving acts *vs.* sense-fulfilling acts. *See* intention/intentionality, meaning-intentions
sensuous presentation. *See* aesthetics
serialism, 17, 120
Shakespeare, 152
shaman. *See* Pythagoras and Pythagoreanism
Shaughnessy, Mary Alice, 171
Shaw, George Bernard, 102
Shepard, Odell, 1
shifters, linguistic, 182–184, 192, 193, 195
  See also "I"
Sievert, John, 171
signs and signification, 127–130, 140, 141, 208, 213, 214, 216, 217
  ambiguity of term "sign," 187
  closed economy of signification, 130

entendre as stripping away indicative significations, 196
See also expression *vs.* indication; index/indication
silence
  "acousmatic silence," 216–218, 222
  anechoic chamber, 161
  call of conscience, 201–205
  Pythagorean silence. *See akousmatikoi* and *mathematikoi*, silence/secrecy of
  in Lacanian psychoanalysis, 216–218, 222
  silent speech, 192, 193, 195, 202, 203, 212, 214, 216–220, 222
  See also call of conscience; drive; Lacan, Jacques and Lacanian theory
sillon fermé. See musique concrète
Simultanism, 75
Singin' in the Rain (Donen and Kelly), 157, 158, 171, 223
slapback, 167, 177
  See also Paul, Les
slips of the tongue, 214
Smalley, Denis, 150
Société académique d'archéologie, sciences et arts, 91
solfège, Schaefferian, 15, 29, 33, 41, 52
  classification of sound universe, 29
Solfège de l'objet sonore (Schaeffer), 15, 32
soliloquization, 190–194, 221
  vs. interlocution, 188
solipsism, 22
Solomos, Makis, 18
sonic body. *See* vocalic body
sonic uniqueness, 151–156
sonorous envelope, 6
sorcery, 80, 82
soul, 55, 70, 101, 114, 183
  as essence of voice, 185, 190–192
  preparation of soul for sound, 110–113
sound
  intrinsic properties, 6, 24, 29, 33, 36, 38, 120, 134–6, 145, 225
  necessary components, 7, 166, 223–4
  See also sound object; source, cause, and effect
"sound fragments." *See* sound object

sound object, 4, 5, 8, 15–41, 135, 147, 156, 202
  vs. acoustical signal, 34
  ahistorical aspects of, 35–36, 40–41
  allure, 32, 120–123, 129, 133, 253n5
  defined, 33
  discovery of, 15–17
  duration, register, and timbre, 29
  eidetic reduction of, 31–36
  grain, 29, 32, 122, 177
  imaginative variation, 32–33
  intersubjectivity of, 21–22, 32, 34
  invariant and essential features, 32
  irreducibility of, 25, 30–31
  modes of listening to, 26–30
  morphological features, 28, 33, 120, 121, 122, 123, 132, 138
  objections to Schaeffer's theory, 36–41
  objectivity of, 32–34
  ontological considerations of, 34, 36–39
  "originary experience," 35–36
  phantasmagoric aspects of, 39–40
  phenomenology as method for discovering, 18–26, 30–33
  reacquaintance, revisiting, re-grasping, rediscovering, 34–35
  reification, 37, 39
  relationship between sound object and acousmatic reduction, 22, 25–26, 30, 36, 37
  relationship between sound object and music theory, 34
  relationship between the sound object and technology, 25, 33, 35
  repetition of, 25, 32, 34
  solfège. See solfège, Schaefferian
  vs. sonic effects, 8
  vs. sonic events, 38
  sonorousness of, 34
  vs. "sound fragments," 16, 17
  types/tokens of, 32–34, 146, 147
  unity of, 21
  See also acousmatic reduction; eidetic reduction; ontology; phenomenology; Schaeffer, Pierre and Schaefferian tradition; source, cause, and effect

sound-on-sound recordings. *See* recordings, overdubbed
sound recording, 77, 79, 132, 183
  law of *record whatever,* 132, 133
  magnetic tape, 17, 24, 30, 40, 119, 124, 131, 166, 171
  See also broadcast media; phonographs/phonography
soundmarks, 1, 151
  defined, 227n1
soundscapes, 2, 7, 8, 124, 131, 151, 152
source, cause, and effect
  acousmate, 74–79, 88–91, 93
  "The Burrow," 142–145, 148, 159, 202
  cause *vs.* source of sound, 7, 142
  central claims regarding, 7–8
  chiasm, 159–161
  counter-reductions, 151–156
  effect, sound as, 4, 16, 23, 132, 223
  false attribution of cause, 168–170
  false attribution of source, 168
  in "key myth," 48, 50, 51
  in Les Paul's work, 168–179
  Moodus noises, 2–4
  natural attitude, 196
  necessary components of all sounds, 7, 166, 223–4
  ontological voice, 202–6
  phantasmagoria, 102, 109–111, 116, 118
  phenomenological treatment, 135–7,
  phenomenological voice, 194–5
  phonographic voice, 185
  Presque rien (Ferrari), 123, 126, 131–2
  psychoanalytic voice, 206–214, 216, 217, 219–221
  in Pythagorean tradition, 64
  in Schaefferian tradition, 4–7, 15, 16, 24–25, 27–30, 33, 36–39, 74–79, 93, 120–123, 137–8, 143,
  spacing of source, cause and effect. *See* spacing (*espacement*)
  uncertainty about source and cause, 7, 142–145, 148–151, 157, 159, 221
  underdetermination of source and cause by effect, 7–9, 145–151, 153, 154, 156, 159–160, 165, 174, 185–186, 194–5, 202–3, 205, 206, 225–226

  See also acousmaticity; *acousmêtre*; nontology; ontology; sound object; underdetermination
spacing *(espacement),* 129, 147, 149, 150, 153, 157, 165, 185, 193–195, 202, 203, 225, 226
  defined, 47, 129, 259n62
  See also ontology; source, cause, and effect; underdetermination
spatiality, 131, 136, 178, 191
  characteristic of sound, 145, 178
  space-less imaginary auditory world, 145–147
speaking
  chatter/inauthentic discourse, 199–201, 204, 215–217
  statement *vs.* enunciation, 214–217
  "who speaks," 165, 166, 195, 202–206, 217
  See also voice
speaking-to-oneself, 188, 190–194, 205, 221
  interlocution *vs.* soliloquization, 188
  See also auto–affection; voice, internal soliloquy
speculation/inference as to causal sources, 4, 24, 202
paranoia, 140, 154, 156, 160, 207
Speech and Phenomena (Derrida), 186, 207
spontaneity, 5, 137, 142, 185–187, 192
Stadthalle, Heidelberg, 105, 106
Steel, Dr., 2
Stein, Erwin, 161
Stephaniensaal, Graz, 104
stereo microphone. *See* microphone
Sterne, Jonathan, 131, 245n7
Stoicism, 64
Straus, Erwin, 136, 138, 151
Strawson, P. F., 145–147, 154
string sounds, 7, 134, 168, 169, 173–176, 224
Stromateis (Clement of Alexandria), 46, 64, 66, 68
subject
  "Cartesian subject," 197
  "I", 185
  split subject, 218, 219

traditional philosophical subject compared with *Dasein*, 197
See also auto-affection; "I"; voice
sunken orchestra pits, 105
superego, 6, 203, 206, 207, 209
See also call of conscience; commands/orders; voice
supernatural effects, 81, 89, 90, 93, 204
supplement
  defined, 113–114
  imaginative supplementation, 8, 9, 15, 36, 98, 101, 108, 124, 145, 148, 165, 166, 172, 216, 224
  supplementary relationship of *physis* and *technê*, 113–118, 165, 222
  tactual supplementation of audition, 145
  See also phantasmagoria; vocalic body
surface. *See* aesthetics
"suspending," 7, 9, 23, 39, 134, 195, 217
  See also acousmatic reduction; bracketing; *epoché*; phenomenology;
"sympathetic hearing," 115
Symphonie pour un homme seul (Schaeffer and Henry), 123
"symphony of noises," 15, 77

tabernacles and consecrated spaces, 68, 69
tape recording, 17, 24, 30, 40, 119, 124, 131, 166, 171
  Les Paulverizer, 172–179, 226
  Nagra recorder, 124
  portability of tape recorders, 124
Taylor, Timothy, 25
technê, 97–119, 126, 129, 165, 167, 171, 175
  dismissal of, 101, 114, 170, 220–221
  and psychoanalytic voice, 220–222, 226
  supplementary relationship between *technê* and *physis*, 113–116, 165, 222
  See also bodily techniques; phantasmagoria; *physis*; supplement
technology, 4, 24, 25, 49, 52
  dubbing, 157, 166–168, 175–179
  essentialist view of, 39, 40
  "live" acousmatic music, 165–179

production, commodity, and technology of music, intercalation of, 99
sound object, relationship with, 33
transmission, inscription, storage, and reproduction of sound, 4
tricks with. *See* trickery
See also broadcast media; phonographs/phonography; tape recording
"The Tell-Tale Heart" (Poe), 159, 160
"Tennessee Waltz," 177
territorial listening, 148, 149, 152
terror. *See* fearful sounds
The Testament of Dr. Mabuse (Lang), 157, 207
Theme and Variations for Violin and Piano (Messiaen), 29
theurgy, 70
"they" and they-self. *See Dasein*
Timaeus, 66
timbre. *See* sound object, morphological features of
time consciousness, 193
Tingley, Katherine 112
tinnitus, physiological or psychic, 159
tokens/types of sound, 32–34, 146, 147
topos, 66, 67, 71, 108
  phoné vs. topos, 209–213
  See also logos; *phoné*; voice
trace. *See* aesthetics
train sounds, 26, 28, 29, 120–123
Traité des objets musicaux (Schaeffer), 4, 6, 9, 15–19, 26, 27, 36, 47–50, 60, 74, 119, 121, 133, 147, 148, 186
Traité historique et critique de l'opinion (de Mairan), 92
Transactions of the Royal Society, 83
transcendence/transcendent effects, 9, 18, 20, 21, 32, 108, 114, 119, 224, 226
  phantasmagoria as production of, 108
  See also phantasmagoria
Treatise on Light (Descartes), 87
Trench, Herbert, 116
trench warfare, 140
trickery, 32, 89, 114, 165–179
  Les Paulverizer, 172–179, 226
  linguistic tricks and philosophical jokes, 180–182
  magic, 62, 71, 79, 80, 82, 83, 117

Truax, Barry, 124, 126
"the truth effect," 45
types/tokens of sound, 32–34, 146, 147

ubiquity. *See acousmêtre*, powers of
uncertainty. *See* source, cause, and effect
　See also certainty; underdetermination
underdetermination, 7, 8, 9, 39, 62,
　　145–150, 156, 157, 159, 165–168,
　　172, 174, 185, 186, 194–5, 202–203,
　　206, 226, 260n62, 263n1
　acousmatic underdetermination
　　compared with theory of sound
　　object, 147–148
　chiasm as function of, 159–161
　counter-reduction, as recoil from,
　　151–156
　imaginative supplementation,
　　encouraged by, 9
　Les Paul's work, 166–168, 174–176
　in purely auditory world, 145–147
　relationship to spacing, 165, 195, 260n62
　of sound in "The Burrow," 147
　of source and cause by sonic effect,
　　8–9, 39, 147–150, 153, 156, 172,
　　174, 176, 185, 194–195, 202–3,
　　260n62, 263n1
　voice, 185, 186, 194–5, 202–3, 206
　See also certainty; spacing; supplement
"unfulfilled" meaning. *See* intention/
　　intentionality
uniqueness of sound, 151–156
unsettledness, 148, 150, 152, 157, 159, 168,
　　170, 171, 177
U.S. Geological Survey, 3

Vancouver, 152
Varèse, Edgard, 120
variation as technique for revealing
　　essence. *See* eidetic reduction
variation, imaginative. *See* eidetic
　　reduction
Variations and Fugue on a Theme of Mozart
　　(Reger), 29
Variations pour une porte et un soupir
　　(Henry), 29
Vasse, Denis, 207

veil, 65–69, 211, 212
　of Isis, 113
　musicians, obscuring, 116
　"nothing" behind, 212
　physical/literal *vs.* allegorical veils, 65, 68
　structural function of veiling, 211
　unveiling, 68, 173, 210, 212, 220
　See also Pythagorean curtain/veil
ventriloquism, 88–90, 150, 177, 218
Venusberg music (Wagner), 98
vespers, 109
vestige. *See* aesthetics
videocentrism, 152, 153
Villiers de L'Isle-Adam, Auguste, 180
De vita pythagorica (Iamblichus). *See On
　　the Pythagorean Way of Life*
vocalic body, 8, 109, 224
　compared with sonic body, 8, 9
　See also phantasmagoria; supplement;
　　underdetermination
voice, 8, 149, 150, 152, 153, 155, 180–222
　acousmatic silence *vs.* acousmatic
　　voice, 217
　acousmaticity of voice, 204, 205
　alogical voice, 210, 214, 217, 222
　animation of, 189
　aphonic voice, 213–218
　atopical voice, 214, 217, 218, 222
　authentic/inauthentic discourse,
　　199–201, 204, 215–217
　authority, voice of, 210
　bearer or source, 209
　and body, 209–213
　chatter, 215, 217
　as cinematic object, 207
　commands, 194, 207, 214, 215, 217, 219
　conscience. *See* call of conscience
　deconstruction, 207
　desire, voice as object of, 213
　dispossession of voice, 217, 218
　divine effect, 210
　drive, voice as object of, 213, 214,
　　217, 222
　ear, circuitry between voice and ear, 192
　enunciation *vs.* statement, 214–217
　epistemological status of, 206
　ethics, 214–217

expropriation of voice, 218
exteriority, alterity, and non-identity, 206, 207
   as guarantee of speaker's presence, 207
His Master's Voice, 213, 215, 217, 219
inscription of phenomenological voice, 189–192
inscription of phonographic voice, 182–186
inscription of recorded sound, 183, 191, 192
inscription of silence, 218
interlocution *vs.* soliloquization, 188
internal soliloquy, 188, 190–195, 221
internal voice inside head, 221
introjected voice of authority, 203
logos *vs. phoné*, 208, 209
   and meaning, 208
meaning-intention, 188, 189
metaphysics of presence, 186
the mouth, 212
obedience, acousmatic voice as the voice of, 213
objet (a), 212-3, 271n86
object, voice as, 6, 206–214, 217, 219, 272n90
omnipresence and omnipotence, 209
ontological voice, 200–203
order, voice as threat to, 209
as permanently phantasmagoric, 220
phoné vs. phoné, 213, 214
political aspects, 215, 216
prescriptive demands, 216
psychoanalytic voice, 206–222
the Said *vs.* the Saying, 152
self-identity of subject, 207
speaking-to-oneself, 192
sonorousness, 191
sound, differentiation from, 208
source, differentiation from, 208
statement *vs.* enunciation, 214–217
stranger in one's self, 207
superego, 6, 203, 206, 207, 209
surplus-meaning, 209
topos *vs. phoné*, 209–213
underdetermination of, 186, 206
unheard voices, 213
uniqueness, 152–156
"who speaks," 165, 166, 195, 202–206, 217
See also acousmatic voice; *akousmata*; call of conscience; expression *vs.* indication; ontological voice; phenomenological voice; phoné; phonographic voice; psychoanalytic voice; speaking-to-oneself
A Voice and Nothing More (Dolar), 11, 48, 208, 209, 214
The Voice in Cinema (Chion), 39, 46, 156, 207
"voix acousmatique," 186

Wackenroder, Wilhelm Heinrich, 79, 101, 102, 105, 108, 114, 116
Wagner, Cosima, 116
Wagner, Richard, 98, 101–103, 105, 108, 114–117, 119, 143, 221
Waksman, Steve, 166
Wangunks, 1, 8
Warburton, Dan, 125
watches and clocks. *See* horology
wax cylinders. *See* phonograms
Westerkamp, Hildegard, 150
"What the GRM brought to music" (Battier), 49
"who speaks," 165, 166, 195, 202–206, 217
Wickenhauser, Richard, 104
Wiener Symphoniker, 117
will-less knowing. *See* aesthetics, Schopenhauer
will-to-truth, 128
Wishart, Trevor, 150
witchcraft, 78, 80, 81, 93
Wittgenstein, Ludwig, 223
The Wizard of Oz, 210
Wolfrum, Philipp, 105
wonder, state of, 4, 23
Wood, Gaby, 181
Wordsworth, William, 183
World Soundscape Project, 151, 152
World's Fair, Brussels, 120, 122
"The World Is Waiting for the Sunrise" (Les Paul and Mary Ford), 169

writing
  inscription and intention, 190
  iterability, 181, 182
  logocentrism, 67
  phonographic voice as form of, 192
  the Scribe (automaton), 181, 182, 184, 185, 192
  voice's privilege over written sign, 185
  writing machines, 181
  See also inscription

X-rays, 113

Xenakis, Iannis, 120

"You Are My Lucky Star," 157

Zahavi, Dan, 23
Zeno, 64
Zeuxis. *See* desire
Zimmerman, Robert, 110
Zischen, 139, 160–1
  See also "The Burrow"; *Pfeifen*
Žižek, Slavoj, 6, 11, 186, 206–211, 217
Zoroaster, 71

CPSIA information can be obtained
at www.ICGtesting.com
Printed in the USA
BVOW03s0158300917
496275BV00002B/7/P

CONWY & AROUND IN 50 BUILDINGS

PETER JOHNSON & CATHERINE JEFFERIS

amberley

A glimpse of an earlier Conwy: 'The Welsh Harp Inn, Conwy' 1809, apparently a late-eighteenth-century building; the probably late medieval building next door, abutting the town wall, seems to be a 'House That Jack Built'. Both buildings are now gone – or hiding behind more modern façades. (© National Library of Wales)

For Alexander, with our love.

First published 2016

Amberley Publishing, The Hill, Stroud
Gloucestershire GL5 4EP

www.amberley-books.com

Copyright © Peter Johnson & Catherine Jefferis, 2016

The right of Peter Johnson & Catherine Jefferis to be identified as the Author of this work has been asserted in accordance with the Copyrights, Designs and Patents Act 1988.

Map contains Ordnance Survey data © Crown copyright and database right [2016]

All rights reserved. No part of this book may be reprinted or reproduced or utilised in any form or by any electronic, mechanical or other means, now known or hereafter invented, including photocopying and recording, or in any information storage or retrieval system, without the permission in writing from the Publishers.

British Library Cataloguing in Publication Data.
A catalogue record for this book is available from the British Library.

ISBN 978 1 4456 6101 8 (print)
ISBN 978 1 4456 6102 5 (ebook)

Origination by Amberley Publishing.
Printed in Great Britain.